CW01521375

Grain Boundaries

Grain Boundaries

Their Microstructure and Chemistry

P. E. J. Flewitt and R. K. Wild
The University of Bristol
Interface Analysis Centre
Bristol
UK

JOHN WILEY & SONS, LTD

Chichester · New York · Weinheim · Brisbane · Singapore · Toronto

Other Wiley Editorial Offices

John Wiley & Sons, Inc., 605 Third Avenue,
New York, NY 10158-0012, USA

WILEY-VCH Verlag GmbH, Pappelallee 3,
D-69469 Weinheim, Germany

John Wiley & Sons Australia, Ltd, 33 Park Road, Milton,
Queensland 4064, Australia

John Wiley & Sons (Asia) Pte Ltd, 2 Clementi Loop #02-01,
Jin Xing Distripark, Singapore 129809

John Wiley & Sons (Canada) Ltd, 22 Worcester Road,
Rexdale, Ontario M9W 1L1, Canada

Library of Congress Cataloging-in-Publication Data

Flewitt, P. E. J.
    Grain boundaries : their microstructure and chemistry / P.E.J. Flewitt and R.K. Wild.
        p.   cm.
    Includes bibliographical references and index.
    ISBN 0-471-97951-1 (cloth : alk. paper)
    1. Grain boundaries. I. Wild, R. K. II. Title.

    TN690.F598   2001
    620.1′1299—dc21   00-043684

British Library Cataloguing in Publication Data

A catalogue record for this book is available from the British Library

ISBN 0-471-97951-1

Typeset in 10/12pt by Techset Composition Ltd., Salisbury, Wiltshire.
Printed and bound in Great Britain by Antony Rowe, Chippenham, Wilts
This book is printed on acid-free paper responsibly manufactured from sustainable forestation, for which at least two trees are planted for each one used for paper production.

Contents

Preface

Over the twentieth century, the understanding of grain boundaries, interphase boundaries and free surfaces has advanced on the basis of the thermodynamic foundations laid down by J. W. Gibbs in 1906 and an appreciation of the microstructure set by Rosenhain in 1911. Certainly, over the intervening period there has been a clear recognition that these discontinuities in engineering materials make a significant contribution to the physical and mechanical properties of materials, spanning the range from polycrystalline and multiphase metals and alloys through to the latest ceramic and electronic materials. The intervening period has seen the development of understanding of the crystallography of grain and interphase boundaries together with the classic book produced by Donald McLean in 1957 entitled simply *Grain Boundaries in Metals*. In the 1960s the rapid advance of transmission electron microscopy produced direct evidence of the structure of these boundaries and, more recently, scanning probe techniques have similarly advanced our understanding of the structure of free surfaces. However, it is in the past 25 years, with the work of Hondros and his coworkers and others that the composition of boundaries and surfaces has been able to be modelled for a range of materials subject to equilibrium and non-equilibrium heat treatment and related conditions. This period has also seen the advent of several high spatial resolution microchemical techniques, including Auger electron spectroscopy and scanning transmission electron microscopy coupled with either energy dispersive X-ray microanalysis or electron energy loss spectrometers, which make it possible to measure local compositions at interfaces where there is coverage of a single layer of atoms of a different elemental species. Moreover, the most recent developments of some of these and other techniques provide scope for extending our knowledge of chemical composition to include the chemical state of atoms residing at grain boundaries, interphase boundaries and free surfaces. As a consequence, at the end of the twentieth century, it seemed appropriate to draw this understanding together and to produce this book which sets out our views in this area. However, we recognise that there is much to be done to couple the understanding of crystal structure and composition and provide predictions of physical and mechanical properties. It will be interesting to see this evolution which we believe will rapidly progress over the first decade of the new century.

As a consequence, this book is directed to an audience of specialist graduates, undergraduates, scientists and engineers to provide a basis for understanding the role of grain and interphase boundaries and surfaces with respect to their local chemical composition. A framework is set to guide the reader through the background to these interfaces, the underlying theory and the techniques available for measuring local compositions to a description of the observed chemistry in a range of materials. The final chapter is devoted to the engineering applications and, in this respect, principally the mechanical properties of metals and alloys are addressed because this reflects the main efforts to date.

P. E. J. Flewitt
R. K. Wild
March 2000

Acknowledgements

We would like to acknowledge the help of our colleagues within the Technology Division of BNFL–Magnox Generation and Dr L. A. Mitchell for encouraging the production of this book. We would like to acknowledge the interaction afforded by contacts within several Universities, in particular Professor A. G. Crocker at Surrey University, Professor R Faulkner at Loughborough University and Professor L. M. Brown at Cambridge University. P. E. J. Flewitt recognises the invaluable collaboration with Dr P. Doig, Mr D. Lonsdale, Dr R. Moskovic and Mr R. A. Stevens over many years. R. K. Wild would like to express his thanks to all at the Interface Analysis Centre at the University of Bristol for their help and encouragement and, in particular, Professor J. Steeds, Professor G. C. Allen, Dr J. Day, Dr K. Hallam, Dr P. Heard, Dr C. Younes, Dr G. Meaden, Mr M. Holt and Mr J. Nicholson. We would like to thank Professor A. G. Crocker (University of Surrey) and Dr A. J. Flewitt (University of Cambridge) for helpful comments on the text.

Quotation

One thing I have learned in a long life: that all science, measured against reality, is primative and childlike—and yet it is the most precious thing we have.

> Albert Einstein
> Creator and Rebel, 1973
> (London: Hart-Davis MacGibbon)

Chapter 1

Boundaries and Interfaces in Materials

1.1 INTRODUCTION

All solid materials are made up of atoms, ions or molecules, and in the crystalline state the vast majority of these are arranged in regular or periodic arrays. In the case of atoms, for example, the precise arrangement is determined by the size and shape of the particular atom and the nature of the bond between adjacent atoms that leads to a configuration, in a single crystal, producing the lowest energy state of the system. As a result, there are many types of crystal structure consistent with the 14 possible lattice arrangements (Bravais lattices), and indeed the same material may have different structures depending upon the temperatures and pressure to which it is subject (Flewitt and Wild, 1994).

When a solid forms from the liquid it does so by the nucleation and growth of small crystallites randomly oriented throughout the solidifying system, and these grow as the liquid solidifies. The crystals continue to grow until they meet with neighbouring crystals and, in general, the individual crystals will not be either of the same orientation or aligned. The boundary where they meet becomes a surface and the mismatch in orientation is accommodated by distortion and gaps in the atomic arrangement. If a solid is polished flat and then lightly etched using a suitable weak acid, the boundaries can be highlighted and observed in an optical or scanning electron microscope (Figure 1.1). This disruption of the perfect coordination of the crystal structure across adjacent grains is similar to the structure of a typical high angle boundary observed in a two dimensional raft of bubbles (Lomer and Nye, 1952) (Figure 1.2). In crystalline materials these zones of misfit are a few nanometers or less in width. These boundaries in solid materials have lower coordination than the adjacent crystals and tend to be areas of mechanical weakness in the body. The energy of the grain boundary will vary as the misorientation between adjacent grains increases, and this is shown schematically for an idealised crystal in Figure 1.3. A major factor determining the physical, mechanical, electrical

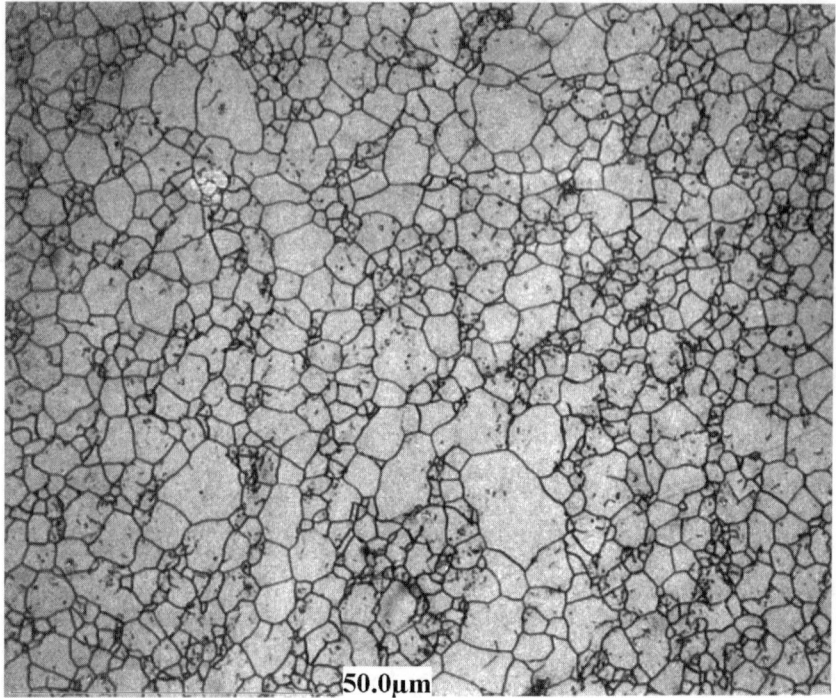

Figure 1.1 Grain boundaries exposed in a metal by polishing and etching

and chemical properties of polycrystalline materials is the grain boundary structure. However, many materials are made up of more than one phase, as shown in Figure 1.4. Each phase in the overall microstructure may have a very different composition and crystal structure, although in some cases these differences may be quite small. One phase will be separated from a second phase by an interface or boundary, known as an interphase boundary (Figure 1.5), which because of its similarity to grain boundaries may be considered to be an extension to the concept of a grain boundary.

As a consequence, in crystalline materials two different internal interfaces are present: homophase boundaries and heterophase boundaries (Kalonji and Cahn, 1982). The first category includes grain boundaries, twin boundaries, domain boundaries, etc., whereas the latter separates crystals of two thermodynamically different phases. These latter boundaries or hetero-interfaces can span a wide range from metal alloy systems where discrete and distinctly different phases are separate crystals, to composite materials where oxides and metals may co-exist. However, in some semiconductor heterostructures the crystal structure and orientation can be the same. In addition there is another group: the metal–ceramic interface (Rühle *et al.*, 1990).

Even in the case of pure elements and compounds, grain boundaries in materials almost always contain additional impurity elements to those in the specification,

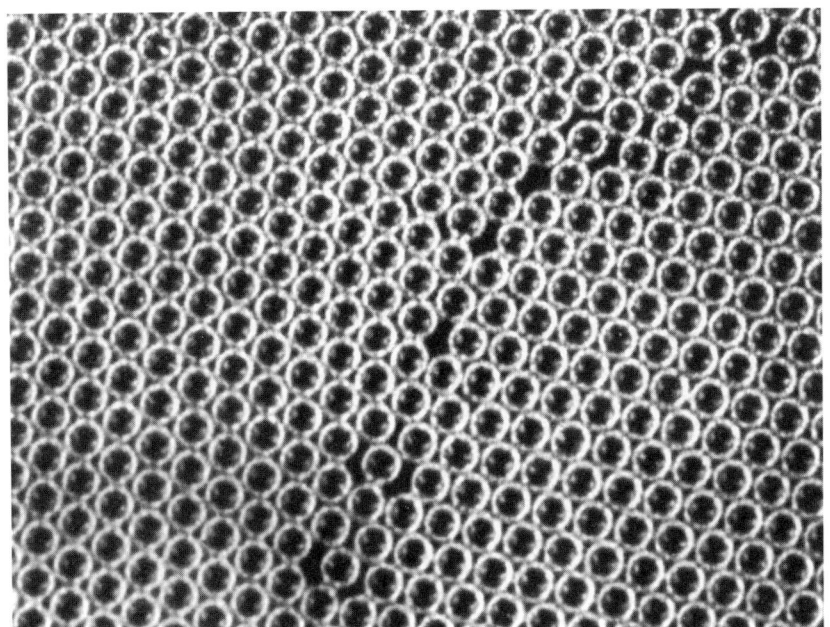

Figure 1.2 High-angle boundary in a two dimensional raft of bubbles (Lomer and Nye (1952))

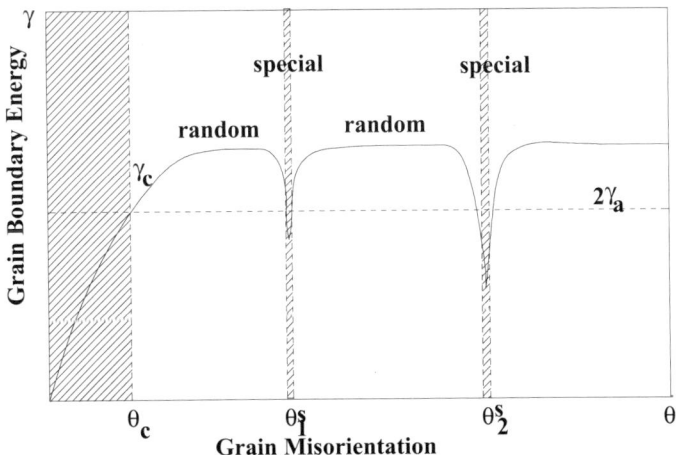

Figure 1.3 Variation in grain boundary energy as a function of misorientation between adjacent grains

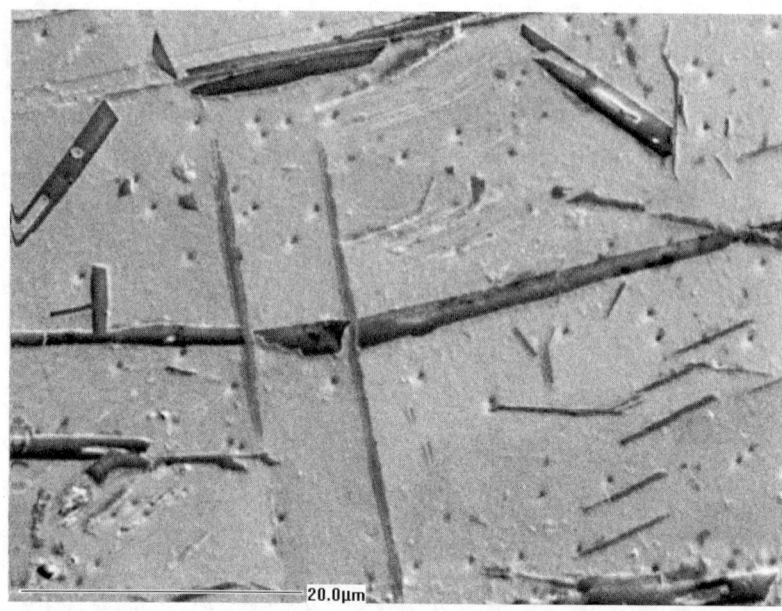

Figure 1.4 Metallographically polished section through a multiphase material with a nickel-base matrix containing aluminium and titanium nitride precipitates

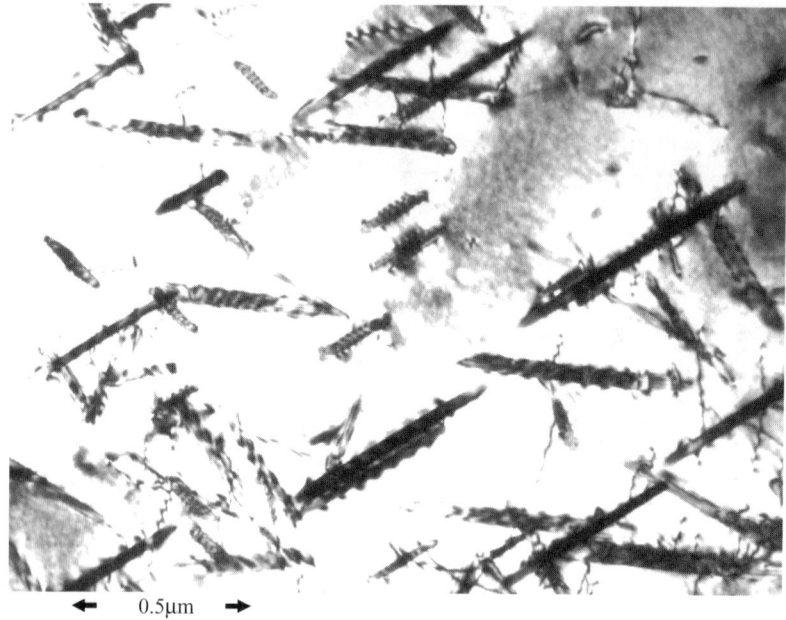

Figure 1.5 Transmission electron micrograph showing the microstructure of a 12% chromium ferritic stainless steel containing retained δ-ferrite

even if only at the part per million or even part per billion level. Moreover, certain engineering materials, usually metal alloys, ceramics or polymeric materials, are produced to contain specific additions of minor elements to achieve the required physical, chemical, electrical and mechanical properties. However, in many of these engineering materials, in particular metal alloys, in addition to the main alloying elements, impurity atoms may also be present. Since the grain boundaries are of lower energy, these atoms can redistribute to the grain boundaries and thereby influence the overall properties of the product. In addition, in alloys which may be either single- or multiphase, the composition is not always uniformly distributed and frequently certain elements will be enhanced or depleted in the region of grain or interphase boundaries. These local changes in chemistry that occur on the nanoscale will result in significant modifications to the bulk properties of the material. Although these local compositions at grain boundaries and other interfaces have been associated with a degradation of the properties of materials used in engineering components and structures, they can by appropriate control be used to produce the desired improvements to the materials.

The mechanical properties of a material are governed by many factors: the strength of the bond between atoms, the strength of the bond between internal surfaces, the ability of atoms to diffuse through the matrix and the internal surfaces, the dislocation motion within the matrix, the influence of second phase particles, etc. When the material fails under the influence of an applied load, it may do so by a variety of modes. Failure may be by cleavage of atom planes, resulting in flat surfaces containing 'river' lines (Figure 1.6a), by ductile tearing, giving a dimpled uneven surface (Figure 1.6b), or by fracture along grain or interphase boundaries, producing smooth regular surfaces (Figure 1.6c). Which type of failure mechanism predominates will be determined by both the conditions and the material properties. The external conditions will include temperature, load, strain rate, etc., while the material properties will include grain boundary energy, bond strength, impurity content, etc. Ashby and coworkers attempted to predict the type of failure mode that would predominate for a wide range of materials (Ashby *et al.*, 1979). He developed a series of diagrams, known as Ashby maps, for various metals, alloys and ceramics, to predict the type of failure that might be expected for any given condition. A typical Ashby map is reproduced in Figure 1.7 for a nickel–chromium alloy (nichrome) for various times to failure and various stresses at 293 K.

It is the purpose of this book to consider the nanochemistry or local composition of grain boundaries and interphase boundaries in solids and to extend this to free surfaces and other interfaces. To achieve this, we will explain the structure and properties of these boundaries and consider the mechanisms that lead to local changes in their nanochemistry. In general, engineering materials are polycrystalline and contain a large number of individual grains, for example a volume of polycrystalline material of $1\,cm^3$ will contain $> 10^6$ grains of $100\,\mu m$ mean diameter; the latter dimension corresponds to a linear array of about 10^{17} atoms. The individual grains have a specific orientation and hence misorientations compared with the surrounding grains, and thereby specific grain boundary types.

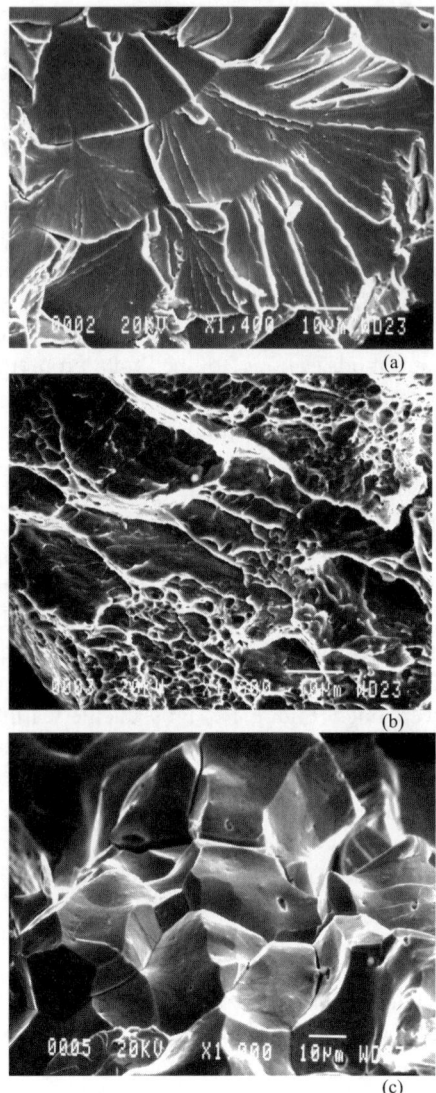

(a)

(b)

(c)

Figure 1.6   Examples of (a) cleavage, (b) ductile and (c) intergranular fracture in iron based alloys

The overall properties of a polycrystalline material are governed by the contribution from all the boundaries rather than a specific individual boundary. Moreover, the local chemistry of each grain boundary will be a function of the misorientation. Therefore, the inter-relationship between the specific local atomic arrangement and composition of a grain boundary on the one hand and the contribution of all boundaries and the overall properties on the other is complex. However, it is an

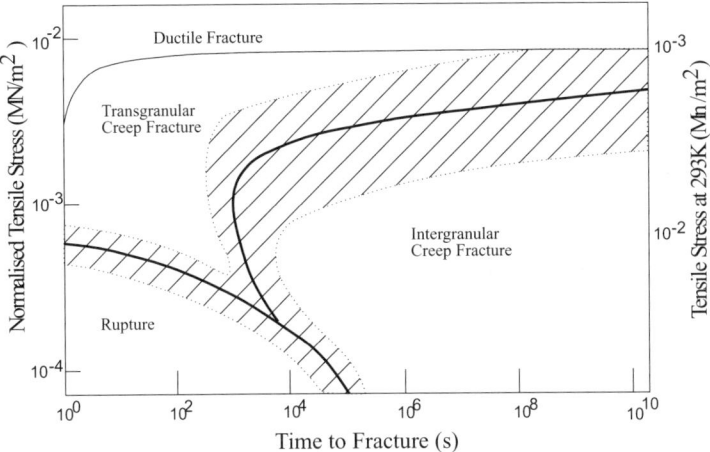

Figure 1.7 Ashby map for nickel–chromium alloy failing at 20 °C at various rates and stresses (Ashby *et al.* (1979))

essential input to the ability to predict the overall properties of materials. In this chapter we introduce some of the underlying concepts which are developed more fully in subsequent chapters of this book.

1.2 PRINCIPLES GOVERNING GRAIN SHAPE AND SIZE

For an ideal system, the grain shape, size and distribution aim to minimise the surface energy and thus reduce the surface area. The ideal shape will minimise the energy for a grain of a given size, but simply increasing the grain size or decreasing the number of grains will reduce the grain boundary area and hence the total surface energy. Grains therefore tend to grow with time, but for this to occur the atoms must be able to migrate from one grain to the next across a grain boundary. This can only occur at relatively high temperatures, typically greater than half the melting point. The rate of grain growth will increase with increasing temperature, and the mobility of a grain boundary, M, takes the form

$$M = M_0 \exp(-Q/RT) \tag{1.1}$$

where M_0 is a constant, Q is the activation energy, R is the gas constant and T is the temperature.

The understanding of the growth, shape and structure of grains has developed from other systems that can be used to model grains. The likely shape of a grain in a polycrystalline material has been gained from observations of bubbles in foams (Weaire, 1994; Weaire and Glazier, 1993; Kumar *et al.*, 1994; Fradkov *et al.*, 1993; Smith, 1954). The equilibrium shapes formed by bubbles are analogous to the shapes of grain boundaries. These have been studied for over a century, with an early paper published by Lord Kelvin (1887), although considerable progress was made about 50 years ago by Smith (1948 and 1954). A bubble will attempt to achieve its minimum energy configuration. For example, consider a bubble in two dimensions where boundaries all have the same energy; here, the angles where three bubbles meet at a point must all be 120°. The bubbles can be considered as polygons and, if all the polygons were hexagons, the structure could be made up with the corner angles all 120°. However, in practice there are polygons with less than and more than six sides. For the corner angles to remain at 120°, those with less than six sides have to be convex and those with more than six sides must be concave (Figure 1.8).

Extending these ideas to three dimensions involves the same principles but is more complicated. Here, only a tetrahedron can shrink and disappear without distorting adjacent structures. When this disappears it leaves four bubbles meeting at a point with the bubble edges at angles of 109.5°, with each edge formed by three faces at 120°. As long ago as 1887, Lord Kelvin claimed to have solved the minimum energy configuration by considering an array of equal-sized polycrystals given by a body centred arrangement of 14-sided polyhedra known as tetrakaide-cahedra; in practice, this polyhedron is only observed very rarely. Smith (1952) showed that the number of polygons per three dimensional cell must outnumber the corners by exactly one and that the average number of edges per polygon tends to $5\frac{1}{7}$ as the number of corners approaches 6. This is the three-dimensional equivalent of the hexagon in two dimensions and explains the frequent occurrence of pentagonal faces in bubbles and grains in materials. Aboav (1970) made empirical observations and Weaire (1974) developed this argument and derived a formula (the Aboav–Weaire law) for the average number of sides m of the neighbouring cells of an n-sided cell such that

$$m = a + \frac{b}{n} \tag{1.2}$$

where a is equal to 5 and b is equal to 6. However, there is a general exact sum rule relating to terms of the distribution function of n:

$$\sum_n f(n)mn = \sum_n f(n)n^2 \tag{1.3}$$

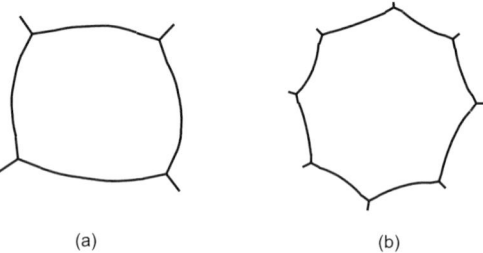

(a)   (b)

Figure 1.8   Polygons with (a) less than and (b) more than six sides but with angles of 120° between the sides

which this does not satisfy and requires the modification by Aboav (1980) when

$$m = 5 - a + \frac{6a + \mu_2}{n}$$ (1.4)

where

$$\mu_2 = \sum (n - 6)^2 f(n)$$

These principles determine the grain shapes in most materials where the grain boundary energies are approximately equal. There are, however, cases where the energies between boundaries may differ. For example, in multiphase alloys there will be interphase boundaries where the energies are not equal. In a two-phase alloy there will be grain corners where two interphase and one single phase meet at a point. Here, the equilibrium angle θ of the included phase is determined by the interfacial free energy, γ_{11} for the monophase and γ_{12} for the duplex phase, where

$$\cos \frac{\theta}{2} = \frac{\gamma_{11}}{2\gamma_{12}}$$ (1.5)

This has important implications for material strength and structure. If θ is less than $60°$, the second phase must spread along all grain edges. If θ is zero, the second phase will penetrate completely between the grains and can result in extremely brittle structures, such as occur with brass in contact with mercury, or it may cause poor corrosion resistance.

With the advent of faster computers with large memory capabilities, it has become possible to search for the ideal theoretical shape that will result in the minimum energy or area. The Kelvin tetrakaidecahedron had 14 faces, but Kusner (1992) has calculated that the average number of faces in bubbles cannot be less than 13.39. Weaire and Phelan (1994) have computer generated a structure with two basic cell structures, one 12-sided and the other 14-sided, which lock together in a unit of eight cells within a repeating simple cubic lattice. A quarter of the cells have 12 slightly curved pentagonal faces, and the other cells have fourteen faces with the extra two faces being hexagons. This gives a structure with an average of 13.5 faces per cell. Figure 1.9a shows the two tetrakaidecahedrons (shapes with 14 faces) first studied by Lord Kelvin, and Figure 1.9b shows the more efficient packing that can be achieved with the Weaire–Phelan structure.

For theoretical modelling, computer simulation of three-dimensional arrangements of grains can be used to generate Voronoi polygons, which are also called Dirichlet domains or Thiesson polygons. Voronoi cells are formed by a growth process that initiates simultaneously from a random distribution of points in space, followed by radial growth at a constant rate. The shapes of these three-dimensional Voronoi cells resemble metallic grains, although their surfaces are planar and not curved to satisfy the requirements for grains described above. These cells fill the volume completely and, unlike an overall arrangement of tetrakaidecahedra, their sizes vary stochastically. It is these aspects of the Voronoi cells that have made them a favoured approximation for grains of materials in computer simulations and, as

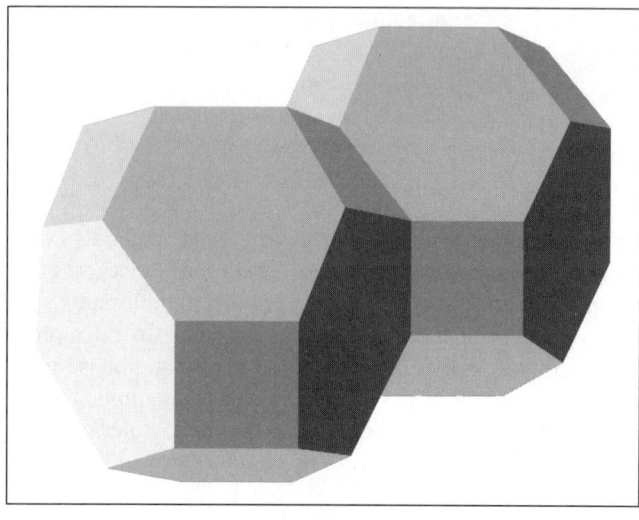

(a)

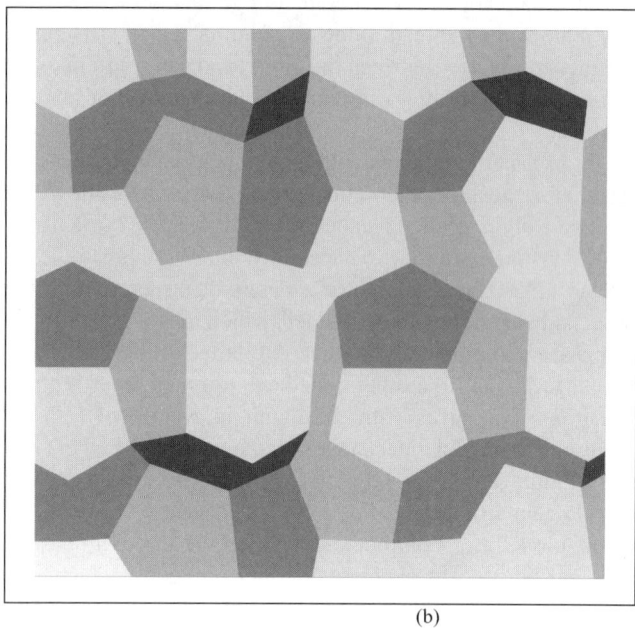

(b)

Figure 1.9 Computer generated structure of grains made up of (a) basic cell structures of 12 and 14 sides to give Kelvin cells (b) with an average of 13.5 faces per cell to give the ideal foam structure decribed by Weaire and Phelan (1996)

such, they have found applications in a variety of disciplines (Thorvaldson, 1992, 1993; Crain 1978; Boots and Murdoch, 1983). More recently, Telley *et al.* (1996) have shown that grain growth can be simulated by a Laguerre (or weighted Voronoi) diagram, entirely determined by a collection of weighted sites. Figure 1.10 shows an example of a Laguerre diagram and its dual Delaunay triangulation. This is a cellular structure generated by a set of weighted points which produces a convex cell made of every point, p, satisfying

$$d^2(p, p_i) - w_i \leqslant d^2(p, p_i) - w_j, \quad 1 \leqslant j \leqslant N \tag{1.6}$$

where $d(p.p_i)$ is the Euclidean distance, w is the weight and p is the point. The Delaunay triangulation is defined as follows. To every cell, L_i, there corresponds a vertex, p_i, with an edge D_{ij} having endpoints p_i and p_j. From this diagram, Telley *et al.* (1996) develop a simple motion equation to simulate grain growth in two dimensions.

The structure that results from modelling using computed Voronoi cells is similar to that found in a material where all the grain boundaries have the same energy. In practice, the grain boundary energy will vary as a function of the angle of mismatch between the crystal lattices in adjacent grains (Figure 1.3). Furthermore, the introduction of trace elements that may segregate to grain boundaries will further modify the grain boundary energy. Therefore, in order to model the grain structure in a real material, the computer simulation must include the energy of the grain boundary. Yang *et al.* (1995) have considered grain growth in situations where there are anisotropic grain boundary energies. They use a Monte Carlo approach to

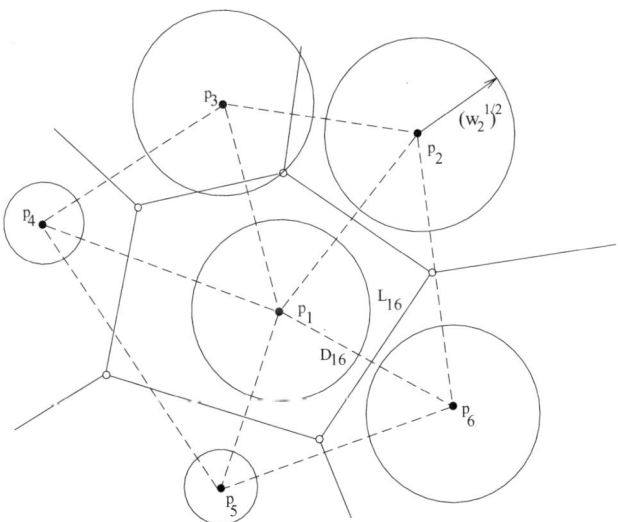

Figure 1.10   Laguerre (solid lines) and weighted Delaunay (dashed lines) construction for a set of weighted points $P = \{(p_i, w_i), 1 \leqslant i \leqslant 7\}$; $D_{16} =$ distance between P_1 and P_6; $L_{16} =$ boundary between P_1 and P_6 (Telley *et al* (1996))

demonstrate that, even in a single-phase polycrystalline material, anisotropic grain boundary energy alone may lead to an anistropic microstructure. One system they have modelled is alumina (Al_2O_3), where the basal planes have a low surface energy and the plane perpendicular to the basal plane has a high surface energy (Iwasa and Bradt, 1984). This leads to a microstructure where there are large tubular grains and the computer generated (Figure 1.9) and observed microstructures (Figure 1.11) give remarkably good agreement.

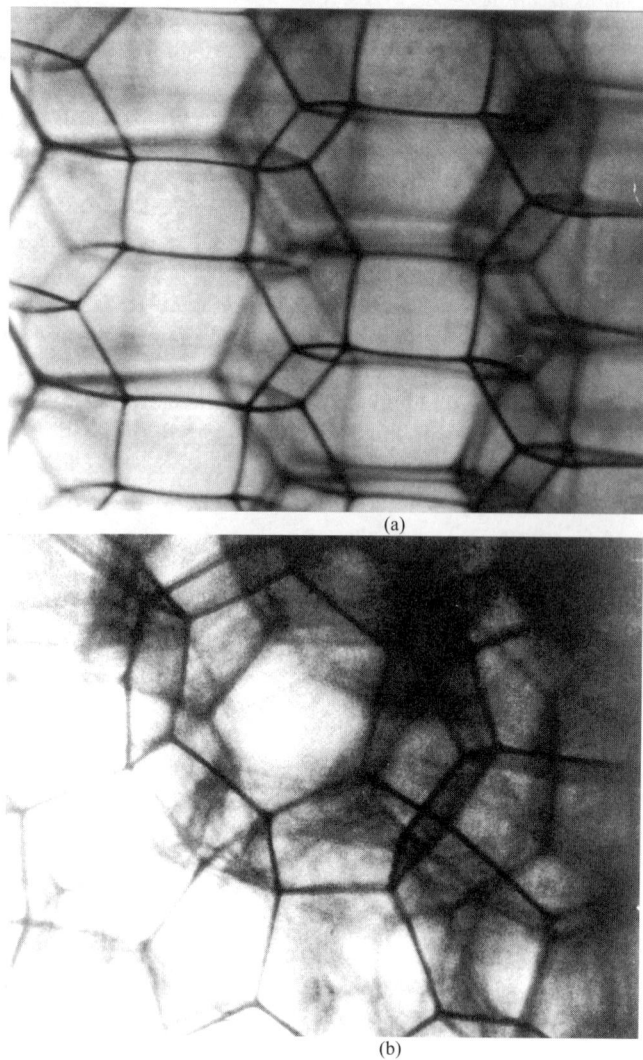

(a)

(b)

Figure 1.11   (a) Foam structure showing a region of Kelvin cells and (b) ideal foam structure corresponding to Figure 1.9b. Reproduced by permission of Taylor and Francis from Weaire and Phelan (1994) *Phil. Mag. Lett.*, **70**, 345

1.3   SIZE AND ORIENTATION OF GRAINS

Grain size is an important parameter which is recognised to have considerable influence on the physical, mechanical and chemical properties of materials. To measure this in a three-dimensional solid is not easy. However, most materials are capable of being mounted, cut and polished to yield a flat surface that can subsequently be interrogated using optical or electron optical techniques. The flat polished surface may be lightly etched to reveal defects such as grain boundaries, second phases, dislocations and inclusions. This produces a two dimensional representation of the grain distribution throughout the material and, by measuring the size of the two-dimensional grains intersecting the surface, a histogram can be constructed of the grain size against grain population. The grain size can be estimated by comparing grains under observation with a standard distribution defined by an ASTM standard built into the graticule of the optical microscope (Gabler, 1956).

In order to understand the relationship of one grain to another it is also necessary to measure individual grain orientation. Grain sizes vary considerably from less than one micrometre to several hundred micrometres; it is only when grains are several millimetres in diameter that they can be individually oriented using classical X-ray diffraction techniques. More usually, to achieve such measurements it is necessary to use a method that has good spatial resolution and yet provides diffraction information. Backscattered electron images recorded in the scanning electron microscope produce diffraction lines that correlate with lattice planes in the grain. Interpretation of these patterns allows the grain orientation to be determined. This effect was observed by Venables and Harland, 1973 and developed to the stage where it could be routinely used to identify crystal orientation in the mid-1980s (Dingley and Baba-Kishi, 1986). It has since been further developed and is now known as orientation imaging microscopy (Adams, 1993; Adams *et al.*, 1993) (see Chapter 4). Fatigue can occur in steels under cyclic stress conditions, and in the secondary electron image (Figure 1.12a), slip bands are visible. Electron backscatter detection measures the crystal orientation, shown schematically in Figure 1.12c, which can be overlayed on to the electron image. Local misorientation is measured and the distribution can be displayed as a histogram (Figure 1.12d), and high- and low-angle grain boundaries may be identified (Figure 1.12b). By overlaying the misorientation map on the electron image, it is possible to correlate high-angle boundaries with changes in direction of the slip bands.

1.4   ENERGY OF POLYCRYSTALLINE MATERIALS

Each grain boundary in a polycrystalline material may be regarded as a sink or source for atoms within the total system. Atoms in the vicinity of a particular grain boundary will have a driving force to diffuse towards that boundary and will diffuse if the temperature is sufficiently elevated. However, in addition to this, the boundary

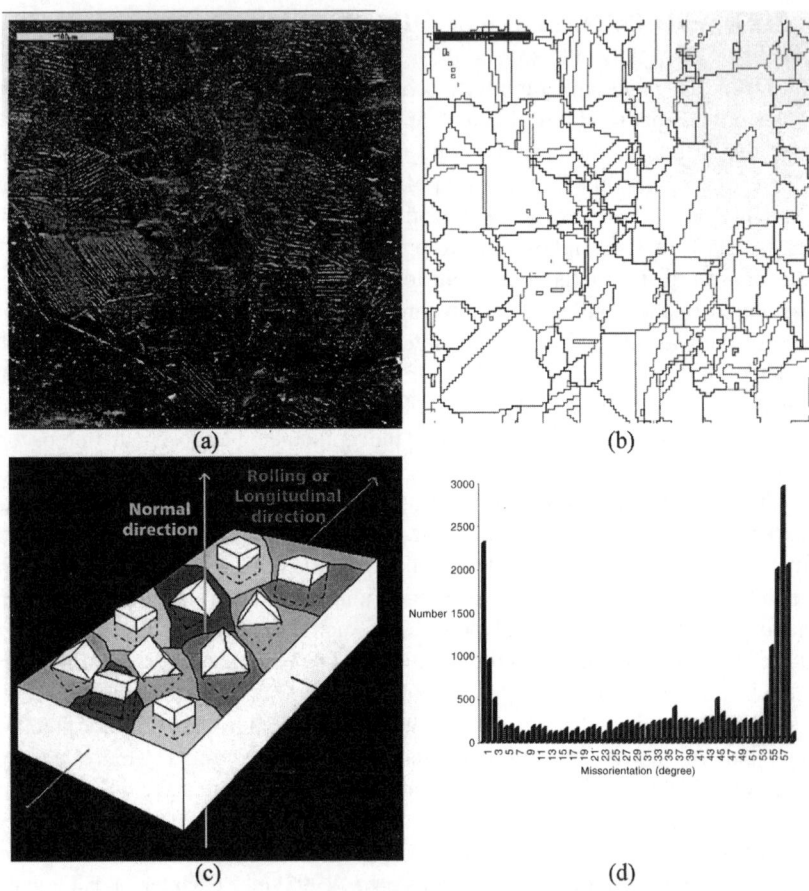

Figure 1.12 Orientation image microscopy (OIM): (a) secondary electron image of steel following fatigue cycles; (b) electron backscatter misorientation map showing high- and low-angle grain boundaries; (c) schematic of EBSD (electron back-scatter diffraction); (d) histogram of grain misorientations. Reproduced by permission of Oxford Instruments

can act as a source of atoms and as a consequence it will achieve an equilibrium condition where creation and absorption of atoms are constant. During this process, the boundary may move from its original location, resulting in the growth of certain grains at the expense of neighbouring grains. The misorientation of the particular boundary will determine the manner in which it may act as a sink or source and migrate. The low-angle grain boundaries can move only by the climb of primary dislocations in the boundary plane which have a Burgers vector component normal to that plane (Hirth and Lothe, 1982). High angle grain boundaries act as sources and sinks by the climb of secondary dislocations in the boundary plane (Chan and Balluffi, 1986). In the case of an interphase boundary the situation is further complicated and the movement of atoms will also depend upon the properties of the phases and the interactions that occur between them.

When considering the energy of a grain or interphase boundary, there are three related quantities that must be taken into account:

(a) surface tension,
(b) surface free energy and
(c) surface stress.

Surface tension, γ_T, is defined as the work, w, required to create a unit area of surface, A, at constant temperature, T, volume, v, and chemical potential, μ. Hence, the surface tension is given by

$$\gamma_T = \left(\frac{dw}{dA}\right)_{T,v,\mu} \tag{1.7}$$

Therefore, the surface tension is related to the number of bonds broken per unit area and to the energy required to break an individual bond. In general, the surface tension will be lowest for those crystallographic planes that contain the highest density of atoms, and, indeed, the surface tension is highly anisotropic. This will have important consequences for grain boundary adhesion.

Surface free energy, γ, is defined as the change in the Helmholtz free energy, τ, of the system per unit area of interface, A. The surface free energy for pure metals is given by

$$\gamma = d\tau/dA \tag{1.8}$$

whereas for alloys at constant v, T and strain it is given by

$$\gamma - \frac{d\tau}{dA} - \sum_i \mu_i \left(\frac{dn_i}{dA}\right) \tag{1.9}$$

where μ_i is the chemical potential and dn_i/dA is the change in the number of atoms of components, i, in the bulk grains owing to change in the grain boundary energy.

Surface stress, σ_{ij}, is defined as the work required to deform a surface. To specify this, it is necessary to define two normal σ_{xx} and σ_{yy} and one shear σ_{xy} stress components which are related to the surface tension, γ_T, by

$$\begin{aligned}
\sigma_{xx} &= \gamma_T + d\gamma/d\varepsilon_{xx} \\
\sigma_{yy} &= \gamma_T + d\gamma/d\varepsilon_{yy} \\
\sigma_{xy} &= d\gamma_T/d\varepsilon_{xy}
\end{aligned} \tag{1.10}$$

where ε_{ij} is the strain. Clearly, if the term $d\gamma/d\varepsilon_{ij}$ is zero in these equations, then the surface stress, σ_{ij}, equals the surface tension, γ.

When two grains come in contact, one may grow at the expense of the other to reduce the total energy of the system. For this to occur there must be a driving force which is generally the difference in chemical potential across the boundary. All other factors being equal, between the two grains the curvature of the boundary can lead to growth, but stored energy, elastic strain, temperature, impurity content, second phase particles and grain boundary orientation all influence the rate of growth. As a consequence, if we consider a grain boundary with a radius of curvature, r, the difference in chemical potential from grain 1 to grain 2 is given by

$$\mu_1 - \mu_2 = \frac{2\sigma_{ij}v}{r} \cong \frac{2\gamma v}{r} \qquad (1.11)$$

where μ_1 and μ_2 are the chemical potentials of grains 1 and 2, v is the specific volume and σ_{ij} is the surface stress which, as in equation (1.10), can be assumed to be equal to the surface tension, γ_T, if $d\gamma/d\varepsilon_{ij}$ is zero. For a boundary of width d_{gb} with a potential difference between the two grains of $\Delta\mu$, the force, F, operating on the boundary is equal to $-\Delta\mu/d_{gb}$. As a consequence, the grain boundary will move with a velocity, v_{gb}, given by

$$v_{gb} = B_M \frac{\Delta\mu}{d_{gb}} \qquad (1.12)$$

where B_M is the atom mobility.

Substituting for $\Delta\mu$ in equation (1.12) from equation (1.11), we obtain

$$v_{gb} = B_M \left(\frac{\Delta\mu}{d_{gb}}\right)^m = \frac{D_B}{kT}\left(\frac{2\gamma v}{d_{gb}r}\right)^m \qquad (1.13)$$

The right hand side is a constant at constant T and therefore $v_{gb} = \text{const}/r^m$. Assuming that r is proportional to the grain dimension D, then the grain boundary velocity is equal to const/D^m which is equal to $dD T/dt$. Integration gives

$$D = kt^n \qquad (1.14)$$

where n is generally less than 0.5.

In the case of a heavily cold worked polycrystalline material, the individual grains will contain a large number of defects, such as dislocations, so that adjacent grains may have different amounts of stored energy. As a result, the various grains will have deformed by different amounts within the polycrystal to give the overall strain for the body. In this situation, one grain may have little or even no stored energy and the adjacent grain may have a substantial amount of stored energy equal to γ_s/N_A where γ_s is the stored energy per mole and N_A is the Avogadro number. In this case, at an appropriate temperature, the grain boundary will move with a velocity

$$v_{gb} = B_M \left(\frac{v_s/N_A}{d_{gb}}\right) \qquad (1.15)$$

Clearly, for this equation the grain boundary velocity varies linearly with stored energy within the grains.

The above considers the movement of a grain boundary in a system subject to plastic strain, but it is possible for grain boundaries to move as a result of the influence of elastic strain. This can be demonstrated by considering the simplest case of a uniaxial strained specimen that contains two grains. In the case of uniaxial strain the energy per unit volume can be shown to be $\sigma^2/2E$, where σ is the stress and E is the Young modulus of elasticity. If the two grains have moduli given by E_1 and E_2, the grain boundary velocity is given by

$$v_{gb} = \frac{B_M}{d_{gb}} \frac{v\sigma^2}{2} \left(\frac{1}{E_1} - \frac{1}{E_2} \right) \tag{1.16}$$

Low concentrations of solute or impurity atoms, which will have different atom mobilities, B_M, in the matrix and grain boundaries, can drastically affect grain boundary motion. For example, an increase in the bulk composition of tin in lead from 1 to 60 ppm can decrease the grain boundary velocity by up to four orders of magnitude. This is due to the segregation of the solute or impurity atoms to grain boundaries, where the tin may be enriched by several orders of magnitude and represent a significant fraction of the grain boundary. Grain boundary segregation will be considered later.

1.5 DIFFUSION

Since a grain boundary is a highly disordered two-dimensional plane between two highly ordered, but misaligned, single crystals, it is an ideal region for preferred transport of atoms. Certainly, grain boundary diffusion is recognised to be orders of magnitude faster than that in the corresponding bulk. However, the grain boundary is a region of modified potential and may act as a driving force for the diffusion of bulk atoms from the grains to the grain boundary. In this section we will confine the consideration to diffusion along grain boundaries.

The dislocation models provide a basis for explicit descriptions of point defect and solute interactions with grain boundaries (see Section 1.6). While there may be some doubt regarding the exact structure of grain boundaries at elevated temperatures where diffusion is significant, analyses have been undertaken to establish grain boundary vacancy and solute atom interactions (King and Smith, 1981; Grovenor *et al.*, 1980; Harmann *et al.* 1976; and Rae, 1981). However, this can lead to the migration of the specific grain boundaries and, indeed, the ordered structure is preserved during this process. This can produce diffusion induced grain boundary migration (DIGM) and recrystallisation (DIR). An early mechanism proposed to describe DIGM invoked an atomistic mechanism: the climb of the grain boundary dislocations as a consequence of the Kirkendall effect on grain diffusion (Smith and King, 1981; Balluffi and Cahn, 1981). The model has the following main features:

(a) interdiffusion, which is confined to the grain or interphase boundaries;
(b) the flux of solute atoms into the boundary and the flux of matrix atoms out of the boundary that are different;
(c) the active sinks and sources for point defects are the source of grain boundary dislocations;

(d) the glide and climb of appropriately oriented grain boundary dislocations, constrained to the boundary plane
(e) the chemical potential driving the alloying process is sufficient to overcome capillary forces and thereby allow breakaway of the grain boundary from its solute atmosphere;
(f) the composition of the region left behind the moving grain boundary depends upon the core structure of the grain boundary dislocations. This leads to specific grain boundaries migrating by different amounts.

The incubation time for DIGM to commence is associated with the nucleation of dislocations which move to produce migration, and the dislocation associated with the original grain boundary position accommodates a rotational mismatch between a sessile and a mobile boundary that is similar to twinning. A series of experiments undertaken by Hillert and Purdy (1978) has been interpreted by a thermodynamic model (Hillert, 1983) where the important feature is that, once an alloyed region is formed by interdiffusion at the grain boundary, the boundary interfaces will have different energies. Hence, a chemical potential is established that allows growth of the region bounded by the lower energy interface. A necessary requirement for this model is the contribution from dislocation misfit to allow migration. These misfit dislocations are also DSC dislocations (see Section 1.6) and will be correctly oriented for grain boundary migration. Diffusion-induced migration and recrystallisation have been observed in a range of systems, including coarsening of hexagonal Ag_2Al precipitates in an aluminium alloy (Laird and Aaronson, 1967), hexagonal Pd_2Si precipitates in a silicon alloy (Cherns *et al.*, 1982), Ni_3Al/copper diffusion couples (Ma *et al.*, 1993) and aluminium during interdiffusion of iron or chromium oxide (Lee *et al.*, 1993; and Backhaus-Ricoult *et al.*, 1996).

Second-phase particles present in the matrix of a polycrystal will have a restraining effect on grain boundary mobility. This is simply addressed by considering the case of a spherical particle of radius r, interacting with a grain boundary (Figure 1.13). Here, the restraining force, F_R, is given by

$$F_R = 2\pi r \cos \phi d_{gb} \cos(\alpha - \phi) \tag{1.17}$$

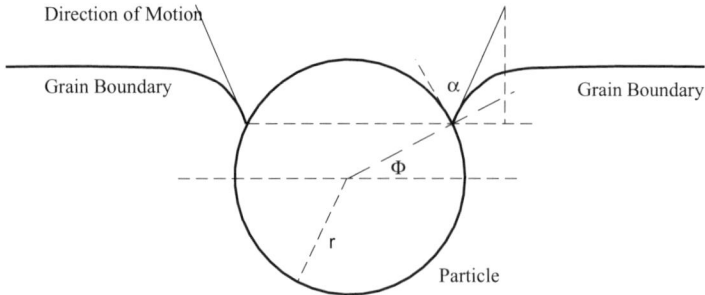

Figure 1.13   Spherical particle of radius r, interacting with a moving grain boundary

where d_{gb} is the grain boundary width, ϕ is the angle between the particle radius at the grain boundary contact and the grain boundary surface, α is the angle between the grain boundary and the particle and the maximum restraining force, F_{max}, becomes

$$F_{max} = \pi r d_{gb}(1 + \cos \alpha) \tag{1.18}$$

1.6 STRUCTURE OF BOUNDARIES AND INTERFACES

As two atoms approach one another, there is an attractive force acting between the atoms that initially draws them closer together, but, as they approach to within a few tenths of a nanometre they then encounter the repulsive force between the positively charged nuclei. The balance between these forces determines the ultimate interatomic spacing and the structure adopted by the atoms in the bulk. In most materials the atoms form in regular arrays determined by the particular bonding arrangement, although in some, such as glasses, there is no long range order in the structure and these are classed as amorphous. The regularity of the array in the crystal structure can be described in terms of symmetry elements (Kelly and Groves, 1970; and Barrett and Massalski, 1986), and these elements determine the directionality of the physical properties of the crystals. The crystal can be defined by a unit cell which is described by three vectors, *a*, *b* and *c*, to give the crystallographic axes. Figure 1.14 shows the 14 possible lattice arrangements (the Bravais lattice) for crystalline materials.

There are five types of interatomic bond as shown schematically in Figure 1.15. These are:

(a) ionic,
(b) covalent,
(c) metallic,
(d) molecular,
(e) hydrogen.

In the case of ionic bonding the atoms either gain or lose an electron by transfer so that their outer electron shell is complete. As a consequence, the atoms are electrically charged, either positively or negatively, and thereby attract atoms of opposite charge. This results in a strong non-directional structure with high melting points. For the covalent bond, pairs of atoms share outer electrons to fill the outer electron shells; these materials have strong directional structures with a high melting point. For the metallic bond, all the atoms collectively share the valence electrons and this results in non-directional structures of high density that are electrically conducting with a wide range of melting points. The molecular bond (Van der Waals bond) arises from the displacement of charge within electrically neutral atoms or molecules, producing a weak attractive force between them. It is analogous to the metallic bond but results in weak, soft crystals with low melting points that are good insulators. The hydrogen bond is weak and mediated by the hydrogen atom. It arises because the hydrogen is a small atom and the charge is easily displaced.

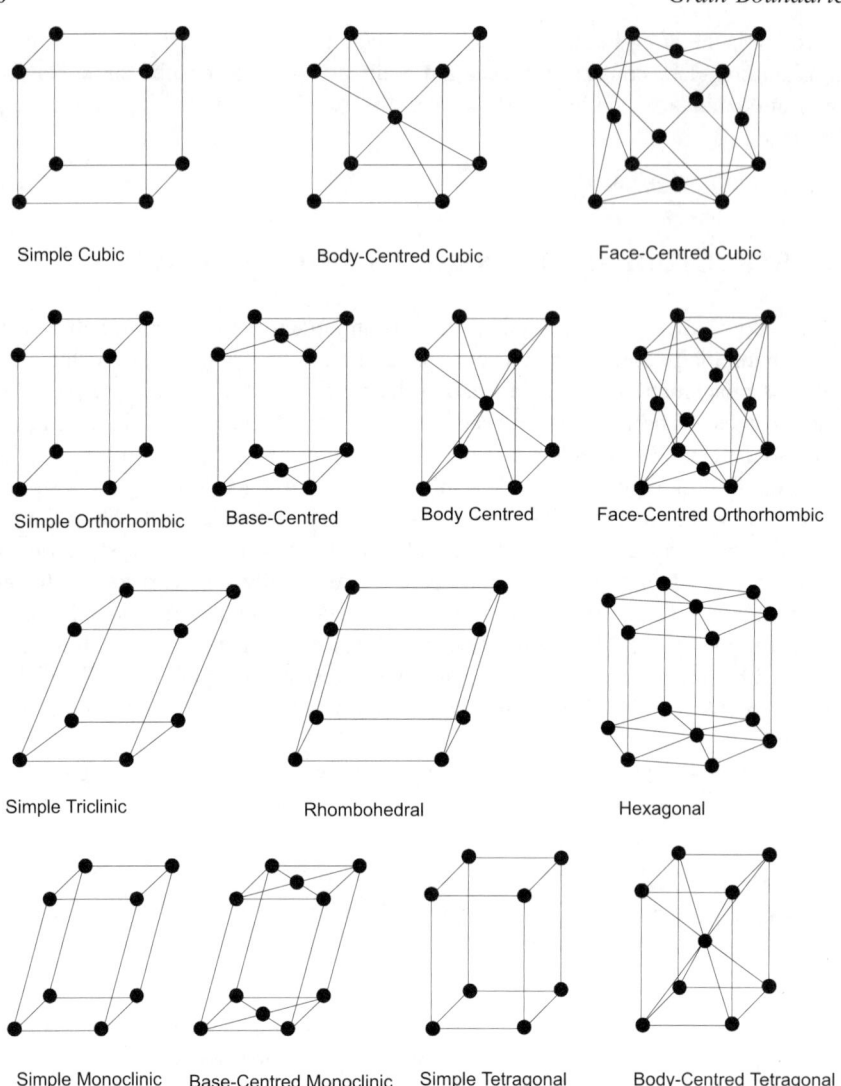

Figure 1.14   The 14 possible Bravais lattice arrangements (the Bravais lattice) for crystalline materials

1.6.1   Dislocation Models

The crystallography of a planar interface depends upon six macroscopic and four microscopic parameters or degrees of freedom: *macroscopic*—three parameters describe rotation of one crystal with respect to another, two describe the orientation of the boundary plane and one describes if the transformation from one crystal to

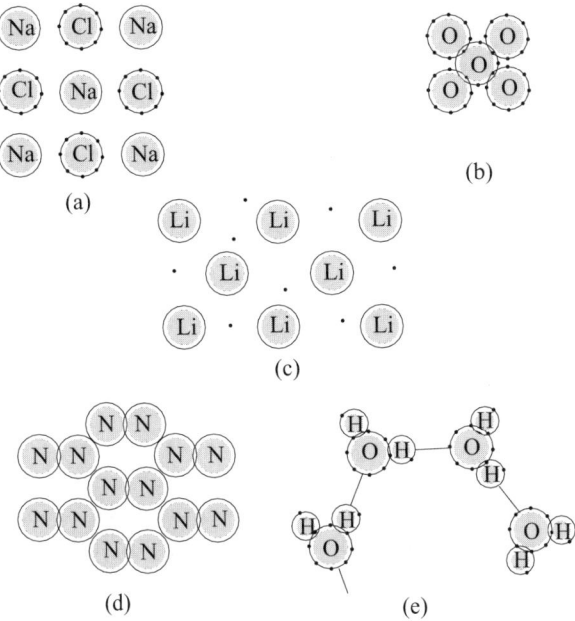

Figure 1.15 The five types of interatomic bond: (a) ionic, (b) covalent, (c) metallic, (d) molecular and (e) hydrogen

another requires an inversion; *microscopic*—three parameters describe the translation of one crystal with respect to another, and one parameter defines the position of the interface.

At the interface, the atoms relax to positions that differ from the adjacent parent crystals and, apart from introducing structural defects, including dislocations, this can have an important effect on the material properties. For example, the conductivity of various ceramics depends upon charged defects that segregate to grain boundaries to induce space charges in adjacent crystals. In the case of grains in a polycrystal, planes may be aligned but tilted through an angle θ (Figure 1.16a), which allows two of the degrees of freedom and is known as a tilt boundary. Alternatively, they may have no tilt but be rotated through an angle α (Figure 1.16b), which is referred to as a twist boundary. In general, all grains will have a tilt and rotational component. If the angle between two grains is small, then the grain boundary may be considered as one continuous crystal containing an array of dislocations. In the case of a simple tilt boundary, a line of edge dislocations provides the boundary. If these dislocations have a spacing d_D and a is the lattice spacing, then the tilt angle, θ, is given by

$$\theta = \frac{a}{d_D} \qquad (1.19)$$

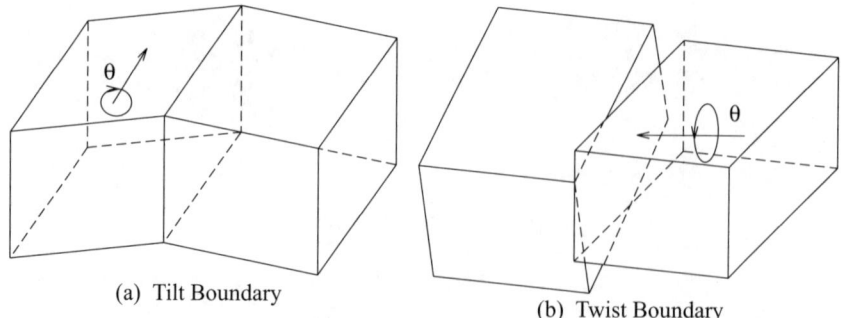

(a) Tilt Boundary

(b) Twist Boundary

Figure 1.16   (a) Tilt and (b) rotational or twist boundaries. In general, grain boundaries all have a tilt and rotational components

For a twist boundary, where the mismatch is due to rotation, then, again for low angles, the two grains may be considered as a single crystal but containing a square grid of screw dislocations (Weertman and Weertman, 1964; Read, 1953).

The orientation of the crystal planes in adjacent grains or phases will, in general, have no relationship unless the material has been treated in some specific manner to allow alignment between them in a special way. In certain materials, for example, the grains are almost all aligned with each other and as a consequence the angle between adjacent grains is small. In the case of second phases precipitating within a matrix there is, often in the early stages, a high degree of matching of atom planes across the interphase boundary and the boundaries are referred to as coherent. As the degree of misfit increases, the boundary may remain fully coherent but at the expense of an increase in the strain resulting from distortion of the atom planes in the precipitate or matrix to accommodate the fit. However, as the mismatch increases further, a point is reached where the strains involved are too great so that the mismatch can then be accommodated by incorporating dislocations. In the asymmetric tilt boundary the mismatch must be accommodated by additional edge dislocations with the Burgers vectors lying along the interface. This type of boundary may occur when there is a considerable mismatch between the crystal planes such that, after a given number of atom planes, the atoms have achieved registry. If a_1 and a_2 are the lattice parameters, and $a_1 > a_2$, then the degree of coherence or disregistry is defined as

$$\delta = \frac{a_1 - a_2}{a_1} \tag{1.20}$$

and this type of boundary will contain dislocations with a spacing d_D given by

$$d_D = \frac{a_1}{\delta} \tag{1.21}$$

In the symmetrical twist boundary, where two planes from adjacent grains are rotated through an angle θ with respect to each other, the boundary consists of coherent cells with screw dislocations at the edge of the cells.

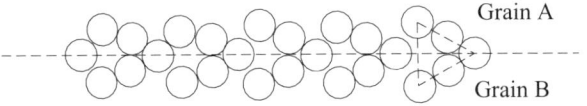

Figure 1.17 Atomic coherence between atoms in grain boundaries at multiple atomic spacings; here five atoms comprise the repeating cell, which is referred to as a Σ5 coincidence site lattice (CSL)

In general, grain and interphase boundaries are aligned in a random fashion and, while all atoms are not coherent in both boundaries, there will be certain alignments where coherence can occur at multiple atomic spacings (Figure 1.17). Here, atoms in the vicinity of the boundary can be considered as repeating units; as illustrated, five atoms comprise the repeating cell. The description of grain boundaries using this approach is known as the coincidence site lattice model (CSL). The lattice is defined by a term Σ, which is always an odd integer, and in which Σ^{-1} of the lattice sites are common to both lattices. In the example shown in Figure 1.17, one-fifth of the lattice sites are coincident and this is referred to as a Σ5 CSL. The CSL model determines the basic periodicity of the atomic structure. Two lattices may be constructed by imagining that lattices A and B on either side of the boundary extend through space and are translated with respect to each other so that atoms of each crystal lattice coincide at one point referred to as the origin.

Bollmann (1970) developed this concept further to define all vector displacements of lattices A and B relative to one another. The approach is called the DSC lattice because it involves a (D)isplacement of lattice B with respect to lattice A by a lattice vector that causes a pattern (S)hift that is (C)omplete. The DSC lattice vector can be determined from vectors connecting atoms of lattice A with atoms of lattice B. These two concepts are complex and the reader is referred to Bollmann (1970) for a complete description.

1.7 COMPOSITION OF BOUNDARIES AND INTERFACE

A grain boundary can be considered as a plane of disorder within a single crystal where the degree of disorder is determined by the angle of misfit between the two sides of the plane. The grain boundary energy will be determined by the type and magnitude of the misfit. In some boundaries, such as twin boundaries, there is perfect alignment of atoms and no increase in dislocation density. In low-angle boundaries a complete description can be given by an array of dislocation, but in high-angle boundaries there is a volume of complete disorder. As a consequence, all these boundaries have increased energy when compared with a perfect single crystal which may be reduced by the addition or substitution of different species of atoms to those comprising the matrix. This is the driving force for alloying atoms, such as molybdenum or manganese or impurity atoms such as phosphorus, sulphur, tin or antimony in steel, to concentrate at grain boundaries when subject to certain heat treatments or external environments. The segregation of elements to grain boundaries will be dealt with in more detail later (see Chapter 2), but the driving force for segregation is to reduce the total energy of the system.

The interface core structure can change locally as a result of changes in temperature, composition or crystal misorientation. Increases in temperature will result in the production of defects, such as point defects, line defects or steps. Molecular dynamic modelling studies have determined the structure factor of the boundary core region as a function of increasing temperature (Ciccotti *et al.*, 1983). However, experimentally there was no evidence for a sudden disordering transition when aluminium grain boundary dislocation structures were heated *in situ* in a transmission electron microscope at temperatures of 0.32–0.96 T_m (Hsieh and Balluffi, 1989). It would appear that increases in temperature influence the short-range order, but the long-range order remains unaffected. Crystal misorientation can also result in a phase change at a grain boundary. As the angle between two grains changes, it will consist of a mixture of units until a certain angle is reached (known as a cusp), and beyond this angle a mixture of different units will apply. It is at the cusp energy that a phase transition may occur.

1.8   ELECTRONIC STRUCTURE

Considerable effort has been directed towards understanding the atomic and electronic structure in a grain boundary (Hashimoto *et al.*, 1984a; Briant and Messner, 1984; Briant, 1990; Cottrell, 1989, 1990; Tang *et al.*, 1993, 1994; Wu *et al.*, 1993). Hashimoto *et al.* (1984b) considered the grain boundary structure with the segregation of phosphorus and boron at different levels. They used the Morse potential to describe the Fe–P and Fe–B systems. A quasidynamic method, similar to molecular dynamics, was then used to simulate the grain boundary structure, from which the binding energy between an impurity atom and the grain boundary was calculated. They concluded that the local environment between phosphorus and iron is similar to that in Fe_3P. Briant and Messmer (1984) and Briant (1990) adopted an approach in which they used a molecular orbital cluster model to predict grain boundary embrittlement of iron, nickel and chromium by antimony. They first assumed that atoms exist as clusters of tetrahedra with the antimony placed at the centre. A self-consistent field method was then used to calculate the molecular orbitals. This approach leads to the conclusion that the antimony acts to embrittle transition metals by drawing electrons from the surrounding metal atoms. Such a transfer of charge leaves fewer electrons to form the metal–metal bonds, so that these bonds are weakened. Tang *et al.* (1993 and 1994) also use molecular orbital cluster calculations to predict the role of boron and sulphur in iron grain boundaries. Here, boron atoms induce less relaxation than sulphur from nearby iron atoms, and the boron forms a relatively strong bonding with the iron while sulphur has very little hybridisation, resulting in a weak bonding state.

The adsorption of hydrogen to surfaces of transition metals and interstitial sites has been explained in terms of orbital electronegativities (Johnson *et al.*, 1977) and by theoretical calculations using a self-consistent field, scattered wave approach (Messmer, 1977). In nickel, the 4s orbital bonds with the hydrogen 1s, whereas in palladium and platinum the 4d orbitals bond with the hydrogen 1s. The theoretical predictions have been confirmed by observing shifts in the peak positions in photoelectron spectra from the free metal surfaces.

Experimental observations of the electronic structure of a grain boundary have recently been made (Brown *et al.*, 1997; Ozkaya *et al.*, 1995; Hallam and Wild, 1995). Most studies of grain boundaries have involved the fracture of a material in ultrahigh vacuum, followed by analysis of that surface by Auger electron spectroscopy or in few cases by X-ray photoelectron spectroscopy. While the former has the potential to probe changes in the electronic structure of a surface, it has rarely been used for this purpose and X-ray photoelectron spectroscopy remains the main technique providing information on the binding energies of the filled electron states (Hallam and Wild, 1995). Unfortunately, the boundary observed by these techniques has been modified by the need to fracture and thereby break bonds at the grain boundary.

Recently, EELS has been used to study grain boundaries in iron (Figure 1.18). The spectrum from a grain boundary that contains phosphorus is compared with a spectrum from a nearby grain (Figure 1.18a). The spectrum from a 'pure' grain boundary that contains no phosphorus is compared with a spectrum from a nearby grain (Figure 1.18b). This reveals that the d orbitals of the iron atoms in a boundary that contains phosphorus have more electrons than the d orbitals of iron atoms in the bulk. Hence, phosphorus atoms in a grain boundary donate electrons to the d bands of the iron. However, in a pure boundary, where there is no phosphorus, the situation is totally different. In this case, the number of electrons in the d bands of the iron atoms stays the same (Figure 1.18b). This led Brown *et al.* (1997) to consider the contribution to the cohesive strength of the grain boundary. In transition metals, such as iron, the d bands can contain 10 electrons. If these form covalent bonds in a linear combination of molecular orbitals, then the d orbitals split into five states below the atomic reference level and five above the so-called bonding and antibonding states. The energy levels in an isolated iron atom are filled to the 3d and 4s states which contain six and two electrons respectively (Figure 1.19a). In this notation the first number ($n = 1, 2, 3, 4$, etc.) denotes the atomic shell (K,L,M,N, etc.) while the letter (s,p,d or f) denotes the subshell and refers to the angular momentum of the orbital state ($\ell = 0, 1, 2, 3$). The 3d state can hold up to 10 electrons. When the atom is excited, say by a fast incident electron, the six electrons in the full 2p state can move into the empty 3d states (arrow). This causes the incident electrons to lose energy. An electron energy loss spectrum can reveal the number of empty 3d states. However, electrons in the 3d state have wave functions that are like standing waves with alternating positive and negative lobes (Figure 1.19b). If lobes of similar sign on neighbouring ions overlap, the electron waves interfere constructively, piling up charge in the region where the electron experiences the attractive potential of both

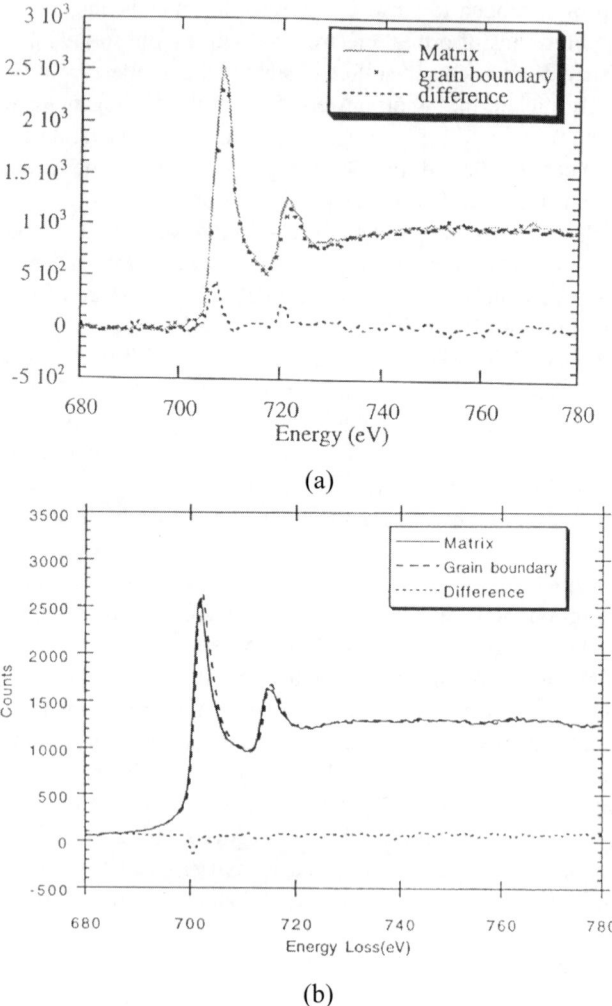

(a)

(b)

Figure 1.18 (a) Electron energy loss spectrum from an iron grain and from a phosphorus-containing grain boundary. The difference between the two spectra indicates that the presence of phosphorus in the grain boundary causes about half an electron per boundary atom to be transferred to the iron. (b) Electron energy loss spectrum from an iron grain and from a pure grain boundary that has no phosphorus. The difference between the two spectra now shows no transfer of electrons to the iron. Reproduced by permission of Ozkaya *et al.* (1995) *J. Microscopy*, **180**, 300

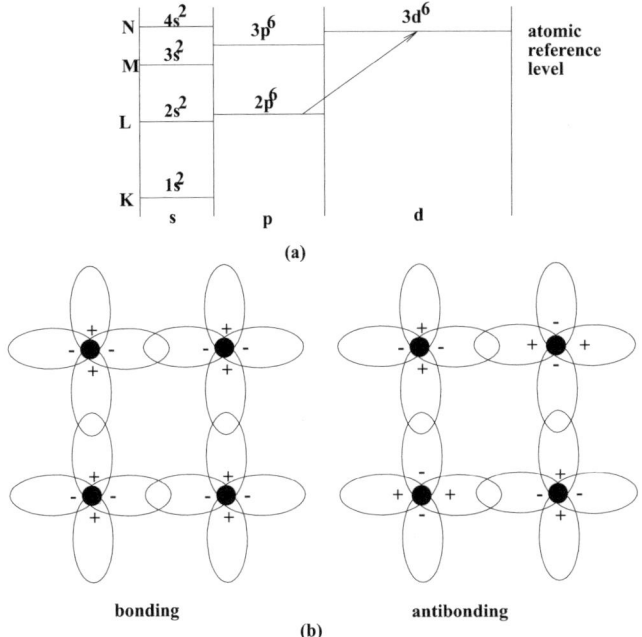

Figure 1.19 (a) The energy levels in an iron atom are filled to the 3d and 4s states which contain six and two electrons respectively. (b) In iron, electrons in the 3d states have wave functions with alternating positive and negative lobes. Lobes of similar sign are attractive and lead to bonding. Those of opposite sign interfere destructively to give antibonding

ions. This lowers the total energy and the configuration is known as 'bonding'. If lobes of opposite sign overlap, they interfere destructively and the resulting configuration is 'antibonding'. Each atomic d state splits into a bonding and antibonding state. The difference in energy between bonding and antibonding states is controlled by the extent to which the wave functions overlap in space. The original atomic level remains roughly unchanged and marks the centre of the array of split levels. Therefore Brown *et al.* (1997) consider maximum bonding occurs when all the bonding states are filled and all the antibonding states are empty. This approach considers the binding energy is proportional to $x(10 - x)$, where x is the number of occupied states (Friedel 1969), so that maximum bonding occurs when $x - 5$, but if the states are totally unoccupied ($x = 0$) or totally full ($x = 10$), then there is no bonding. This simple model gives equal weight to each of the 10 states, but it predicts the cohesive properties of transition metals and their alloys. Since the electron energy loss spectrum of a grain boundary measures the density of unoccupied states, i.e. $(10 - x)$, EELS can be used to predict cohesion at a boundary. Muller (1996) showed that this approach leads to a semi-quantitative measure of the boundary cohesion. If an iron atom in a phosphorus-containing grain boundary has Z

nearest neighbours, then the binding energy of a single iron atom is proportional to $\beta Z^{1/2}x(10 - x)$, where β is the value of the overlap integral that controls the splitting of the state. For a grain boundary in pure iron when broken it becomes an external surface, which changes both the number of nearest neighbours and the electronic binding. According to the experimental results described above, the presence of phosphorus does not appear to affect the surface binding: the phosphorus seems to be chemically unbound, being attached weakly to the surface by Van der Waals forces. However, at the boundary, phosphorus causes the iron to gain about half an electron, over and above the six it had as a pure metal; x has increased from 6 to 6.5, which weakens the boundary. Knowing the boundary energy per iron atom, it has been estimated that the presence of phosphorus in a grain boundary reduces the boundary cohesion by 0.04 eV per atom (about 10 %).

It is important in these calculations to know the bonding of phosphorus, both at the internal boundary and at the free surface that is exposed. Active radicals in the environment, when the sample is fractured, can markedly alter the energy required to propagate an intergranular crack because of changes in the bond energies. Indeed, such environmental effects are known to play an important role in stress corrosion cracking. It is clear that electron theory provides a connection between the experimental observations of electronic structure at a grain boundary and suscept-ibility to fracture.

1.9   STRENGTH OF GRAIN BOUNDARIES

Fracture along grain boundaries has been observed in many different materials spanning metals, alloys and refractory materials. The parameters that are important when analysing the fracture interaction process, which involves separation along interfaces, are the cohesive properties, the work to separate the interface and the maximum force that is necessary to separate a unit area of the interface (Rice and Thomas, 1974). The propensity to intergranular fracture is associated with the chemical composition of the grain boundaries (Lee *et al.*, 1984; Komeda and McMahon, 1981) which, in turn, can dramatically influence properties such as ductility and strength. As a consequence, it is important to be able to explain the processes leading to these changes in cohesion of the grain boundaries, and this has been strengthened by the ability to measure the composition and chemical state of atoms at the grain boundaries and interfaces (see Chapter 4). However, it has been recognised that in the brittle fracture regime, when cleavage fracture predominates, a proportion of intergranular fracture will accompany the cleavage owing simply to geometrical requirements. Certainly, when propagating from one grain to the next, the crack has a choice between either cleavage or brittle intergranular fracture. The mismatch between two misoriented cleavage planes can be accommodated in a number of different ways:

(a) it can propagate on several parallel planes forming a large number of small cleavage steps,
(b) the mismatch can be bridged by ductile tearing or
(c) intergranular cracking may occur in the mismatch region.

The intergranular cracking generated between misoriented cleavage planes in adjacent grains can be regarded as geometrically necessary. Moreover, this type of intergranular cracking has to be distinguished from that arising from the cohesive energy considerations above. As shown by theoretical modelling (Smith *et al.*, 1997), there is a minimum proportion of such geometrically necessary benign intergranular cracks which may vary with the temperature at which the brittle fracture occurs. There are two main approaches that have been adopted to explain grain boundary cohesion, one based on thermodynamic arguments and the other at the atomic and electronic scale invoking modern quantum theory. The thermodynamic approach is attractive for adsorption induced brittle fracture since it considers specifically the effect of interfacial cohesion as a result of segregation of solute impurities to interfaces, and the degree of segregation is determined by the kinetics of the transport process. Here, the essential relationship describing the ideal work of fracture per unit area, ϕ, is given by

$$\phi = 2\gamma_s - \gamma_p \tag{1.22}$$

where γ_s is the energy per unit area of created fracture surface and γ_p is the energy per unit area of the prior grain boundary. This has been the starting point for the underlying theory that has been developed by Seah (1976 and 1980), Rice and Thomas (1974), Hirth (1980), and Asano (1980). Here, for fast low-temperature fracture, γ is assumed not to be an equilibrium value so that for unit area of grain boundary

$$d\gamma_{gb} = VdP - SdT - \sum_i \Gamma_b^i d\mu_i \tag{1.23}$$

where v is the specific volume and S the entropy of the interface region, P is the pressure across the interface, T is the temperature and Γ^i is the quantity of species i with chemical potential μ_i per unit area of interface. For a binary system A–B (where B is the solute) at constant T and P, this gives for the Gibbs–Duhem relationship

$$d\gamma_{gb} = \{[X^B/(1 - X^B)]\Gamma_{gb}^A - \Gamma_{gb}^B\}d\mu^B \tag{1.24}$$

where X^B is the solute molar fraction and, if Henry law is obeyed since $d\mu^B = RT \, d\ln\alpha$, where α is the solute activity then $d\mu^B = RT \, d\ln X^B$. Therefore, for a dilute solution, substituting in equation (1.23) gives

$$\gamma_{gb}^A = \gamma v_{gb}^{A_0} - RT\Gamma_{gb}^B \tag{1.25}$$

and for the surface

$$\gamma_s^A = \gamma_s^{A_0} - RT\Gamma_s^B \tag{1.26}$$

where $\gamma_{gb}^{A_0}$ and $\gamma_s^{A_0}$ are the grain boundary and surface energies for the pure A system. At fracture $\Gamma_s = \frac{1}{2}\Gamma_{gb}$, so from equations (1.22), (1.25) and (1.26)

$$\phi = 2\gamma_s^{A_0} - 2\gamma_{gb}^{A_0} \tag{1.27}$$

for a given material irrespective of the amount of segregation. However, as considered by Hirth (1980), an additional term has to be added to equation (1.27). In the Γ versus μ diagram (Figure 1.20), curve (a) shows the equilibrium excess of B atoms at the grain boundary, Γ_{gb}, as a function of the chemical potential, whereas curve (b) shows the surface excess for the fracture surface, $2\Gamma_s$.

In the case of slow fracture at higher temperatures, if equilibrium is maintained the system starts at position K on curve (a) and moves to position L on curve (b). Hence, the level of segregation increases in fracture so that, at constant μ, equation (1.22) gives

$$\phi = 2\gamma_s^L - \gamma_{gb}^K$$
$$= 2\gamma_s^{A_0} - \gamma_{gb}^{A_0} - RT(2\Gamma_s^L - \Gamma_{gb}^K) \tag{1.28}$$

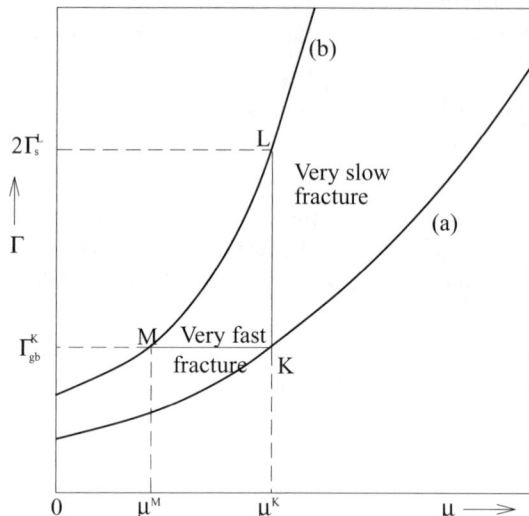

Figure 1.20 Dependence of Γ on μ for solute atoms at sites on (a) the grain boundary and (b) the fracture surfaces (after Seah, 1980)

Therefore, there is a decrease in ϕ as a result of solute segregation to grain boundaries in slow high-temperature fracture. At lower temperatures, fast fracture starts at K and moves to M with the total amount of segregation constant so that

$$
\begin{aligned}
\phi &= 2\gamma_s^M - \gamma_{gb}^K \\
&= 2\gamma_s^{A_0} - \gamma_{gb}^{A_0} - RT(2\Gamma_s^M - \Gamma_{gb}^K) \\
&= 2\gamma_s^{A_0} - \gamma_{gb}^{A_0}(\text{const. } \mu)
\end{aligned}
\tag{1.29}
$$

However, in this change the segregated solute atoms release energy:

$$
\int_{\mu^M}^{\mu^K} \Gamma \, d\mu
$$

so that

$$
\begin{aligned}
\phi &= 2\gamma_s^{A_0} - \gamma_{gb}^{A_0} - \int_{\mu^M}^{\mu^K} (2\Gamma_s^M - \Gamma_{gb}^K) \, d\mu \\
&= 2\gamma_s^{A_0} - \gamma_{gb}^{A_0} - (\mu^K - \mu^M)\Gamma_{gb}
\end{aligned}
\tag{1.30}
$$

This result was first observed by Rice (1976). It shows that the maximum cohesive force for a boundary separating continuously over its entire area is directly related to the cohesive energy but not to the magnitude of the lattice trapping term.

In the case of the atomic or electronic scale models, the first principles electronic structure calculations have been limited because of the complexity of the structure of grain boundaries (section 1.8). As a consequence, it has been more usual to consider polyhedral atomic cluster models where only the segregated atom and a limited number of grain boundary atoms are included in the local environment used in these calculations (Briant and Messmer, 1980, 1982 and 1984). The technique used for calculating the molecular orbitals is the self consistent field $x\alpha$ scattered wave theory. This is a method that has been applied to many local bonding problems such as chemisorption and amorphous metals. Three types of interactions can be obtained for such calculations: the local orbital energy, the one-electron molecular orbital wave function and the valence electron density of the atoms in the cluster. The results of these calculations indicate that a possible mechanism for impurity atom segregation embrittlement is the formation of strong bonds between the impurity atom and the immediate neighbouring atoms. Here, the electrons participating in these bonds come from surrounding metal–metal bonds, causing them to be weakened. By quantifying the strength of the embrittlement in terms of the amount of charge drawn to the impurity atom, Briant and Messmer were able to rank the relative embrittling potencies of certain impurity atoms at grain boundaries in a given parent lattice to describe how the cohesive energy changes. Table 1.1 shows the results for three clusters in α-iron and two clusters in nickel. Here the results for B in an Ni grain boundary show that, in addition to no change in the charge in the metal–metal bonds across the boundary, a covalent-like bond is formed

Table 1.1   Cluster composition for tetrahedral clusters (after Messmer and Briant, 1982)

Host metal	Impurity	Experimentally determined effect of impurities
Fe	S	Strong embrittler
Fe	P	Moderate to weak embrittler
Fe	C	Cohesive enhancer
Ni	S	Strong embrittler
Ni	B	Cohesive enhancer

between the solute atom and the atoms in the metal. However, the impurity atom–parent bonds are favoured at the expense of the metal–metal bonds when S segregates to the grain boundary. Similar results have been obtained by Wang and Zhao (1994). Hashimoto *et al.* (1984b) used an approach based on cluster calculations using atom positions determined from atomic relaxations considered on large grain boundary cells with approximate pair potentials. For the case of B and P in $\Sigma5$ and $\Sigma9$ grain boundaries in α-iron, the results indicate that the grain boundary comprises an alternating structure of strong and weak bonds. The addition of B tends to strengthen the boundary by replacing the weak Fe–Fe bonds with stronger ones arising from the B atoms. The addition of P atoms, on the other hand, acts to rearrange the atoms into clusters, such as Fe_3P, and weaken the boundary by drawing charge from strong bonds responsible for cohesion.

1.10   REFERENCES

Aboav, D. A. (1970) *Metall*, **3**, 383.
Aboav, D. A. (1980) *Metall.* **13**, 43.
Adams, B. L. (1993) *Mater. Sci. Eng.*, **A166**, 59.
Adams, B. L., Wright, S. I. and Kunze, K. (1993) *Metall. Trans. A*, **24A**, 819.
Asano, R. J. (1980) *Phil. Trans. R. Soc. (Lond.) A*, **295**, 151.
Ashby, M. F., Gandhi, C. and Taplin, D. M. R. (1979) *Acta Metall.*, **27**, 699.
Backhaus-Ricoult, M., Peyrot-Chabrol, A. and Hagége, S. (1996) *Mater. Sci. Forum*, **207–209**, 169.
Balluffi, R. W. and Cahn, J. W. (1981) *Acta Metall.*, **29**, 493.
Barrett, C. S. and Massalski, T. B. (1986) *Structure of Metals*, Third edition, New York: McGraw-Hill.
Bollmann, W. (1970) *Crystal Defects and Crystalline Interfaces*, New York: Springer-Verlag.
Boots, B. N. and Murdoch, D. J. (1983) *Computers and Geoscience*, **9**, 351.
Briant, C. L. (1990) *Metall. Trans. A*, **21A**, 2339.
Briant, C. L. and Messmer R. P. (1980) *Phil. Mag.*, **42**, 569.
Briant, C. L. and Messmer, R. P. (1982) *Acta Metall.*, **30**, 1811.
Briant, C. L. and Messmer, R. P. (1984) *Acta Metall.*, **32**, 2043.
Brown, L. M., Allen, G. C. and Flewitt, P. E. J. (1997) *Phys. World*, **45**, May.
Chan, S. W. and Balluffi, R. W. (1986) *Acta Metall.*, **34**, 2191.
Cherns, D., Smith, D. A., Krakow, W. and Batson, P. E. (1982) *Phil. Mag.*, **A45**, 107.
Ciccotti, G., Guillope, M. and Pontikis, V. (1983) *Phys. Rev. B*, **27**, 5576.

Cottrell, A. H. (1989) *Mater. Sci. Technol.*, **5**, 1165.
Cottrell, A. H. (1990) *Mater. Sci. Technol.*, **6**, 974.
Crain, I. K. (1978) *Comput. Geosci.*, **4**, 131.
Dingley, D. J. and Baba-Kishi, K. (1986) *Scanning Electron Microscopy*, Part 2, 383.
Flewitt, P. E. J. and Wild, R. K. (1994) *Physical Methods for Microstructural Characterisation of Materials*, Bristol: Institute of Physics Publishing.
Fradkov, V. E., Magnasco, M. O., Udler, D. and Weaire, D. (1993) *Phil. Mag. Lett.*, **67**, 203.
Friedel, J. (1969) in *The Physics of Metals*, ed. J. M. Ziman, Cambridge: Cambridge University Press.
Gabler, F. (1956) *Mikroscopie*, **11**, 2.
Grovenor, C. M. R., Rae, C. M. F. and Smith, D. A. (1980) in *Grain Boundary Structure and Kinetics*, ed. R. W. Balluffi, Metals Park, OH: ASM, p. 331.
Hallam, K. R. and Wild, R. K. (1995) *Surf. Interface Analysis*, **23**, 133.
Harmann, G., Gleiter, H. and Bäro, G. (1976) *Acta Metall.*, **24**, 353.
Hashimoto, M., Ishida, Y., Yamamoto, R. and Doyama, M. (1984a) *Acta. Metall.*, **32**, 1.
Hashimoto, M., Ishida, Y., Wakayama, S., Yamamoto, R., Doyama, M. and Fujiwara, T. (1984b) *Acta Metall.*, **32**, 13.
Hillert, M. (1983) *Scr. Metall.*, **17**, 237.
Hillert, M. and Purdy, G. (1978) *Acta Metall.*, **26**, 533.
Hirth, J. P. (1980) *Phil. Trans. R. Soc. (Lond.) A*, **295**, 139.
Hirth, J. P. and Lothe, J. (1982) *Theory of Dislocations*, second edition, New York: Wiley.
Hsieh, T. E. and Balluffi, R. W. (1989) *Acta Metall.*, **37**, 1637.
Iwasa, M. and Bradt, R. C. (1984) in *Structure and Properties of MgO and Al_2O_3 Ceramics*, ed. W. D. Kingery, Westerville, OH: Am. Ceram. Soc., p. 767.
Johnson, K. H., Yang, C. Y., Vvedensky, D., Messmer, R. P. and Salahub, D. R. (1977) *Proc. of Seminar of Materials Sci. Div. of Am. Soc. for Metals*, eds W. C. Johnson and J. M. Blakely, October 1977, p. 25.
Kalonji, G. and Cahn, J. W. (1982) *J. Phys. (France)*, **43(Cb)**, 25.
Kelly, A. and Groves, G. W. (1970) *Crystallography and Crystal Defects*, London: Longmans.
Lord Kelvin (1887) *Phil. Mag.*, **24**, 503.
King, A. H. and Smith, D. A. (1981) *Radiat. Effects*, **54**, 169.
Komeda, P. J. and McMahon, C. J. (1981) *Metall. Trans. A*, **12A**, 31.
Kumar, S., Kurtz, S. K. and Weaire, D. (1994) *Phil. Mag.*, **69**, 431.
Kusner, R. (1992) *Proc. Roy. Soc.*, **439**, 683.
Laird, C. and Aaronson, H. I. (1967) *Acta Metall.*, **15**, 73.
Lee, D. Y., Barrera, E. V., Stark, W. P. and Marcus, H. L. (1984) *Metall. Trans. A*, **15A**, 1415.
Lee, H. Y., Kang, S. J. L. and Yoon, D. Y. (1993) *Acta Metall. Mater.*, **41**, 2497.
Lomer, W. M. and Nye, J. F. (1952) *Proc. R. Soc.*, **212**, 576.
Ma, C. Y., Gust, W., Fontnelle, R. A. and Predel, B. (1993) *Mater. Sci. Forum*, **126–128**, 317.
Messmer, R. P. (1977) *Modern Theoretical Chemistry*, Vol. 8, ed. G. A. Segal, Plenum Press, New York.
Messmer, R. P. and Briant, C. L. (1982) *Acta Metall.*, **30**, 457.
Muller, D. A. (1996) PhD thesis, Cornell University.
Ozkaya, D., Yuan, J., Brown, L. M. and Flewitt, P. E. J. (1995) *J. Microscopy*, **180**, 300.
Rae, C. M. F. (1981) *Phil. Mag.*, **44A**, 1395.
Read Jr, W. T. (1953) *Dislocations in Crystals*, New York: McGraw-Hill.
Rice, J. R. (1976) *Effect of Hydrogen on the Behaviour of Materials*, eds A. W. Thompson and I. M. Rennstein, Norfolk: AIME, p. 455.
Rice, J. R. and Thomas, R. (1974) *Phil. Mag.*, **29**, 73.
Rühle, M., Evans, A. G., Ashby, M. F. and Hirth, J. P. (1990) *Metal–Ceramic Interfaces*, Oxford: Pergamon Press.
Seah, M. P. (1976) *Proc. R. Soc. (Lond.) A*, **349**, 535.

Seah, M. P. (1980) *Acta Metall.*, **28**, 955.

Smith, C. S. (1948) *Trans. AIME*, **175**, 15.

Smith, C. S. (1952) in *Metal Interfaces*, Cleveland: Am. Soc. for Metals, p. 65.

Smith, C. S. (1954) *Trans. Chalmers Univ. Technol.*, Gothenburg, Sweden, Vol. 152.

Smith, D. A. and King, A. H. (1981) *Phil. Mag.*, **A44**, 333.

Smith, G. E., Crocker, A. G. and Flewitt, P. E. J. (1997) *Damage and Failure of Interfaces*, ed. P. Rossmanith, Rotterdam: Balema Press. p. 229.

Tang, S., Freeman, A. J. and Olson, G. B. (1993) *Phys. Rev. B*, **47**, 2441.

Tang, S., Freeman, A. J. and Olson, G. B. (1994) *Phys. Rev. B*, **50**, 1.

Telley, H., Liebling, D. M. and Mocellin, A. (1996) *Phil. Mag. B*, **73**, 395.

Thorvaldsen, A. (1992) *Mater. Sci. Forum*, **94–96**, 307.

Thorvaldsen, A. (1993) *J. Appl. Phys.*, **73**, 7831.

Venables, J. A. and Harland, C. J. (1973) *Phil. Mag.*, **27**, 1193.

Wang, C. Y. and Zhao, D. L. (1994) *Mater. Res. Soc. Symp. Proc.*, **318**, 571.

Weaire, D. (1974) *Metall.*, **7**, 157.

Weaire, D. (1994) *Phil. Mag.*, **69**, 99.

Weaire, D. and Glazier, J. A. (1993) *Phil. Mag.*, **68**, 363.

Weaire, D. and Phelan, R. (1994) *Phil. Mag.*, **69**, 107.

Weertman, J. and Weertman, J. R. (1964) *Elementary Dislocation Theory*, New York: Macmillan.

Wu, R., Freeman. A. J. and Olson, G. B. (1993) *Phys. Rev. B*, **47**, 6855.

Yang, W., Chen, L.-Q., and Messing, G. L. (1995) *Mater. Sci. Eng.*, **A195**, 179.

Chapter 2
Grain Boundary Composition

2.1 INTRODUCTION

Grain boundaries in materials are regions of transition in between adjacent, but similar, crystals and as such they are distinct from the bulk crystal both in structure and chemical composition. The interphase boundary is an extension of this since it separates crystals of different structure and composition. The classical thermodynamics of both fluid and solid surfaces developed by Gibbs (1957) over a hundred years ago considers a range of topics including adsorption, thermal, mechanical and surface effects, external surfaces, homogeneous and heterogeneous nucleation and internal surfaces. The latter case represents a fundamental difference between solid and liquid surfaces. Moreover, it is recognised that alloying and impurity elements are often redistributed within a solid material to external surfaces, internal surfaces, such as grain boundaries, special boundaries, including twins or stacking faults, and interphase boundaries (Joshi, 1978). There are several possible driving forces for elemental segregation to these surfaces, or interfaces, one simply described by Gibbs (1957) as adsorption which results in a reduction in the surface, grain boundary or interfacial free energy. In this chapter we examine the theories based upon equilibrium thermodynamics and consider how solute and impurity element equilibrium segregation to surfaces, grain boundaries or interphase boundaries can be quantitatively estimated in a solid material. This approach is extended to non-equilibrium segregation which occurs with a different driving force. The driving forces for thermally induced non-equilibrium segregation arise when solute or impurity atoms interact with vacancies where the thermal regime is insufficient to achieve equilibrium. However, in addition, there are potential driving forces where excess interstitial atoms are created by non-equilibrium processes such as neutron irradiation and external stress. The reader is recommended to consider the reviews of Seah and Hondros (1973 and 1977) Briant and Banerji (1978), Westbrook (1969), Cabané and Cabané (1991) and Lejcek and Hofmann (1995).

To understand the origin of segregation it is important to recognise that for an atomic description of a condensed material it is necessary to know the mean environment associated with each atom, the interatomic spacing and the number and nature of the next nearest and at least second nearest neighbour atoms. At a given temperature, equilibrium is achieved when the level of arrangement of the atoms is such that for a solid solution a compromise is reached to give a random distribution of the various atom types on the lattice sites. Hence, a particular atom will reside at a site where the neighbouring atom environment is energetically favoured. However, when an atom is close to a surface, a grain boundary or an interface, the local atomic environment cannot be optimised in the same way as in a bulk crystal. The formation of lattice defects produces local modifications to the crystal structure, including revised interatomic distances, which lead to local modifications to the composition. It is this modification to the composition that enables segregation of a given atom type within a polycrystalline material and, as such, it is a necessary requirement to achieve equilibrium at the grain boundaries and, indeed, in the total system.

Hence, the driving force for equilibrium segregation of elements to surfaces, grain boundaries or interfaces in a material is the minimisation of the interfacial free energy. This phenomenon is observed, for example, in ferritic steels where changes in grain boundary composition arise from the equilibrium segregation of Group IV to VI impurity atoms and metallic alloying elements, Figure 3.6, to the prior austenite grain boundaries brought about by heat treatment within a critical temperature range (see Chapter 3). Alternatively, non-equilibrium segregations to grain boundaries have been observed in a range of binary and complex alloy systems following cooling below a critical rate. In general, non-equilibrium solute or impurity atom segregation arises from diffusion to, and decomposition at, grain boundaries of vacancy–atom complexes. Under these circumstances, changes in the equilibrium vacancy concentration with temperature provide the driving force for diffusion to the grain boundary sinks, so that the extent of the segregation depends upon the excess vacancy concentration and the strength of the vacancy–atom bond. However, in addition, other non-equilibrium processes such as neutron irradiation can induce crystal lattice defects that can lead to the diffusion of interstitial–atom complexes to grain boundaries, producing non-equilibrium segregation. The extent of both equilibrium and non-equilibrium segregation to grain boundaries can be either enhanced or retarded by cosegregation with other impurity or alloying elements. These processes arise when segregating elements have interacted significantly with each other, but unfortunately the exact role of these interactions is not well understood or is often incorrectly measured. As a consequence, evidence in the literature is often conflicting with respect to the conditions that lead to segregation of specific alloying or impurity elements in more complex alloy systems such as those used commercially.

2.2   EQUILIBRIUM SEGREGATION

Three main approaches have been developed to describe equilibrium segregation to surfaces, grain boundaries and interfaces. The first is based upon the change in interfacial free energy with the bulk composition known as the Gibbs adsorption isotherm (Gibbs, 1957). The second accounts for the distribution of individual components between the boundary and the bulk and is described by the thermo-dynamics of a regular system. Deviations from the regular behaviour are considered as modifications to the segregation free energy, whereby the concentration dependent excess segregation reflects mutual interactions in the system for both the grain boundary and the bulk. The third is based on the reaction equilibrium concept where the free energy of the surface, grain boundary or interface is considered to be independent of concentration, and this affects ternary additions and their contribu-tion to the activity of the individual component atoms.

2.2.1 Thermodynamics of an Ideal Binary System

For macroscopic systems subject to a uniform pressure and containing a planar boundary or interface, it is difficult to establish a model that can be used to describe both experimental observations and theoretical calculations. Figure 2.1a shows a model that describes an interface where composition is not homogeneous in all directions. When equilibrium is achieved it is assumed that composition is constant in a direction parallel to the interface, I, and the interface has no thickness. This approach defines the dividing surface used by Gibbs (1957), which is given in Figure 2.1b.

Segregation to a surface can be described by formal Gibbsian thermodynamic theory based upon the change in the free energy at an interface in a multicomponent system at constant volume and temperature (Gibbs, 1957). Using the adsorption theorem for a binary alloy system containing A and B type atoms, the surface energy, γ, is given by

$$d\gamma = -H_S\,dT - C_A d\mu_A - C_B\,d\mu_B \qquad (2.1)$$

where H_S is the specific surface excess enthalpy, C_A and C_B are the surface excess concentrations of atoms A and B and μ_A and μ_B are the corresponding chemical potentials, and T is the temperature. Equation (2.1) provides a relationship between the surface composition given indirectly by the excess surface concentration and the bulk composition expressed by the chemical potentials and the temperature. Unfortunately, equation (2.1) is difficult to use because it is necessary to know the surface energy for the particular alloy system and the variation with temperature and bulk composition.

McLean (1957) developed a classical model to describe either solute or impurity atom segregation to grain boundaries in binary alloy systems using the adsorption analogue. The model adopted considers a grain boundary or interface and a perturbed zone about this (Figure 2.1c). The grain boundary is represented by a zone in the material where the elastic distortion of the atomic sites more readily accommodates solute or impurity atoms compared with the adjacent bulk lattice. This model considers P solute atoms distributed randomly within N lattice sites and

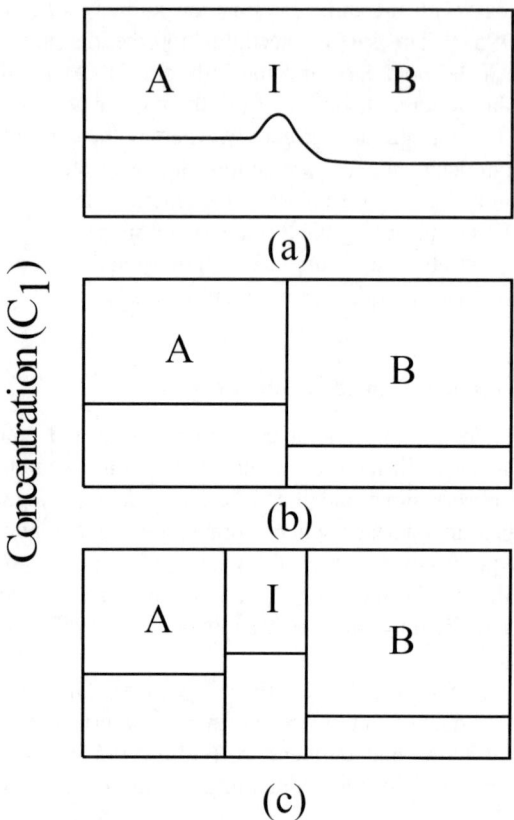

Figure 2.1 Schematic diagrams showing interface models for the change in concentration of components A and B along the *X* axis which is perpendicular to the interface I. (a) represents a real distribution, (b) is that assumed in the Gibbs thermodynamic model, while (c) is that assumed in subsequent models such as that adopted by McLean (1957)

p solute atoms distributed within n grain boundary sites where each have a free energy e. The total free energy, G, due to the presence of solute atoms is given by

$$G = pe + Pe_L - kT \ln \omega \tag{2.2}$$

where e_L is the interaction energy of the solute atom within the lattice, k is the Boltzmann constant, T is the temperature and ω is the atomic interaction energy. Thus, the term $kT \ln \omega$ is the configurational enthalpy which results from the difference in the arrangement of solute atoms in the bulk compared with the grain boundary. The arrangement most likely to minimise the free energy of the system is given when

$$\frac{p}{n-p} = \frac{P}{N-P} \exp\left(\frac{e_L - e}{kT}\right) \tag{2.3}$$

This can be expressed simply in a notation where concentrations, X, are mole fractions such that

$$\frac{X_b}{X_{b0} - X_b} = \frac{X_c}{1 - X_c} \exp\left(-\frac{\gamma_1}{RT}\right) \tag{2.4}$$

where X_c is the bulk concentration, X_b is the grain boundary concentration, X_{b0} is the saturation concentration of X_b, R is the gas constant and $\gamma_1 [= (e_L - e)N_A]$ is the grain boundary adsorption energy, where N_A is the Avogadro number. If the grain boundary is covered by a fraction of B atoms, θ_B, such that $\theta_B = X_b/X_{b0}$, then for a dilute alloy this reduces to

$$\theta_B = \frac{KX_c}{1 + KX_c} \tag{2.5}$$

where $K = (\gamma_1/RT)$. This has the same form as the saturation adsorption equation developed earlier by Langmuir (1918) to describe adsorption behaviour for solid and gaseous systems. As a consequence, equation (2.4) is referred to as the Langmuir–McLean segregation equation. Although this equation has the form of the Langmuir equation for adsorption at the free surface of a solid, there are fundamental differences that have to be recognised. For example, the environmental constraints on atoms at grain boundaries are different from free surfaces which have an empty half space and as such the coordination number for the atoms will be significantly different compared with a grain boundary which differs from, but approximates more closely to, a bulk crystal.

Equation (2.4) predicts that segregation of a second atom species to grain boundaries in a polycrystalline material will increase as the solute content is raised and the temperature is decreased. Moreover, since the model assumes a given number of adsorption sites at a grain boundary, the segregation will approach a saturation concentration, X_{b0}; for a typical binary system this varies from about half to complete monolayer coverage. At lower temperatures it is difficult for the segregating atom to diffuse to the grain boundary, despite the fact that the driving force is large. As a consequence, the kinetics of the overall process is important and McLean addressed this to give the time dependency for this segregation process. Unfortunately, the predictions from the model, equation (2.4), do not completely agree with the experimental observation with respect to the change in grain boundary concentration with temperature; it does not decrease as rapidly with increasing temperature. Certainly, a single value for the grain boundary adsorption energy, γ_1, would not be expected to account for the interaction of solute atoms with all potential occupancy sites at a grain boundary. For example, for a simple tilt grain boundary (Figure 2.2), using only the concept of misfit, solute atoms will have a

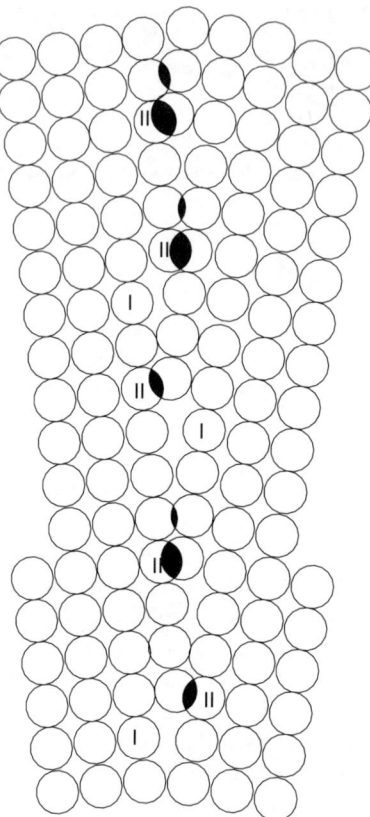

Figure 2.2   Schematic diagram of an unrelaxed tilt boundary. Sites labelled I are located in regions of hydrostatic tension and will attract oversize solute atoms, while type II sites are located in regions of compression and will repel oversize solute atoms. Between these two extremes there is a spectrum of sites with a range of interaction energies for oversize solute atoms (after White and Coghlan, 1977)

larger boundary energy for sites I than for sites II and III. Furthermore, locations in the region of site II, where the crystal lattice is subject to hydrostatic compression, will have a negative boundary energy and therefore will tend to reject solute atoms. Thus, a solute or impurity atom will not have the same boundary energy for each site on a given grain boundary and indeed will vary with boundary orientation.

Seah and Hondros (1977) applied the multilayer gas adsorption theory of Brunauer, Emmett and Teller (BET) (1938) to overcome some of these limitations of the Langmuir–McLean model. They invoked the grain boundary analogue of the truncated isotherm derived by Brunauer *et al.* (1940). This statistical mechanics approach assumes the constraints of the Langmuir–McLean model by considering an array of identical absorption sites with no interaction between the absorbed atoms. Thus

$$\frac{X_{gb}}{X_{gb0} - X_{gb}} = \frac{X_c}{X_{c0}} \exp\left(-\frac{\gamma}{RT}\right) \tag{2.6}$$

where X_{c0} is the solute atom concentration at the limit of solid solubility. For atoms with a low solubility:

$$X_{c0} = \exp\left(-\frac{\gamma_L}{RT}\right) \tag{2.7}$$

where γ_L is the free energy of the solution. Thus, equation (2.6) becomes

$$\frac{X_{gb}}{X_{gb0} - X_{gb}} = \frac{X_c}{X_{c0}} \exp\left(-\frac{\gamma_L}{RT}\right) \tag{2.8}$$

Equation (2.8) has the same form as the Langmuir–McLean equation (2.4), derived for the dilute approximation. Here $\gamma = \gamma_1 - \gamma_L$, which is the difference in free energy of the first and subsequent condensing layers. For a system containing highly active surface species, such as S and Sn in α-Fe (Seah and Hondros, 1973), γ_1 will be just greater than γ_L. Hence, the temperature dependence of sulphur segregation conforms to the truncated BET formulation for monolayer segregation. Hondros and Seah (1977a) used this approach to explain their results for grain boundary segregation in a series of iron- and nickel-base alloy systems. Multilayer segregation has also been detected in other systems such as Te in Fe, Sn in Fe, P in W and Bi in Cu, as summarised by Balluffi (1979). As temperature is reduced, the amount of segregation of S in α-iron increases rapidly, with γ_1 being only a few kJ mol^{-1} greater than γ_L. However, when the temperature reaches a value where the bulk sulphur concentration exceeds the equilibrium solubility limit, there is a rapid change in the segregation behaviour due to precipitation. Thus

$$\frac{X_{gb}}{X_{gb0} - X_{gb}} = \exp\left(-\frac{\gamma}{RT}\right) \tag{2.9}$$

Since $X_c = X_{c0}$ at temperatures below that at which precipites form, segregation will continue to rise with decreasing temperature, but more slowly, towards the saturation limit X_{gb0} of about one monolayer.

Unfortunately, this approach still neglects any structural change at the grain boundary and the chemical interaction between atoms. The zero order quasi-chemical approximation or regular solution model, which assumes a constant boundary energy between nearest neighbour atom pairs, is accommodated in the Fowler and Guggenheim isotherm (1939). In this approach, the grain boundary adsorption analogue includes an additional term to describe the free energy of adsorption when compared with the Langmuir–McLean model, such that

$$\frac{X_{gb}}{X_{gb0} - X_{gb}} = X_c \exp\left(-\frac{\gamma_1 - Z_C \gamma_A \frac{X_b}{X_{b0}}}{RT}\right) \tag{2.10}$$

where Z_C is the coordination number for adsorbed atoms and γ_A is the interaction energy between neighbouring atoms. The free energy term depends upon both the probability that a neighbouring atom site is occupied and the value of the interaction energy. During the early stages of atom adsorption, X_{gb} approaches zero and equation (2.10) reduces to the Langmuir–McLean form, equation (2.4). Moreover, if γ_A is zero so that there is no interaction between atoms, then equation (2.10) will also approximate to equation (2.4), whereas if γ_A is either positive or negative, the additional atoms are either mutually repulsed or attracted respectively. Attractive interactions increase the total adsorption energy term in equation (2.10) so that the adsorption isotherm tends towards a steeper version of the Langmuir–McLean isotherm. As the temperature falls, γ_A becomes more negative and segregation to the grain boundaries rises more sharply than predicted by equation (2.4). This transition behaviour is shown in Figure 2.3a, together with measurements for Se and Te in α-iron. For $Z_C\gamma_A < -4RT$ the curves become S-shaped, whereas for the final curve the segregation does not follow this form but moves discontinuously from A to B as a result of the difference in energy between high and low atomic coverage of the grain boundary (Seah, 1980).

Since there is no obvious reason why grain boundary segregation should conform to either the saturation coverage or the Langmuir–McLean adsorption type model, Seah and Hondros (1977) examined the differences by considering experimental observations of segregation of tin to grain boundaries in α-Fe. In this system, depending upon the bulk composition and the solid solubility limit, the Sn concentration at a grain boundary can increase to several equivalent monatomic layers (Figure 2.3b). To explain these results, Seah and Hondros (1977) invoked a number of classical multilayer gas adsorption theories and showed that the BET theory fits the observations. The difference between multilayer free surface adsorption and multilayer grain boundary segregation is that, in the case of the free surface adsorption, segregating atoms coexist with vacant sites, whereas segregated atoms mix with matrix atoms at the grain boundary. In the analogue of the BET formulation, the pressure, P, and the quantity adsorbed, v, are replaced by X_c and X_{gb} respectively, and the solute concentration for precipitation, X_{c0}, is introduced to give the grain boundary analogue:

$$\frac{X_{gb0}}{X_{gb}}\left(\frac{X_c}{X_{c0} - X_c}\right) = \frac{1}{K'} + \frac{K' - 1}{K'}\left(\frac{X_c}{X_{c0}}\right) \tag{2.11}$$

where $K' = \exp(-\gamma/RT)$ and X_{gb0} is the coverage by one monolayer. Thus, atoms that have a free energy of adsorption γ_1 in the first layer and γ_L in successive layers give $\gamma = \gamma_1 - \gamma_L$. No lateral interactions in the layer are accommodated in the model, and each layer can be incomplete before the next has started. Applying this approach to the data of Figure 2.3, Hondros and Seah (1977b) confirmed the analogue form for BET behaviour by plotting (X_c/X_{c0}) and $[X_c/X_{gb}(X_{c0} - X_c)]$, to show good correlation, thereby proving that Sn segregation in α-Fe follows this theory. Although the BET theory was first applied in this analogue form, it has been

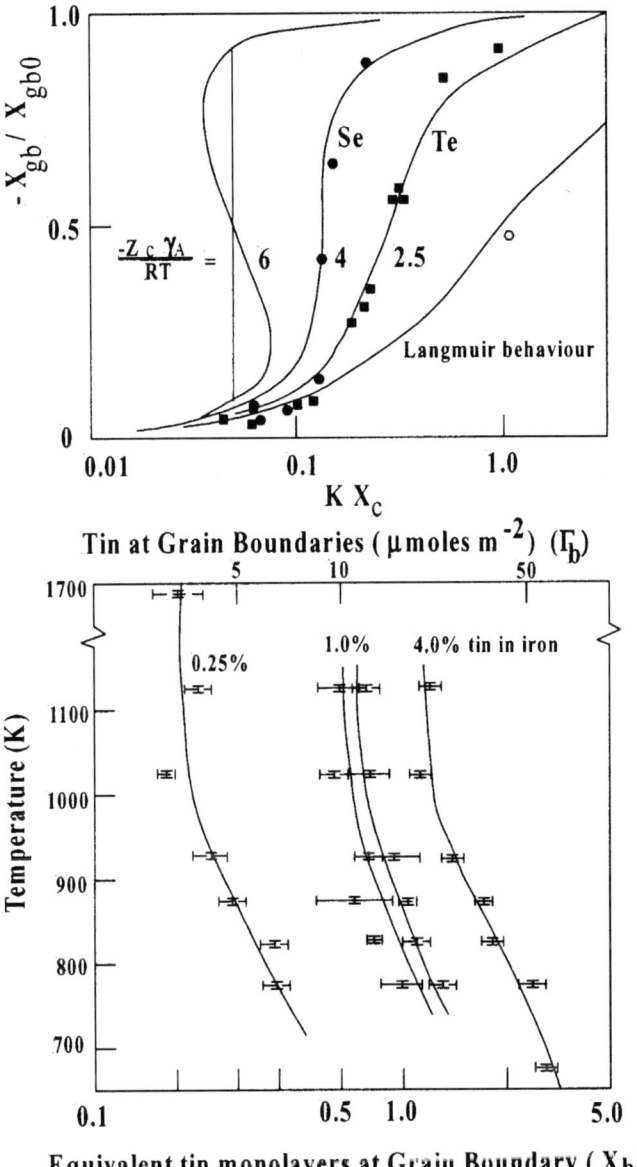

Figure 2.3 (a) Grain boundary Fowler-Guggenheim isotherms with the experimental results for Se and Te in iron (after Hondros and Seah, 1977a). (b) The isothermal segregation of tin in iron measured by Auger electron spectroscopy (after Seah and Hondros, 1973)

derived subsequently using the Langmuir–McLean approach with the assumptions of the original BET model.

At dilute levels, the truncated BET and complete BET theory are very similar, giving

$$\frac{X_{gb}}{X_{gb0}} = \frac{X_c}{X_{c0}} \exp\left(-\frac{\gamma}{RT}\right) \tag{2.12}$$

At low grain boundary or interfacial enrichment values, when $X_{gb} \ll X_{gb0}$, the truncated BET model given by equation (2.8), can be written as an enrichment ratio, β, such that

$$\beta = K^1 X_{c0} \tag{2.13}$$

where $K^1[= \exp(\gamma_1/RT)]$ has values in the range 1.8–10.8 for a wide variety of binary systems. Using equation (2.13), the grain boundary enrichment ratio for solute atoms B in a matrix A can be predicted to within an order of magnitude using a knowledge of the bulk solubility in the matrix. This led Seah and Hondros (1973)

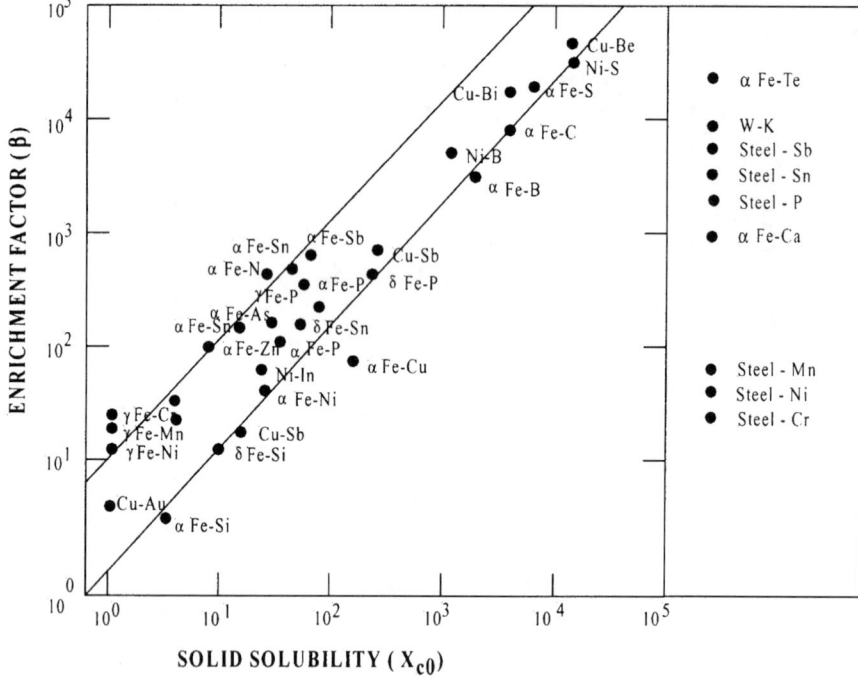

Figure 2.4 Correlation between measured grain boundary enrichment ratio β, and solid solubility X_{co} (after Seah and Hondros, 1973). The solid solubility of the solute atoms in the systems given on the right hand side of the figure is unknown (after Lejcek and Hofmann, 1995)

to plot Figure 2.4 which shows the measured grain boundary enrichment ratio, β, correlated with the solid solubility, X_{c0}, for a range of iron-, nickel- and copper-base binary alloys. On the right hand side of this figure are shown systems where the solid solubility of solutes is unknown (Lejcek and Hofmann, 1995). The values for the parameter K^1 for a given binary system depend upon temperature, the bulk concentration of the segregant and also the presence of a third element. Certainly, K^1 decreases with increasing temperature and bulk composition owing to changes in activity arising from increasing interactions between the segregating atoms and the need for these atoms to occupy sites at the grain boundaries which are of a higher energy (Hofmann, 1987; Seah, 1976). Irrespective of these limitations, this approach allows the relative contribution to segregation to be determined by making use of the appropriate binary phase diagram to establish an enrichment factor to identify elements that would segregate to a surface, grain boundary or interface. However, the value of the thermodynamic parameters for grain boundary segregation will be modified if a first order phase transformation occurs in the presence of a first-order phase transformation in the matrix. Similar changes can also arise for second-order phase transitions, for example changes in the segregation free energy for some impurity elements have been observed near to the Curie temperature in ferromagnetic materials (Hofmann, 1990).

As an alternative, we can consider the equilibrium developed between free solute atoms and grain boundary sites when atoms are adsorbed during segregation. This approach is similar to that adopted by Graham (1953) to characterise adsorption of gases on to a solid surface. For this condition to be reached, there is an adsorption equilibrium between free solute atoms in the bulk vacant boundary sites and occupied sites. The equilibrium constant for this reaction, K, is given by

$$K = \exp -\frac{\gamma_1}{RT} = \frac{a_{xgb}}{a_x a_{gb}} \tag{2.14}$$

where a_x is the activity of solute atoms, a_{gb} are available grain boundary sites and a_{xgb} is the activity of sites occupied by solute atoms. For dilute solutions, the activity coefficients are unity and all sites are identical. By setting a_{xgb} equal to a_{gb}, the equilibrium constant reduces equation (2.14) to

$$K = \frac{\theta_{gb}}{X_c(1 - \theta_{gb})} \tag{2.15}$$

In the case of a dilute system this is equivalent to the Langmuir–McLean segregation equation (2.5):

$$\theta_b = \frac{KX_c}{1 + KX_c} \tag{2.16}$$

The benefit associated with this approach lies in the fact that, if Langmuir–McLean adsorption is obeyed, the equilibrium coverage K^1 should be invariant. As argued by Graham (1953) for the case of free surface adsorption, K varies with θ_{gb} so that the departure from the ideal case provides information on the adsorption process. Briant (1990 and 1991) has used a similar approach that gives a phenomenological description of grain boundary segregation. Certainly, changes in bulk activities and solid solubility are responsible for the segregation of a particular atom type to a grain boundary.

Cahn and Hilliard (1959) derived the maximum excess solute atom concentration at a grain boundary C_{gb}. The approach adopted introduced the concept of an interfacial layer, but using the Gibbs adsorption theorem at constant pressure and temperature

$$-d\sigma = C_1 \, d\mu_1 + C_2 \, d\mu_2$$

combined with the Gibbs–Duhem equation

$$X_1 \, d\mu_1 + X_2 \, d\mu_2 = 0 \tag{2.17}$$

where γ is the surface energy, C is the total excess of solute atoms per unit area μ_1 and μ_2 are the chemical potential of components 1 and 2 respectively and X_1 and X_2 are the atomic fractions of components 1 and 2. If these two equations are differentiated with respect to the concentration, eliminating $d\mu_1/dx$ and using $X_2 = (1 - X_1) - X$, then

$$\left. \begin{array}{l} -\dfrac{d\gamma}{dx} = C_{gb} \dfrac{d\mu_1}{dx} \\[2ex] C_{gb} = C_2 - \dfrac{X}{1-X} C_1 \end{array} \right\} \tag{2.18}$$

The measured quantity of segregation at a given atomic fraction of solute atoms X_0 is denoted by C_{gb} and the upper limit C_{gb}^0 is estimated by Cahn (1977) to be

$$C_{gb}^0 = \gamma_1 \{ kT[1 + \ln(X_e/X_0)] \} \tag{2.19}$$

where γ_1 is the surface energy of component 1 and X_e is the equilibrium solubility limit of component 2 via solution 1. For phosphorus in α-Fe at a temperature of 1273 K, Cahn and Hilliard (1959) found that $C_{P(Fe)}^0 = 0.19 \times 10^{16}$ atoms cm^{-2}, assuming that $X_0 = 0.002$, $X_e = 0.02$ and $\gamma_{Fe} = 850$ mJ m^{-2}. This corresponds to grain boundary coverage of phosphorus for a layer 0.3 nm thick. This model places no limit on to the concentration profile normal to the grain boundary plane, and hence solute atoms may extend for some distance about the grain boundary. For a single element segregating to a grain boundary, the model provides a correct order of magnitude where the upper limit corresponds to the surface energy for the segregated atoms approaching zero. However, the surface energy cannot decrease below ~ 400 J m^{-2} and, in addition, this model applies only to binary alloy systems.

2.2.2 Thermodynamics of a Multicomponent System

The high concentration of segregated species at grain boundaries, combined with the fact that, in practice, alloy systems usually contain many elements, means that the chemical interaction between atoms cannot be accommodated by the simple Langmuir–McLean model. It is necessary to have a theory that allows for interaction between all alloying elements present that can potentially segregate. Examples are Cr and Ni in a steel and impurity elements such as P, Sn, Sb, As, etc., which cannot be incorporated into the term X_{c0} as above. Using the approach described by McLean (1957) to develop the Langmuir–McLean equation (2.4) and the various derivatives described, Guttman (1980) proposed a regular ternary solution model. In this model there is a preferred interaction between the impurity and major alloying atoms present in the particular system. For a regular ternary system, such as α-Fe containing both alloying elements and impurity atoms, segregation can be described by

$$\frac{X_{bi}}{X_{bi}^0} = X_{\mu i} \exp\left(\frac{\Delta G_i}{RT}\right) \bigg/ 1 + \sum_{j=1}^{2} X_{\mu i}\left[\exp\left(\frac{\Delta G_i}{RT}\right) - 1\right] \tag{2.20}$$

where the free energy terms ΔG_i are given by

$$\left.\begin{aligned}\Delta G_1 &= \Delta G_1^0 + \alpha_{12} X_{gb2} \\ \Delta G_2 &= \Delta G_2^0 + \alpha_{12} X_{gb1}\end{aligned}\right\} \tag{2.21}$$

and X_{gb1} and X_{gb2} are the grain boundary molar fractions of monolayer segregation for the impurity and alloying elements, X_{bi}^0 is the saturation value, X_{c1} and X_{c2} are the respective bulk molar concentrations and ΔG_1^0, and ΔG_2^0 are the free energies of the impurity and alloy elements in the respective binary alloy systems. Moreover, $\alpha_{12}' = \alpha_{12} - \alpha_{10} - \alpha_{20}$, where the α_{ij} are the interaction coefficients $[= -\Delta H^0(/N_i/N_j)$ of a regular solution, ΔH_{ij}^0 are the enthalpies (gram atom) and N_i and N_j are the mole fractions of alloying and impurity elements] are the changes in nearest neighbour molar bond energies for the relevant impurity/alloy atoms. As a consequence, equation (2.20) allows both competitive and non-competitive segregation to an interface or a grain boundary to be modelled for the alloying and impurity elements. For a surface active impurity element, the segregation can be enhanced by the presence of an alloying element if $\alpha_{ij} > 0$, (Figure 2.5a). In the competitive case this is valid at the higher temperatures (Figure 2.5b). Similarly, as the concentration of an alloying element is increased, in Figure 2.6 there is a corresponding increase in the segregation of both elements at a given temperature. This model demonstrates that, with preferential attraction between specific atoms in an alloy system, synergistic equilibrium cosegregation of these atoms to interfaces or grain boundaries can occur.

2.2.3 Thermodynamic Parameters for Segregation

The phenomenological theories described above require a knowledge of the segregation and interaction parameters if they are to be used quantitatively for comparison with experimentally measured values. In particular, it is important to have values for the enthalpy, ΔH_1^0, and the entropy, ΔS_1^0, and methods have been developed to derive values of these parameters from a knowledge of atomic bonding and structural factors at interfaces or grain boundaries as described in Sections 1.6 to

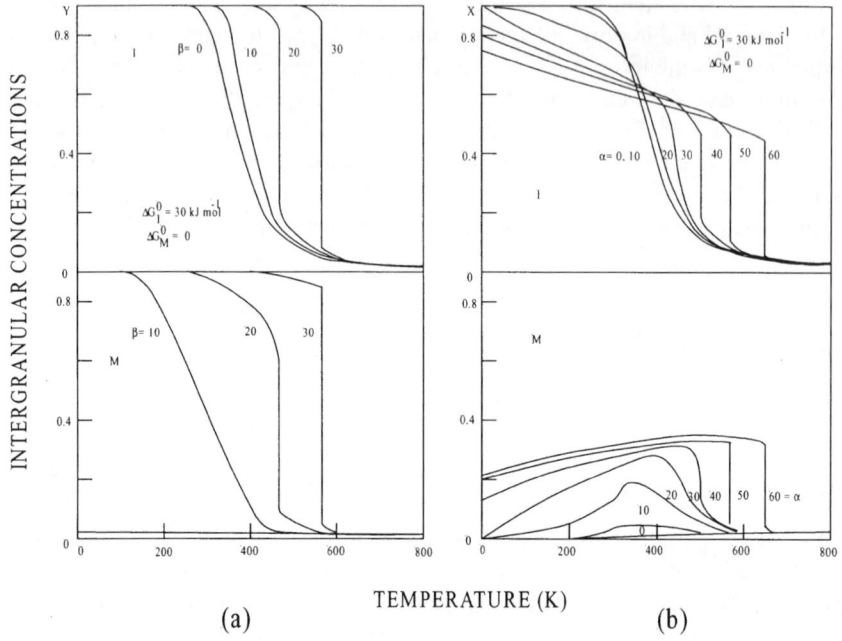

Figure 2.5 Calculated temperature dependence of grain boundary concentrations of solutes I (impurity) and M (alloy) in regular ternary solutions for the indicated values of the interaction coefficients α and β (kJ mol^{-1}): (a) no competition between I and M. (b) competition between I and M. $X_I^B = 0.01$ %, $X_M^B = 2$ % (after Guttmann, 1980)

1.8. We will now briefly consider the underlying methods relevant to segregation, and in particular, to grain boundaries. However, it has to be recognised that the majority of the theories predict thermodynamic parameters of segregation for solute to free surfaces, but the similarity between these and grain boundaries means that the same principles can be applied (Lejcek and Hofmann, 1995).

The approach based upon the bond breaking model developed by Williams and Nason (1974) considers heats of formation and mixing to predict the enthalpy and the entropy for interfacial segregation (Wynblatt and Ku, 1977 and 1979). It includes the chemical contribution at the interface ΔH_I^c, and the elastic strain energy relaxation, ΔH_I^e; the latter can be changed by relaxation of the electronic density of states (Medema, 1978). The enthalpy for interfacial segregation, ΔH_I^0, is given by

$$\Delta H_I^0 = \Delta H_I^c + \Delta H_I^e \tag{2.22}$$

where ΔH_I^c is given by the energy change that occurs if an atom, M, is replaced by an atom, I, that segregates to the boundary. For a regular solution

$$\Delta H_I^c = (\gamma_I - \gamma_M)A - [Z_Z \Delta H^M/(Z_Z X_2 X_M)][Z_L(X_2 - X_M) + Z_P(X_I - 1/2)] \tag{2.23}$$

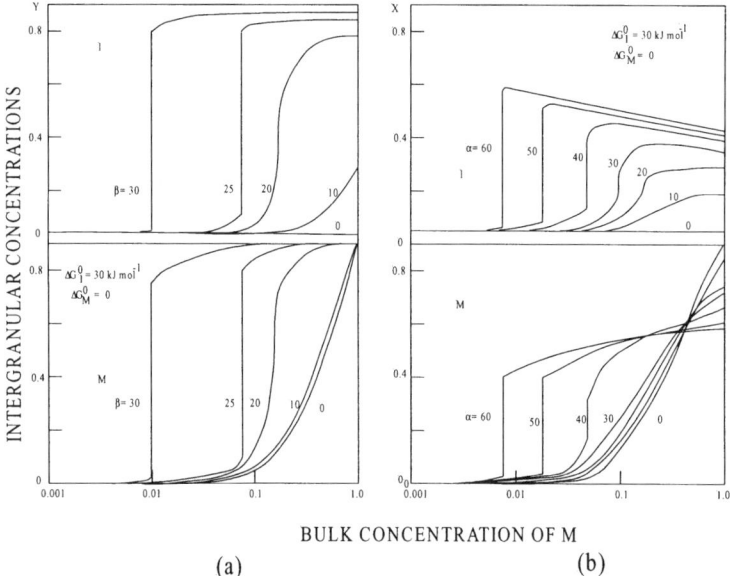

BULK CONCENTRATION OF M

(a)   (b)

Figure 2.6 Calculated grain boundary concentrations of I (impurity) and M (alloy) as a function of the bulk M concentration in regular ternary solutions for the indicated values of α and β (kJ mol^{-1}): (a) no competition; (b) competition. $T = 573$ K, $\Delta G_I^0 = 30$ kJ mol^{-1}, $\Delta G_M^0 = 0$, $X_I^B = 0.01$ %. The same curves hold for any set of proportional values of T, ΔG_I^0, α and β (after Guttmann, 1980)

where γ_I and γ_M are the interfacial energies of the individual components, A is the area per atom, ΔH^M is the enthalpy of mixing for the M–I atom system, Z_Z is the coordination number and Z_p and Z_L are the perpendicular and the lateral coordination numbers respectively. By contrast, the elastic strain enthalpy, ΔH_I^e, arising from mismatch between solute, I, and solvent, M, atoms is given by (Langmuir, 1918 and Medema, 1978)

$$\Delta H_I^e = [24\pi B\mu r_I r_M (r_I - r_M)^2]/(3Br_I + 4\mu r_M) \tag{2.24}$$

where B is the bulk modulus of the solute, μ is the shear modulus of the parent and r_I and r_M are the radii of the solute and parent respectively. Since in metals $\gamma_I \approx \gamma_M$, then ΔH^M has a more significant contribution. Hence, if $\Delta H^M < 0$, the segregation enthalpy is reduced, giving rise to ordering and oscillations of concentration (Sundaraman and Wynblatt, 1975, and Viefhaus and Rusenberg, 1978), whereas if $\Delta H^M > 0$, segregation is enhanced, leading to multilayer segregation. This concept has been extended to ternary alloy systems by Hoffmann and Wynblatt (1989).

The segregation entropy, ΔS_{I}^0, has been described by Wynblatt and Ku (1979) as

$$\Delta S_{\text{I}}^0 = (S_{\text{I}} - S_{\text{M}}) - \left(\frac{2\Delta S_{\text{M}}}{ZX_1 X_{\text{M}}}\right)[Z_{\text{L}}(X_{\text{I}}^{\text{p}} - X_{\text{M}}) + Z_{\text{F}}(X_i - 1/2)]$$

$$- \left(\frac{d}{dt}\right)\left[\frac{24\pi B\mu r_{\text{I}} r_{\text{M}}(r_{\text{I}} - r_{\text{M}})^2}{3Br_i - 4\mu r_{\text{M}}}\right] \tag{2.25}$$

where ΔS_{M} is the excess entropy of mixing for the alloy and S_{I} and S_{M} are the specific interfacial entropies for the pure components. Seah and Lea (1975) and Hondros and Seah (1983) considered that ΔS_{I}^0 consists of three contributions:

(a) the vibrational entropy,
(b) an anharmonic contribution and
(c) site multiplicity.

The latter two are small compared with vibrational entropy and may be neglected. The vibrational contribution, $\Delta S_{\text{I}}^{\text{V}}$, is given by the change in the Debye temperature for a solute in the matrix and the interface (Seah and Lea, 1975; Hondros and Seah, 1983; Ewing, 1971; Kumar, 1981) where

$$\Delta S_{\text{I}}^{\text{V}} = 3R[1 - \ln(kT/h\nu_{\text{E}})] \tag{2.26}$$

where $kT \gg h\nu_{\text{E}}$, ν_{E} is the Einstein frequency and $h\nu_{\text{E}} = k\theta_{\text{E}}$, where θ_{E} is the Einstein temperature ($= 0.775\theta_{\text{D}}$, the Debye temperature). Hence

$$\Delta S_{\text{I}}^{\text{V}} = 3R\ln(\theta_{\text{D}}\theta_{\text{D}}^*) \tag{2.27}$$

where θ_{D}^* is the Debye temperature for the solute atom positioned at a distorted grain boundary site.

In the case of transition metal alloys, interfacial segregation can be predicted by considering three contributions:

(a) the heat of solution,
(b) the difference in the interfacial energies of the pure metals and
(c) the elastic mismatch energy.

Medema (1978) established for an M–I system that

$$X_{\text{I}}^\phi/X_{\text{I}} = \exp\left(\left[f\Delta H_{\text{MI}}^{\text{S}} - g(H_{\text{I}}^\phi - H_{\text{M}}^\phi)\nu_{\text{E}}^{2/3}\right]/RT\right) \tag{2.28}$$

where $\Delta H_{\text{MI}}^{\text{S}}$ is the heat of solution of I in M, H_{I}^ϕ and H_{M}^ϕ are the interfacial enthalpies and f and g are constants.

By contrast, a bond breaking model has been used to extend a quasichemical formulation for chemical composition at the surfaces of non-regular solutions (Bozzolo *et al.*, 1992). This considers nearest neighbour atom interactions, the different behaviour of individual atom layers parallel to the interface, the equal relaxation of bonds and the enthalpy of segregation, ΔH_I^0, in the interface for a binary alloy system. The enthalpy of segregation is given by

$$\Delta H_\mathrm{I}^0 = [(\varepsilon_{ij} - \varepsilon_{\mathrm{MM}})/2][Z_\mathrm{Z} - (Z_\mathrm{L} + Z_\mathrm{P})(I + \alpha_0)] \tag{2.29}$$

where ε_{ij} are the bond enthalpies for the ij nearest neighbours, Z_Z, Z_L and Z_P are as defined above and α_0 is a relaxation parameter. The segregation entropy can also be expressed in this form.

More recently, Bozzolo *et al.* (1992, 1993 and 1994) and Rodriguez *et al.* (1994) calculated properties of an alloy, including heats of formation, surface energies and heat of segregation for substitutional impurity atom additions. This approach defines the energy of segregation as the difference between the heat of formation of a semi-infinite crystal, M, with an impurity atom, I, at a lattice site on a plane parallel to an interface or grain boundary and the same structure but without the impurity atom that is now distributed within the matrix. Thus, the energy of segregation, $\Delta E_\mathrm{I}^\mathrm{S}$, is given by

$$\Delta E_\mathrm{I}^\mathrm{S} = \Delta E_\mathrm{I}^{\mathrm{ST}} + \Delta E_\mathrm{I}^\mathrm{c} \tag{2.30}$$

where $\Delta E_\mathrm{I}^{\mathrm{ST}}$ is the strain energy which can be evaluated from

$$\Delta E^{\mathrm{ST}} = \gamma_{\mathrm{I}\phi}^\mathrm{S} - \gamma_{\mathrm{Ib}}^\mathrm{S} - \gamma_{\mathrm{M}\phi}^\mathrm{S} \tag{2.31}$$

where $\gamma_\mathrm{I}^\mathrm{S}$ is the strain energy of the i atom ($i = \mathrm{IM}$) at the interface, ϕ, or within the parent matrix b. Moreover the chemical energy, $\Delta E_\mathrm{I}^\mathrm{c}$, is given by

$$\Delta E_\mathrm{I}^\mathrm{c} = \sum_q g_{\mathrm{Mq}} \left(f_q^\phi \gamma_{\mathrm{Mq}}^{\mathrm{Ic}} + g_q^\phi \gamma_{\mathrm{Mq}}^{\mathrm{IIc}} \right) - N_1 \gamma_{\mathrm{Mb}}^{\mathrm{Ic}} - N_2 \gamma_{\mathrm{Mb}}^{\mathrm{IIc}} + g_{\mathrm{I}\phi} E_{\mathrm{I}\phi}^\mathrm{c} - g_{\mathrm{Ib}} \gamma_{\mathrm{Ib}}^\mathrm{c} \tag{2.32}$$

where E_I^c are the chemical energies of atoms i at the interface ϕ or in the parent b, $\gamma_{\mathrm{Mq}}^{\mathrm{Ic}}$ and $\gamma_{\mathrm{Mq}}^{\mathrm{IIc}}$ are the chemical energies between different atoms for nearest and next nearest neighbours at the qth layer, N_1 and N_2 are the total number of nearest and next nearest neighbours and f_q^ϕ and g_q^ϕ are for the segregated atom the numbers of nearest and next nearest neighbours in the layer q. These equations are for the unrelaxed thermodynamic state and using zero temperature Monte Carlo calculations allow selected atoms to be subject to unconstrained relaxations so that relaxed segregation energies and configuration energies are obtained. The advantage of this model is that simple approximate expressions are derived to evaluate the segregation and values for the driving forces.

2.2.4 Kinetics of Segregation

As referred to previously, the driving force for segregation of specific atoms to a grain boundary or interface increases with decreasing temperature. However, as the temperature is reduced, diffusion becomes slower and as a consequence the rate at which the grain boundary concentration increases becomes progressively reduced. In his classical work, McLean (1957) addressed this problem by using diffusion theory applied to a simple binary alloy system. Here, equilibrium between the concentration of the segregating species in the grain, X_1, and at the boundary, X_{gb}, is reached at a temperature T_1 and the ratio of these concentrations is X_{gb}/X_1 which is defined as the equilibrium concentration ratio α_1. If the system is cooled to a temperature T_2, where the new equilibrium concentration ratio is $\alpha_2(\alpha_2 > \alpha_1)$, and then held at this new temperature, the new equilibrium concentration at the grain boundary is achieved, $\alpha_2 X_1$. Making assumptions with respect to the width of a grain boundary and the grain size, McLean simplified the problem to linear flow in a semi-infinite body so that the Fick diffusion equations could be applied:

$$D\left(\frac{\partial^2 X}{\partial x^2}\right) = \frac{\partial x}{\partial t} \tag{2.33}$$

where t is the time, x is the distance and D is the diffusion coefficient. If the boundary is located at a position $x = 0$, then, at any time after rapid cooling from temperature T_1 to T_2, the concentration at $x = 0$ is $X = X_{gb}/\alpha_2$. Hence, at the grain boundary or interface the condition is

$$D\left(\frac{\partial X}{\partial x}\right)_{x=0} = 0.5 d_{gb}\left(\frac{\partial X_{gb}}{\partial t}\right) = 0.5\alpha_2 \, \mathrm{d}\left(\frac{\partial X}{\partial t}\right)_{x=0} \tag{2.34}$$

where d_{gb} is the thickness of the boundary and the factor of 0.5 accommodates diffusion to a given grain boundary from the two adjacent grains.

Using standard simplifying procedures for solving diffusion equations based on the Laplace transform (Carlslaw and Jeager, 1947) gives an error function solution:

$$\frac{X_x - X_0}{X_{gb} - X_0} = 1 - \exp\left(\frac{4Dt}{\alpha^2 d_{gb}^2}\right)\mathrm{erf}\left(\frac{4Dt}{\alpha^2 d_{gb}^2}\right)^{1/2} \tag{2.35}$$

where D is the diffusion coefficient for the segregating atom in the matrix, t is the time, X_0 is the segregating atom species concentration at time $t = 0$, X_x is the grain boundary concentration after time t, and X_{gb} is the concentration on the grain boundary after infinite time. Clearly, the lattice concentration is not reduced as the grain boundary concentration increases; Figure 2.7 shows the predicted change in grain boundary concentration with time from the initial concentration to the final equilibrium concentration. This analytical solution assumes that the ratio α remains constant with change in concentrations at the grain boundary. This is true at vanishingly low concentrations of the segregant, but at higher values differences do emerge that affect the predicted results. More recently, computer based numerical solutions to the analytical solution have been developed. An example is the finite difference procedure described by Beeré and Buswell (1994) which offers flexibility in the method for calculating equilibrium segregations. Figure 2.8 shows the results obtained from finite difference calculations for α-iron containing 10^{-4} wt % phosphorous when heat treated for 4.8×10^4 h at a temperature of 665 K followed by a change in temperature to 633 K for a further 4.8×10^4 h. Also shown in this figure is the concentration obtained from the analytical method, equation (2.35), for a constant temperature of 665 K (dashed line); the agreement between the analytical

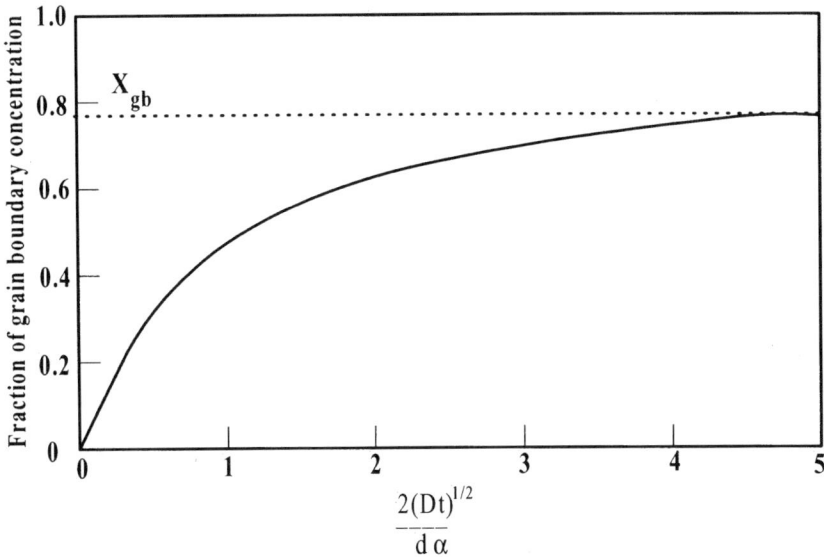

Figure 2.7 Predicted change in grain boundary concentration with time from the initial concentration to the final equilibrium concentration

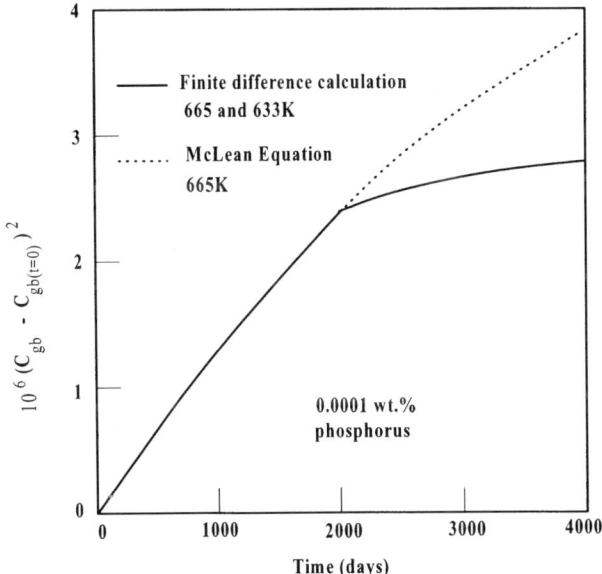

Figure 2.8 Results of a finite difference calculation for α-iron containing 10^{-4} wt % P when heat treated at a temperature of 665 K followed by a period at 633 K: there is a change in slope on reducing the temperature. Also shown by the dotted line is the result of the analytical solution for a constant temperature of 665 K

and numerical method is extremely good. In addition, Figure 2.9 shows the full time and temperature predictions for impurity phosphorous segregation in a steel containing 0.4 wt % C, 1.26 wt % Ni and 0.77 wt % Cr, with a bulk concentration of 0.015 wt % P. The kinetics for this calculation undertaken by Seah (1977) were assumed to be the same as those described above. However, to accommodate interaction between nickel and phosphorous, a modified version of the Guttman segregation theory described in Section 2.2.2 was used with an experimentally determined value for the interaction coefficient between these two elements.

In some cases the volume diffusion coefficients are insufficient to account for the time at temperature to achieve saturation of the grain boundaries. To accommodate this, Militzer *et al.* (1992) proposed a new mechanism to explain the rapid surface segregation of sulphur in an iron–6 at % silicon alloy. The model includes three contributions:

(a) volume diffusion,
(b) dislocation diffusion and
(c) dislocation enrichment

to give the following relationship

$$\frac{C^{\phi}(t) - C^{\phi}(0)}{C^{\phi}(\infty) - C^{\phi}(0)} = \frac{4C^{L}}{d_{gb}}\left(\frac{Dt}{\pi}\right)^{1/2} + \frac{4}{d_{gb}}\left(\frac{r_0}{d}\right)^2\left(\frac{D^d t}{\pi}\right)^{1/2}\left[C^d + \frac{\pi(C^d - C)Dt}{r_0^2 \ln(d/r_0)}\right]$$

$$(2.36)$$

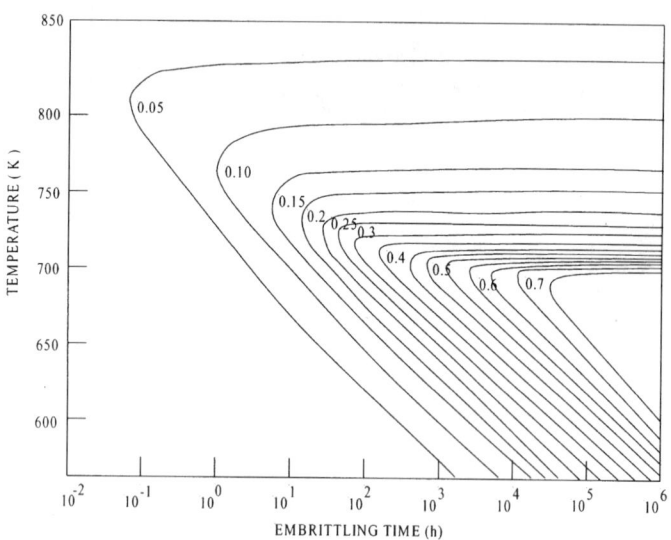

Figure 2.9   Kinetics of grain boundary segregation of phosphorus in SAE 3140 over a wide temperature range, showing existence of 'nose'; numbers on curves show segregation increase in monolayers (after Seah and Hondros, 1977)

where r_0 is the radius of a dislocation, d is the average distance between dislocations, D^d is the dislocation diffusion coefficient, C^d is the impurity concentration in the dislocation, $C^\phi(t)$ is the concentration at the grain boundary after time t, D is the volume diffusion coefficient, d_{gb} is the grain boundary thickness and C^L is the solubility. This relationship suffers from some weaknesses, for example the dependence of $C^\phi(t)$ on D^d suggests that the diffusion fluxes from the bulk into the dislocation line and along the dislocation are equally important; a weakness of the model.

2.3 NON-EQUILIBRIUM SEGREGATION

2.3.1 Thermally Induced Segregation

Non-equilibrium segregation to grain boundaries arising from thermal effects was first reported by Aust *et al.* (1967) and Anthony (1969). It is considered to result from the formation of atom–vacancy complexes within the matrix (Figure 2.10a), where there is substantial misfit between the segregating and the matrix atoms. As for equilibrium processes, the segregating atom can be either solute or an impurity addition to the system. If a solid state system is subjected to a thermal cycle where non-equilibrium conditions are developed, such as cooling rapidly from a high temperature, a supersaturation of vacancies will be developed within the crystal. If, for example, the system is a polycrystal, the grain boundaries act as efficient sinks for vacancies so that the excess vacancy concentration will be reduced by migration to these regions during the period of cooling. However, the interior of the grains retains the excess of vacancies at a concentration corresponding to the higher temperature. The associated gradient in concentration allows the vacancies to diffuse down this gradient towards the grain boundaries and any solute or impurity atoms complexed with the vacancies will diffuse towards the boundary. This will lead to an accumulation of the segregant at, and within, regions of a few nanometres about the grain boundary plane. This is a non-equilibrium condition and the excess concentration of segregated atoms at the grain boundary can be reversed by subsequent heat treatment to achieve the corresponding equilibrium concentration.

The various attempts to quantify this process have been reviewed by Faulkner (1981, 1987). A key feature of these models is that the atom–vacancy complex migrating to the grain boundary moves faster than the individual atoms; this maintains the concentration profile of the complex at a constant value across the complete grain boundary region. Two main approaches have been adopted for predicting the magnitude and width of non-equilibrium segregation profiles:

(a) approximation and numerical solutions using diffusion rate theory, solute drag models, and

(b) the mathematically rigorous application of formal rate theory (Karlsson, 1988 and Lidiard, 1999).

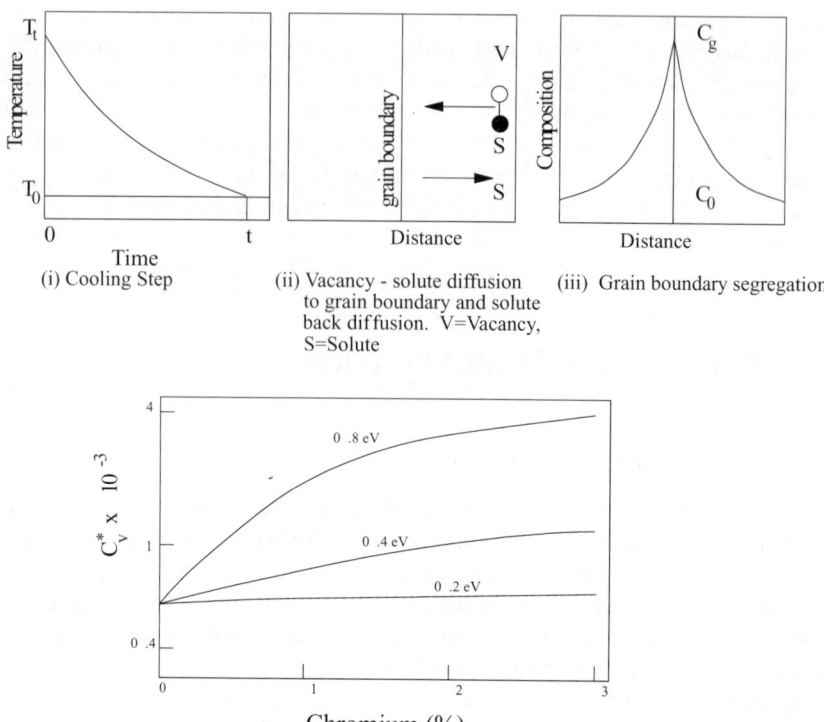

Figure 2.10 (a) Schematic diagram showing (i) a typical cooling profile that leads to (ii) solute–vacancy (s–v) complexes diffusing to a boundary where there will be some back diffusion of solute atoms to the grain interior which leads to (iii) the enrichment of the grain boundary in solute atoms. (b) Calculated variation in equilibrium vacancy concentration, C_v^*, with solute content in γ-Fe–Cr alloys. Values for the different vacancy–solute binding energies, E_{vs}, selected are shown

APPROXIMATE AND NUMERICAL SOLUTIONS

An analysis described by Bercovici *et al.* (1970) includes two main contributions:

(a) the variable size of the diffusion field as time progresses during the period of cooling from the elevated temperature and
(b) blocking of atom-vacancy complexes, caused by the presence of free impurity atoms which oppose the flux of these complexes.

The latter tends to reduce the impurity concentration gradient. The number of atom–vacancy complexes that reach the grain boundary, N_c, is given by

$$N_c = \int_0^{t_f} d_i(t)(\mathrm{d}C_g - \mathrm{d}C_{gb}) \tag{2.37}$$

where $d_i(t)$ is the impurity diffusion distance at time t, C_g is the complex concentration within the grain, C_{gb} is the complex concentration at the grain boundary and t_f is the time required to cool from the higher temperature to the final temperature. This approach, although limited by the integral having to be solved numerically and there is a need to estimate the period of cooling, does have the advantage of offering a physically realistic and practical solution. Here, models have been developed by Doig and Flewitt (1981) and Adetunji *et al.* (1991).

For pure metals the energy of a vacancy is independent of position in the crystal lattice so the equilibrium number of vacancies may be expressed in terms of the vacancy formation energy. Thus, the equilibrium fraction of lattice sites occupied by single vacancies in a pure metal, C_v^*, at temperature T is given by

$$C_v^* = S \exp(-\gamma_v / kT) \tag{2.38}$$

where γ_v is the energy for vacancy formation and S is the entropy, and k is the Boltzmann constant. In the case of γ-iron, for example, the melting temperature $T_m \sim 1800\,\text{K}$, the vacancy formation energy $\gamma_v \sim 1.4\,\text{eV}$, and the entropy A is about 4. Hence, equation (2.38) provides a measure of the equilibrium vacancy concentration (Damask and Dienes, 1966; Schlapink, 1965). By comparison, for a binary alloy system, the energy of a vacancy is position dependent since various lattice sites contain different numbers of constituent atoms surrounding a particular vacancy site (Flynn, 1972). For a binary fcc alloy containing a small concentration of substitutional solute atoms, $C_s(0 < C_s < 0.1)$, the equilibrium vacancy concentration is given by

$$C_v^* = C_v \left[1 - ZC_s + ZC_s \exp\left(\frac{\gamma_{vs}}{kT}\right) \right] \tag{2.39}$$

where Z is the coordination number, which is 12 for an fcc lattice, and γ_{vs} is the vacancy–solute binding energy.

One of the more difficult parameters to establish experimentally for a simple binary alloy system is the binding energy between a solute atom and a vacancy (Howard and Lidiard, 1965). Various theoretical models have been proposed (Demangeat, 1974; Faulkner, 1985, 1981) to allow this parameter to be evaluated. A simple approach proposed by Hasiguti (1967) for fcc alloys provides an estimate:

$$\gamma_{vs} = \gamma_0 + (Z_s - Z_m)\gamma_z + \gamma_d(d - d_m)/d_m \tag{2.40}$$

where γ_0, γ_z and γ_d are constants with dimensions of energy, Z_s and Z_m are the valencies and d and d_m are the atomic diameter of solute and parent atoms respectively. Typical values for γ_{vs} in dilute fcc solutions are of the order of 0.3 eV: for Cr in α-Fe, d_{Cr} and $d_{\alpha\text{-Fe}}$ are 0.168 and 0.152 nm and γ_0 and γ_d are 0.3 and 0.6 eV. Since the valency of Cr equals that of Fe, then the second term reduces to zero to give $\gamma_{vs} = 0.36\,\text{eV}$. The sensitivity of equilibrium vacancy concentration to small changes in the concentration of solute and the value of γ_{vs} is shown in Figure 2.10b where such concentrations are plotted for a range of energies, γ_{vs}, for chromium solute concentrations of up to 3 at. %. A more realistic approach to this problem has been developed by Faulkner (1985 and 1981) using the strain fields associated with the specific solute atom–vacancy complex arrangements. For this, both static and dynamic models allow the energy changes arising in the crystal lattice to be derived. Binding energies for impurity-vacancy (E_{IV}) and impurity interstitial (E_{II}) have been derived from input data based upon atomic size (Table 2.1) (Faulkner *et al.*, 1996). The approach, however, is valid only for metallic systems where there is no substantial directionality in the electronic contributions due to bonding. In ionic and covalent materials these latter contributions can exceed the value of the strain field.

When a pure metal or alloy is cooled from a high temperature, the equilibrium concentration of vacancies at that temperature attempt to adjust to the equilibrium value at the lower temperature. In polycrystalline materials this occurs by diffusion

Table 2.1 Data on impurity–point defect binding energies in several alloy matrices (Faulkner *et al.*, 1996)

Impurity element	Atomic radius (nm)	Ferritic steel		Austenitic steel		Nickel base alloy	
		E_{IV} (eV)	E_{II} (eV)	E_{IV} (eV)	E_{II} (eV)	E_{IV} (eV)	E_{II} (eV)
S	0.104	0.421	0.78	0.46	0.90	0.39	0.77
P	0.109	0.36	0.57	0.41	0.69	0.33	0.58
Si	0.117	0.20	0.26	0.27	0.38	0.19	0.30
As	0.121	0.094	0.12	0.17	0.23	0.10	0.16
Ni	0.1246	0.016	0.0072	0.073	0.10	—	—
Cr	0.128	0.12	−0.096	0.036	−0.012	0.10	−0.049
Cr[a]	0.132	—	—	0.18	−0.14	—	—
Mo	0.136	0.38	−0.31	0.30	−0.26	0.33	−0.27
Mo[a]	0.140	—	—	0.43	0.37	—	—
W	0.137	0.41	−0.33	0.33	−0.29	0.36	0.29
W[a]	0.141	—	—	0.46	−0.40	—	—
Sb	0.141	0.54	−0.43	0.46	−0.40	0.48	−0.39
Sn	0.141	0.54	−0.43	0.46	−0.40	0.48	−0.39
Nb	0.143	0.60	−0.43	0.52	−0.45	0.54	−0.44
Ti	0.147	0.72	−0.54	0.65	−0.55	0.66	−0.52

Values used to obtain binding energies

	Ferritic steel	Austenitic steel	Nickel base alloy
Matrix atom radius	0.1241 nm	0.1269 nm	0.1246 nm
Surface energy/unit area of matrix	~ 1.9 J m^{-2}	~ 1.9 J m^{-2}	~ 1.73 J m^{-2}
Shear modulus of matrix	~ 8.1 × 10^4 MN m^{-2}	~ 8.1 × 10^4 MN m^{-2}	~ 7.7 × 10^4 MN m^{-2}

[a]These values were chosen on the basis of the atomic radius of the metallic structure being dependent on the coordination number. The coordination number of Cr, Mo or W with a bcc structure is 8 yet that of γ-Fe with an fcc structure is 12. It may be imagined that the atomic radii of Cr, Mo or W increase by about 3 % as each acts as a solute atom in the γ-Fe matrix because the increase in interatomic spacing on going from coordination 8 to coordination 12 is about 3 %.

of excess vacancies to strong sinks such as grain boundaries and free surfaces. For cooling under Newtonian conditions, the change in temperature with time is given by (Doig and Flewitt, 1981).

$$T_0 = T_i \exp(-\phi \Delta t) \tag{2.41}$$

where T_i is the initial temperature, Δt is a time interval for cooling, and ϕ is a rate parameter. Holding at a temperature T_i will produce an equilibrium vacancy concentration given by equations 2.38 and 2.39 (Ewing, 1971, and Kumar, 1981), and cooling at a rate described by equation (2.41) induces diffusion of vacancies to the sinks, to relieve the excess concentration. The vacancy profile developed by cooling from temperature T_i to T_a may be evaluated numerically by considering diffusion to occur isothermally for sequential periods of time, Δt, at temperatures defined by equation (2.41), (Adetunji *et al.*, 1991) (Figure 2.11a). The vacancy concentration profile is described by the diffusion equation for semi-infinite solids (Crank, 1975):

$$\frac{C_v(x) - C_v(n)}{C_v(0) - C_v(n)} = \operatorname{erf} \frac{x}{2} (D_v \Delta t)^{1/2} \tag{2.42}$$

where D_v is the vacancy diffusion coefficient $(= D_0 \exp \gamma_v / kTn)$ and n is equal to 1. The spatial extent of this initial vacancy profile is given by the approximation

$$x_n = 2(D_v \Delta t)^{1/2} \tag{2.43}$$

The vacancy profile during a complete cooling cycle may be numerically evaluated for all time intervals, and the form of the profile at the grain boundary is given by the locus of the coordinates. Here, the total vacancy flux, I_v, is the shaded area shown in Figure 2.11a where

$$I_v = \sum_{n=1}^{} \frac{x_n}{2} (C_{v(n-1)} - C_{v(n)}) \tag{2.44}$$

Thus, a vacancy profile and a total vacancy flux for a given start temperature and cooling rate may be determined. For polycrystalline materials it is necessary to include the grain size for evaluating the vacancy segregations in order to limit any profile extending over distances greater than half a grain diameter. Using values typical for γ-iron, $T_m = 1800 \text{ K}$, $\gamma_v = 1.3 \text{ eV}$, $Q_v = 2.6 \text{ eV}$ and a grain size of $2 \times 10^3 \text{ nm}$, the vacancy profiles developed during cooling at a rate $\phi = 1$ are shown in Figure 2.11b for a range of initial annealing temperatures, T_i, over the range T_m to $0.5T_m$. Here, increasing the cooling rate decreases the spatial extent of the vacancy profile from the grain boundary, whereas increasing the initial temperature, T_i, Ti, increases the width of the profile.

It is these vacancy fluxes generated during cooling that produce non-equilibrium segregation of solute atoms because of coupled vacancy–solute pair migration (Adetunji *et al.*, 1991; Hanneman and Anthony, 1969; Floreen and Westbrook, 1969; Aust *et al.*, 1968; Anthony, 1969, 1975). The process is driven thermo-dynamically by the decrease in free energy associated with the annihilation of excess

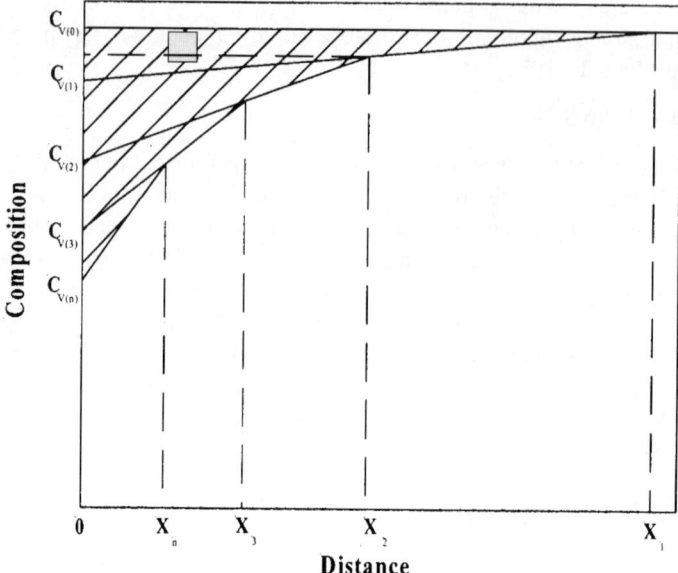

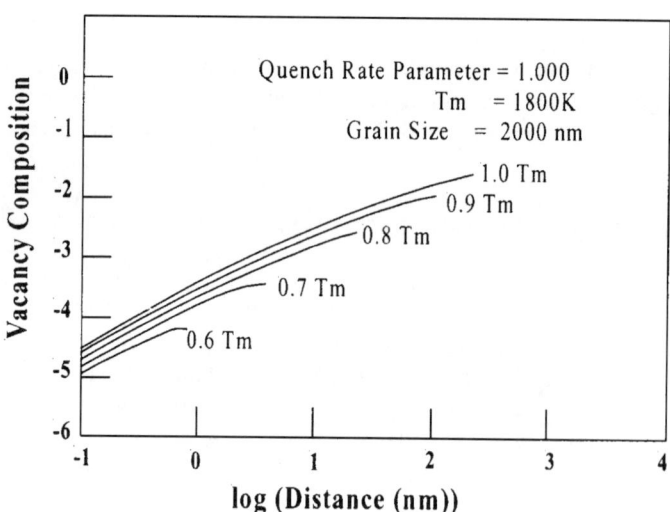

Figure 2.11 (a) Schematic diagram showing the progressive development of a vacancy concentration profile at a grain boundary ($C_{v(0)} \equiv C^*_{v(T_i)}$). (b) Calculated vacancy concentration with distance from a grain boundary (position 0) over the annealing temperature range T_m to $0.3\,T_m$. Based on γ-iron [$T_m = 1800\,\mathrm{K}$; $E_v = 1.3$ eV; $Q_v = 2.6\,\mathrm{eV}$ and grain size 2×10^3 nm], $\phi = 1$ (after Doig and Flewitt, 1981)

vacancies at grain boundary sinks. The overall process does not require the system to be in equilibrium, although the local vacancy–solute pairing reaction is balanced since the pairing depends upon short range diffusion in small volume elements. This can be considered instantaneous relative to the long range diffusion necessary to establish overall solute equilibrium. As a consequence, a system containing vacancy–solute pairs in equilibrium with free solute atoms and free vacancies can be expressed by the reaction

Vacancy − solute pair ⇌ free vacancy + free solute

such that

$$C_{vs} = (C_v^* - C_{vs}) + (C_s - C_{vs}) \tag{2.45}$$

As discussed by Howard and Lidiard (1965), the mass action relationship for this reaction is

$$[C_{vs}]/[C_v^* - C_{vs}][C_s - C_{vs}] = K \tag{2.46}$$

The mass action constant, K, is given by $K = Z_C \exp(\gamma_{vs}/kT)$, where, as discussed above, Z_C is the coordination number. If $C_s \gg C_{vs}$, then equation (2.46) reduces to

$$C_{vs} = KC_s C_v^*/(1 + KC_s) \tag{2.47}$$

The vacancy–solute binding energy γ_{vs} is considered constant so that the resultant gradients for vacancies and solute atoms will be opposite, leading to grain boundary segregation. Usually, a vacancy will move at least one solute atom to the grain boundary sink. Thus inequality in vacancy–solute gradients suggests that the magnitude of each profile may be different. Using the above analysis for vacancy flux together with equation (2.45), it is possible to evaluate numerically the solute flux and resulting profile about a grain boundary. Concomitant with the vacancy–solute complex diffusion transporting solute atoms to a grain boundary is a reverse solute diffusion process that arises from the thermodynamic driving force due to the presence of the non-equilibrium concentration of decoupled solute atoms at, and adjacent to, a grain boundary. Such a process will result in a time dependent reduction in the solute profile by reverse diffusion. This is accommodated using a knowledge of the local solute and vacancy compositions adjacent to the grain boundaries and their relaxation to equilibrium with time (Anthony, 1975).

The contributions from the input variables for an iron–chromium binary alloy are shown in Figures 2.12a and b. In Figure 2.12a the variation in grain boundary solute composition with initial heat treatment temperature is shown for a range of solute–vacancy binding energies but a fixed cooling rate ($\phi = 10^{-1}$) typical of a water quench. Clearly, enhanced grain boundary enrichment is predicted by increasing either the vacancy–solute binding energy for an initial heat treatment temperature or vice versa. However, as shown in Figure 2.12b, the width of the predicted segregation profile developed about a grain boundary is correspondingly increased with decreasing cooling rate. Figure 2.12c shows a predicted tin composition profile in a 2.25 % Cr–1 % Mo ferritic steel containing 0.08 wt % Sn following a heat

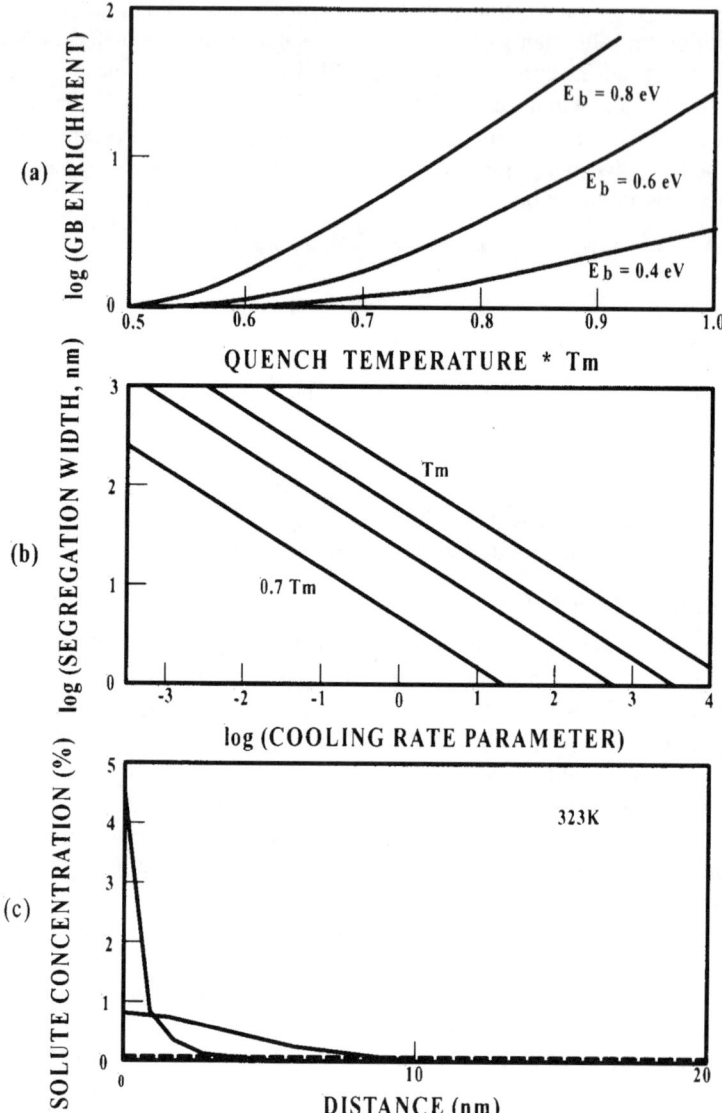

Figure 2.12 Influence of quench temperature, cooling condition, and vacancy–solute binding energy on the grain boundary composition profile for an iron–chromium alloy: (a) effect of quench temperature and binding energy on the grain boundary composition; (b) effect of cooling rate parameter and quench temperature on the solute profile width; (c) calculated solute segregation profiles about austenite grain boundaries following heat treatment at 1323 K and water quench; 2.25Cr–1Mo–0.08Sn, (wt. %) $E = 0.75\,\mathrm{eV}$, $\phi = 1\,\mathrm{s}^{-1}$, tin segregation

treatment at a temperature of 1323 K. Also given is a value for the calculated grain boundary composition that would be measured using STEM-EDS microanalysis with a relatively large analysing probe. The calculated and calculated measured value difference accounts for the microanalysis conditions and is based on the predicted profile.

An alternative but closely related model to that proposed by Doig and Flewitt (1981) is given by Faulkner (1985). This analyses the thermodynamic equilibrium concentration of solute atoms at a grain boundary predicted from equilibrium concentrations estimated from the temperature, T_i, at the start of the heat treatment cycle, and an intermediate temperature of approximately half the melting temperature for the system, $T_{0.5}$. At this latter temperature it is assumed that all further diffusion is negligible so that the predicted grain boundary composition, C_{gb}, is then given by

$$C_{gb} = C_{gb} \exp[(\gamma_b - \gamma_f)/kT_i] - [(\gamma_b - \gamma_f)/kT_{0.5}] \tag{2.48}$$

where γ_b and γ_f are the vacancy–solute complex binding and vacancy formation energies respectively. The kinetic evaluation is made from the starting temperature T_i and subsequent cooling which is converted to an effective time, t^*, using an approach due to Eyre and Maher (1970) where

$$t^* = (K^* k T_i^2)/\phi\gamma_a \tag{2.49}$$

where K^* is a constant assumed to be 0.01 and γ_a is the average activation energy for self-diffusion and impurity diffusion in the adjacent matrix. The overall solution adopts Fick's second law of diffusion for appropriate boundary conditions. This model also accommodates reverse diffusion of the non-equilibrium grain boundary concentration. Hence, the critical time t_c is given by

$$t_c = [d_g^2 \ln(D_c/D_1)]/[4K(D_c - D_1)] \tag{2.50}$$

where d_g is the grain size and K is a constant equal to 0.05. Predictions of this model for Mo and B segregation in a ferritic steel for known effects of cooling rates and start temperature are given in Figure 2.13. Similar models have been developed by Xu and Song (1989), Song *et al.* (1989) and Song (1995).

FORMAL RATE THEORY

A rigorous model to describe non-equilibrium segregation uses coupled diffusion in equations again based upon Fick's second law of diffusion (2.33) where

$$\left.\begin{aligned} \frac{\partial C_v}{\partial t} &= D_v \frac{\partial^2 C_v}{\partial x^2} \\ \frac{\partial C_s}{\partial t} &= D_B \frac{\partial^2 C_s}{\partial x^2} \\ \frac{\partial C_{vs}}{\partial t} &= D_{vs} \frac{\partial^2 C_{vs}}{\partial x^2} \end{aligned}\right\} \tag{2.51}$$

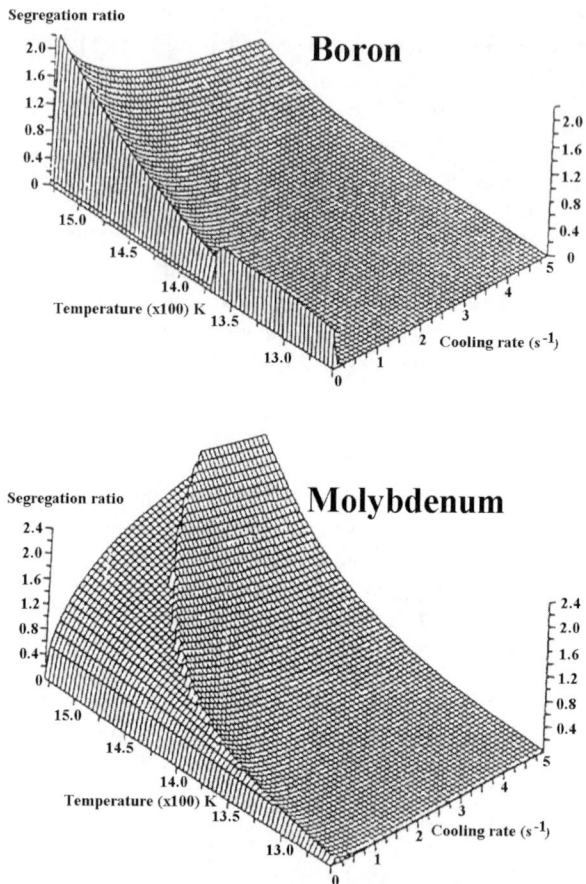

Figure 2.13 Isometric plots of grain boundary composition with respect to start temperature and cooling rate, showing the segregation intensity, F, for Mo and B in a ferritic–martensitic steel. Reproduced by permission of Institute of Materials from Faulkner, R. (1985) *Mater. Sci. Technol.*, **1**, 442

where C_v is the concentration of free vacancies, C_s is the concentration of solute atoms and C_{vs} is the concentration of the complex. These three equations are coupled by assuming that local equilibrium is maintained between the three species within each volume element as given by the form of equation (2.47), i.e. $C_{vs} = C_v C_s K^* \exp \gamma_{vs}/RT$, where K^* contains geometrical and entropy terms. The boundary conditions at the grain boundary are given by combining this equation with the thermal equilibrium concentration of vacancies and a measure of trapping for solute atoms and complexes. Since the diffusion coefficients and the boundary conditions vary with time, exact solutions for these coupled diffusion equations do not exist, at present, for the general case. Therefore, methods such as a finite difference have to be adopted to calculate the amount of segregation and the associated profiles. These calculations have been applied to the segregation of impurity boron in austenitic stainless steel. Figure 2.14 shows the general form of the concentration profile that develops at the grain boundary when cooling at a given rate, $50\,K\,s^{-1}$, from a temperature 1523 K to temperatures of 1273 and 273 K

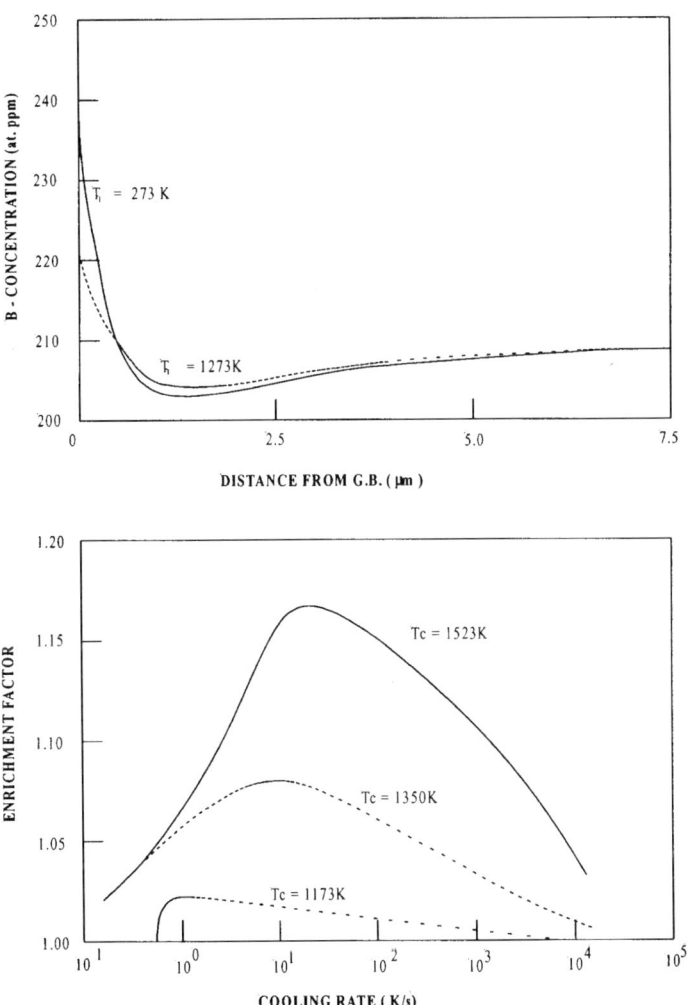

Figure 2.14 (a) Simulated segregation profiles for boron (206 at. ppm) in austenite after cooling at $50\,K\,s^{-1}$ from 1523 K down to 1273 K and 273 K respectively. (b) Calculated enrichment factors as a function of cooling rate for starting temperatures of 1523 K, 1348 K and 1173 K. Austenite with 206 at. ppm boron and a grain size of $150\,\mu m$ (after Karlsson, 1988)

respectively. The diffusion coefficients $(m\,s^{-1})$ of boron and the boron-vacancy complex in austenite were estimated to be

$$D_B = 2 \times 10^{-7} \exp[1.15/RT]$$
$$D_{vB} = 2 \times 10^{-6} \exp[1.15/RT] \tag{2.52}$$

where the activation energy is in eV.

Most of the segregation of the boron takes place during the first phase of the cooling when the mobility of the atoms is high. Here, the enhancement at the grain boundary is given as an enrichment factor (the concentration at the boundary divided by that in the bulk) that is strongly dependent upon the start temperature from which the specimen is cooled (Figure 2.14b).

Generally, a disadvantage of these rate methods is that it is not easy to accommodate the effect of the diffusion process and microstructural parameters on the finally developed segregation profile at a grain boundary. There are cases where these parameters have been examined, for example Bernardini *et al.* (1982) considered the influence of impurity segregations, such as tin to the grain boundaries in α-iron, with respect to the grain boundary diffusion coefficient. Similar thermodynamic arguments are adopted to those we have described to relate the influence of various elements on grain boundary diffusivity in terms of the boundary energy and modifications arising from the presence of an impurity atom.

2.3.2   Neutron Irradiation-induced Segregation

An alternative to the non-equilibrium thermal condition described in Section 2.3.1 is that arising as a result of neutron irradiation. Under conditions of irradiation with fast neutrons, substantial disturbance of the lattice of crystalline materials occurs and this leads to residual defects within the crystal. Figure 2.15 is a time series from a molecular dynamic model for α-iron subject to an incident flux of neutrons with an energy of 10 keV (Gao *et al.*, 1996, 1998) which shows the formation of vacancy–interstitial pairs; Frenkel defects. It is the excess concentration of these defects that provides the necessary driving force for the segregation of solute or impurity atoms to the grain boundaries within a polycrystalline material. Diffusion of point defect–impurity atom complexes then occurs down the concentration gradients created around grain boundaries. Again, as with the thermal non-equilibrium, rate theory (English *et al.*, 1990; Johnson and Lam, 1976) or diffusional solute drag or segregation models (Okamoto and Wiedersich, 1974) can be applied. During neutron irradiation it is assumed that the important point defects are interstitial atoms, because the interstitial–solute binding energies are generally an order of magnitude larger than for the vacancy–solute complex. The main factors that control the magnitude of neutron irradiation induced segregation are the relative diffusion rates of the complexed and free solute atoms in the matrix, with additional factors contributing to irradiation-enhanced diffusion and the interstitial–atom binding energy. As a rule for these complexes, strong binding occurs only with a negative misfit between the solute atom and the matrix atom. This is unlike the vacancy–atom complexes which form provided a misfit exists, either positive or negative. Neutron irradiation induced segregation theory has been developed by several workers (Johnson and Lam, 1976; Okamoto and Wiedersich, 1974; English *et al.*, 1990; Simonen *et al.*, 1989; Murphy 1989a and b; Faulkner *et al.*, 1993, 1994).

Rate theory approaches have been applied to systems that contain low (< 1 at. %) and higher (1–10 at. %) concentrations of solute atoms. Generally, the approach is based upon the fluxes of all the atomic species being defined by a series of partial differential equations which are solved simultaneously to provide the final solute distribution. The kinetic theory of diffusion in dilute alloys, modelled by Murphy (1989a and b), may be used to study diffusion under thermal conditions (Allnott *et al.*, 1983; Lidiard, 1955). In the dilute alloy (< 1 at. %) the solute atoms are

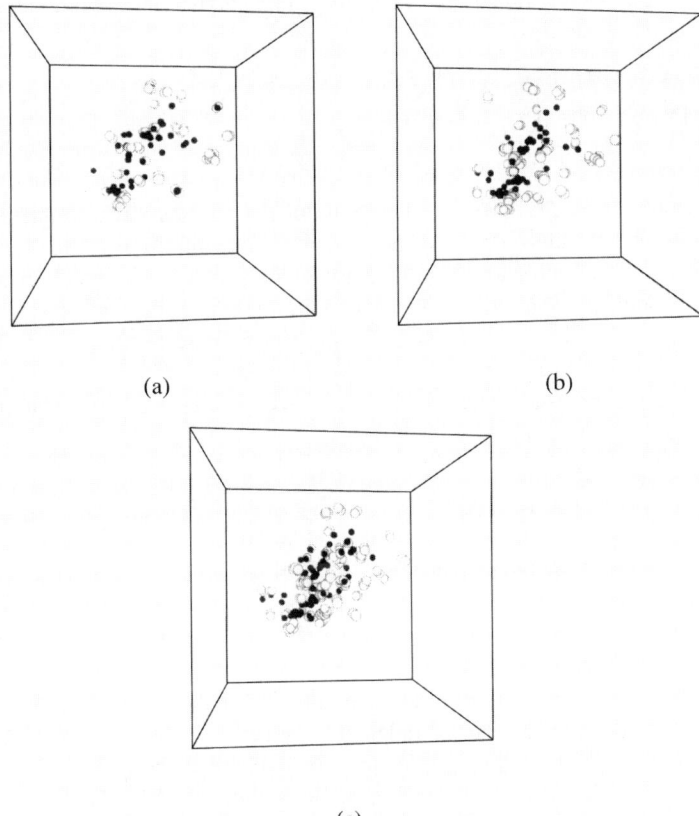

(a)　　　　　　　　　　　(b)

(c)

Figure 2.15 Computer-generated visualisations showing the final damage states of vacant sites (small spheres) and interstitial atoms (large spheres) of typical 10 keV cascades for values of the electron–phonon coupling strength, equal to (a) 0.17×10^{-13}, (b) 4.8×10^{-13} and (c) 79×10^{-13} kg s^{-1}. The block size is $(50a_0)^3$. Reproduced by permission of Institute of Physics Publishing from Gao, F. *et al.* (1998) *Mater. Sci. Eng.*, **6**, 543

considered to be singular so that any clusters of two or more atoms are neglected. In irradiated material the total concentration of vacancies, interstitials and solute atoms, denoted by C_v, C_I and C_s respectively, are given by

$$\frac{\partial C_v}{\partial t} = K - \alpha C_I C_v - L_v - \nabla I_v \tag{2.53}$$

$$\frac{\partial C_I}{\partial t} = K - \alpha C_I C_v - L_I - \nabla I_I \tag{2.54}$$

$$\frac{\partial C_s}{\partial t} = -\nabla I_s \tag{2.55}$$

where I_v, I_I and I_s are the fluxes of vacancies, interstitials and solute atoms, K is the production rate for vacancies and interstitials, ∇ is the recombination coefficient and L_v and L_I are the dose rates for vacancies and interstitials to dislocation and/or grain boundary sinks. These fluxes of vacancies, interstitials and solute atoms are described by the equations

$$I_v = -D_v \nabla C_v - D_{sv}^v C_s \nabla C_v - D_{vs}^v C_v \nabla C_s \tag{2.56}$$

$$I_I = -D_I \nabla C_I - D_{sI}^I C_s \nabla C_I - D_{Is}^I C_I \nabla C_s \tag{2.57}$$

$$I_s = -(D_{vs}^s C_v + D_{Is}^s C_I) \nabla C_s - D_{sv}^s C_s \nabla C_v - D_{sI}^s C_s \nabla C_I \tag{2.58}$$

where the coefficients D_{sv}^s, etc., are functions of the various jump frequencies for vacancies and interstitials and C_v, C_I and C_s denote concentrations of free vacancies, interstitials and solute atoms that exclude association with defect–solute pairs. The first subscript indicates that the coefficients are multiplied by the vacancy, interstitial or solute atom concentrations, and the second subscript indicates that the coefficients are multiplied by their respective concentration gradient.

Some of the vacancies and interstitials will be lost to sinks within the material, and the loss L_v and L_I respectively is given by

$$\left. \begin{array}{l} L_v = D_v^{\text{eff}} C_v k_v^2 \\ L_I = D_I^{\text{eff}} C_I k_I^2 \end{array} \right\} \tag{2.59}$$

where the effective diffusion coefficients, D_v^{eff} and D_I^{eff}, are

$$\left. \begin{array}{l} D_v^{\text{eff}} = D_v + D_{sv}^v C_s \\ D_I^{\text{eff}} = D_I + D_{sI}^I C_s \end{array} \right\} \tag{2.60}$$

This model has been applied to α-iron containing a low concentration of phosphorus to derive the segregation to grain boundaries for specified neutron doses, dose rates and temperatures. For this, the coefficient α is assumed to be a constant and the recombination coefficient is considered to be proportional to the effective diffusion coefficient for the interstitials, D_I^{eff}. However, such a change has little effect on the calculated segregation of the phosphorus since the bulk concentration of this element is small and the diffusion of the interstitials may be considered to be only weakly perturbed by their presence. In these calculations, the boundary condition assumed at a grain boundary is

$$I_v = I_I = I_s = 0 \tag{2.61}$$

so that there is no loss to the grain interior. The grain boundary is assumed to occupy about 10 atom spacings to encompass the diffusion profile and the absorption of vacancies and interstitials is limited so that at $x = 0$. The flux of vacancies and interstitials, I_v and I_I, from the surface of the material is given by

$$\left. \begin{array}{l} I_v = D_v^{\mathrm{eff}}[C_v - C_v^e]/\mu_v + N_v v \\ I_I = D_I^{\mathrm{eff}}[C_I - C_I^e]/\mu_I + N_I v \end{array} \right\} \tag{2.62}$$

where c_v^e and c_I^e are the equilibrium concentrations of free vacancies and interstitials at a free surface μ_v and μ_I are rate limiting parameters for the absorption of vacancies and interstitals, and v is the velocity of the receding surface. The second term in equation (2.62) describes the movement of the free surface that is caused by sputtering, where μ is the velocity of the receding surface and μ_v and μ_I are rate limiting parameters for the absorption of vacancies and interstitials at the free surface. The flux of solute atoms across the free surface is zero, except for the loss of solute atoms by sputtering, so that the boundary condition at the free surface ($x = 0$) is $I_s = N_s C$, where any preferential loss of a given atom species of solute is neglected. Lidiard (1999) has extended the model of English *et al.* (1990) to address radiation induced transport of solute atoms in a bcc lattice. As a result of the more open arrangement of atoms in the bcc lattice compared with those for the fcc lattice the model does not accommodate second nearest neighbour interactions. Here, the form of the rate equations for defects and solute atoms remains independent of the crystal structure where differences between the bcc and the fcc lattice are relationships for the flux and interactions that describe the concentration of the defect–solute complexes. An example of the predictions for these rate theory calculations are given in Figure 2.16 for the segregation of Ni and P to the grain boundaries in α-iron and γ-iron over a range of temperatures for a given neutron flux.

Using an alternative approach, Faulkner *et al.* (1994) extended the point defect based non-equilibrium solute drag segregation model described in Section 2.3.1. The major variables considered include the diffusion coefficients for solute and interstitial–solute complexes, microstructural parameters such as grain size and dislocation density, neutron dose, dose rate, interstitial formation, binding energies with either the solute or impurity atoms and temperature. The maximum segregation to the grain boundary C_{gbI} is given by

$$C_{\mathrm{gbI}} = C_g \frac{\gamma_b^I}{\gamma_f^I} \left[1 + \frac{B_R G^*}{K D_i k_d^2} \exp\left(\frac{\gamma_f^I}{kT} \right) \right] \tag{2.63}$$

where C_g is the solute concentration in the matrix, γ_b^I is the solute–interstitial binding energy, γ_f^I is the interstitial formation energy, D_i is the interstitial diffusion coefficient, K is a geometrical constant, G^* is the point defect production rate which is proportional to the neutron dose rate, k is the Boltzmann constant, T is temperature, B_R is a dose rate correction factor and k_d is the grain interior sink strength for an interstitial. The latter has the form

$$k_d = \{(K_Z \rho_0)^{1/2}[6/d_g + (K_Z \rho_0)^{1/2}]\}^{1/2} \tag{2.64}$$

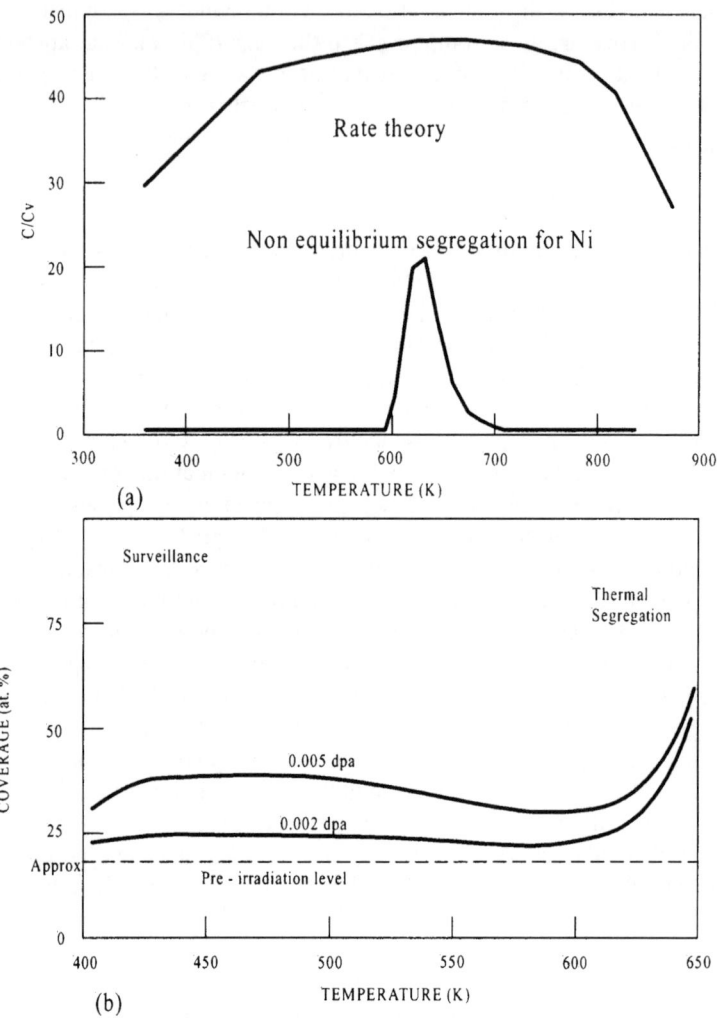

Figure 2.16 Rate theory predictions. (a) Comparison of solute drag and rate theory predictions for the temperature dependence of Ni segregation in austenite. Dose 1 dpa. Dose rate 1×10^{-6} dpa s^{-1}. (b) Calculated temperature dependence of the grain boundary coverage for surveillance irradiations (for a bulk concentration of 0.04 wt % phosphorus) at both 2×10^{-12} and 5×10^{-12} dpa s^{-1} after 32 years. Note that thermal segregation commences at ~ 623 K

where ρ_0 is the dislocation density, K_Z is a bias parameter which defines preferred interaction between interstitials and dislocations compared with vacancies and dislocation and d_g is the grain size. However, only a small proportion of the point defects created during neutron irradiation undergo long-range migration, and this will reduce the absolute magnitude of the segregation. Hence, B is scaled on the basis of experimental observation (Naundorf *et al.*, 1992) so that the segregation process is given by

$$D_{\mathrm{I}} \frac{\partial^2 C_{\mathrm{c}}(x, t)}{\partial x^2} = \frac{\partial C_{\mathrm{c}}(x, t)}{\partial t} \tag{2.65}$$

where $C_{\mathrm{c}}(x, t)$ is the interstitial–solute atom complex concentration and D_{I} is the diffusion coefficient for the complex in the matrix. Here, C_{c} is given by (Faulkner *et al.*, 1994)

$$C_{\mathrm{c}} = C_{\mathrm{i}} \left[K_{\mathrm{c}} K \exp\left(\frac{\gamma_{\mathrm{b}}^{\mathrm{I}} - \gamma_{\mathrm{f}}^{\mathrm{I}}}{kT} \right) + \frac{k_{\mathrm{c}} G^*}{D k_{\mathrm{d}}^2} \exp\left(\frac{\gamma_{\mathrm{b}}^{\mathrm{I}}}{kT} \right) \right] \tag{2.66}$$

where C_{i} is the solute concentration and K_{c} is a geometrical constant. Substituting equation (2.65) gives

$$D_{\mathrm{I}} \frac{\partial^2 C_{\mathrm{i}}(x, t)}{\partial x^2} = \frac{\partial C_{\mathrm{i}}(x, t)}{\partial t} \tag{2.67}$$

This equation has a special form since the diffusion coefficient is associated with a complex, whereas the concentration is associated with the solute atom and therefore it describes dragging solute atoms to grain boundaries during neutron irradiation. Equation (2.67) is solved for the following boundary conditions:

$$\left. \begin{aligned} C &= C_{\mathrm{gb}}(t)/\alpha \\ D_{\mathrm{I}} \left(\frac{\partial C}{\partial x} \right)_{x=0} &= \frac{d_{\mathrm{gb}}}{2} \frac{\partial C_{\mathrm{gb}}(t)}{\partial t} \\ &= \frac{1}{2} \alpha d_{\mathrm{gb}} \left(\frac{\partial C}{\partial t} \right)_{x=0} \end{aligned} \right\} \tag{2.68}$$

where d_{gb} is the width of the grain boundary segregated layer, $\alpha = C_{\mathrm{gbl}}/C_{\mathrm{g}}$ and t is the time. From equation (2.67) we obtain

$$\frac{C_{\mathrm{gb}}(t) - C_{\mathrm{g}}}{C_{\mathrm{gbl}} - C_{\mathrm{g}}} = 1 - \exp\left(\frac{4 D_{\mathrm{I}} t}{\alpha^2 d_{\mathrm{gb}}^2} \right) \mathrm{erfc}\left(2 \frac{(D_{\mathrm{I}} t)^{1/2}}{\alpha d_{\mathrm{gb}}} \right) \tag{2.69}$$

which is an interstitial kinetic relationship for neutron irradiation induced segregation to a grain boundary—the amount of segregation as a function of time at a given neutron irradiation temperature. The critical time, t_{c}, is given by

$$t_{\mathrm{c}} = \frac{K d_{\mathrm{g}}^2 \ln(D_{\mathrm{I}}/D_{\mathrm{s}})}{4(D_{\mathrm{I}} - D_{\mathrm{s}})} \tag{2.70}$$

where K is a numerical constant ~ 0.05, dg is the grain size, and D_{s} is the diffusion coefficient of the solute in the matrix. This approach allows the prediction of Ni segregation in γ-iron (Faulkner, 1985) and silicon segregation in α-iron (Faulkner, 1989) (Figures 2.17a and b respectively).

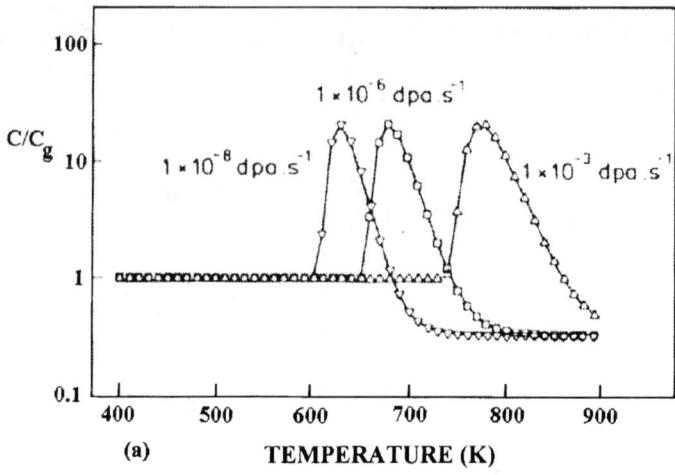

(a)

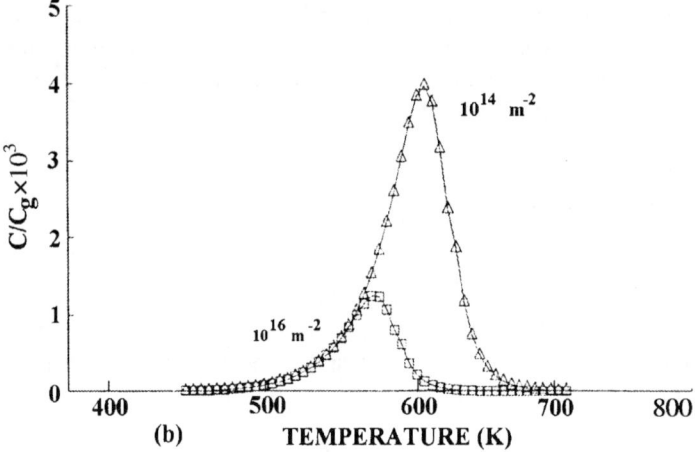

(b)

Figure 2.17 Model predictions of radiation induced segregation for (a) nickel in γ-Fe as a function of irradiation temperature and dose rate. Reproduced by permission of Institute of Physics Publishing from Faulkner, R. (1985) *Mater. Sci. Technol.* **1**, 442. (b) Silicon in ferritic steel showing effect of prior dislocation density. Dose 1 dpa, dose rate 10^{-5} dpa s^{-1} (after Faulkner, R. 1989)

2.3.3 Stress-driven Segregation

When subject to a uniaxial tensile stress, a body increases length in the direction of the applied stress. Since vacancies are mobile in a crystal lattice at temperatures above $\sim 0.4T_{\mathrm{m}}$, this extension can occur in a polycrystalline material by the emission of vacancies from grain boundaries transverse to the tensile stress which are then absorbed into grain boundaries oriented parallel to the direction of the stress. In the Herring–Nabarro model (Herring, 1950, and Nabarro, 1948) this vacancy flow is across grains, whereas in the Coble model (Coble, 1963, and Burton, 1977) it is along grain boundaries. In effect, a similar mechanism could result in the movement of over- or undersize solute or impurity atoms. Moreover, as considered previously, the migrating vacancies could drag solute atoms, in a vacancy–solute complex, to the grain boundaries if the binding energy is appropriate.

Shimoda and Nakamura (1981a and b) suggested that the effect of the applied stress is twofold, affecting:

(a) the limits of grain boundary segregation by modifying the diffusion rate of solute atoms within the grain interior and
(b) the ability of the grain boundaries to absorb solute atoms.

The latter is a thermodynamic contribution where the applied stress is considered to change the free energy towards a more stable system and thereby modify the equilibrium segregation of the solute atoms. Rauh and Bullough (1985) and Rauh *et al.* (1989) have developed a theoretical model that shows how misfitting solute atoms diffuse to local grain boundaries under the influence of a crack tip stress field for both mode I and mixed-mode loads. The model considers a long, semi-infinite crack within an isotropic elastic body, occupying the negative *x*-region of the *x*–*z* plane (Figure 2.18) so that the crack tip coincides with the *z* axis of the orthogonal Cartesian system *x,y,z*. An unfractured grain boundary ahead of the crack occupies the positive *x* region. Here, both crack surfaces and the grain boundary are potential sinks for migrating point defects. Under mixed-mode loading and applied uniaxial

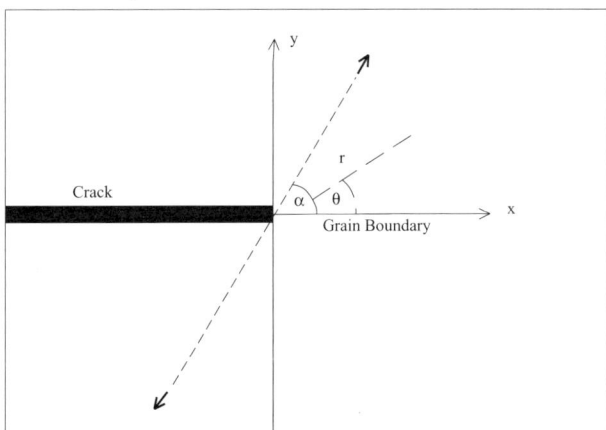

Figure 2.18 Long straight semi-infinite crack and the grain boundary in front of the crack. The loading by an applied uniaxial tension (bold arrows) perpendicular to the crack-top and inclined at an angle α to the grain boundary half-plane as well as the definition of cylindrical polar coordinates (r, θ) are indicated (after Rauh *et al.*, 1989)

stress inclined at an angle α to the grain boundary, the crack tip elastic stress field is characterised by the stress intensity factor, σ, which is related to the intensity factors σ_I and σ_{II} for the opening mode ($\alpha = \pi/2$) and the slip mode ($\alpha = 0$), where

$$\left.\begin{array}{l} \sigma_I = \sigma \sin^2 \alpha \\ \sigma_{II} = \tfrac{1}{2}\sigma \sin(2\alpha) \end{array}\right\} \tag{2.71}$$

The stress field that results from these two crack tip fields interacts with any nearby point defect at a position defined in cylindrical polar coordinates (r, θ, z) relative to the tip, with an energy

$$\gamma_\alpha(r, \theta) = A \sin \alpha \sin\left(\frac{\theta}{2} - \alpha\right)/r^{1/2} \tag{2.72}$$

Here, $-\pi < \theta < \pi$ and A is given by

$$A = \left(\frac{2}{9\pi}\right)^{1/2} (1 + v)K\Delta v \tag{2.73}$$

where v is the Poisson ratio of the elastic body and Δv is the relaxation volume of the point defect. The point defect flux, I_α, due to a field of force $-\nabla\gamma_\alpha$, in the drift approximation is given by

$$I_\alpha = -\frac{D}{kT}C_\alpha\nabla\gamma_\alpha \tag{2.74}$$

where D is the point defect diffusion coefficient, k is the Boltzmann constant, T is the temperature and C_α is the point defect volume concentration. Since

$$\frac{\partial C_x}{\partial t} + \nabla I_x = 0 \tag{2.75}$$

then combining equations (2.74) and (2.75) gives

$$\frac{\partial C_x}{\partial t} = \frac{D}{kT}\nabla\gamma_x\nabla C_x \tag{2.76}$$

which has the explicit form

$$\frac{\partial C_x}{\partial t} = \left(-\frac{AD \sin \alpha}{2}kTr^{3/2}\right)\left[\sin\left(\frac{\theta}{2} - \alpha\right)\frac{\partial C_x}{\partial r} - \cos\left(\frac{\theta}{2} - \alpha\right)\frac{1}{r}\frac{\partial C_x}{\partial \theta}\right] \tag{2.77}$$

Rauh and Bullough (1985) solved this first order hyperbolic equation for the time dependent point defect concentration around the crack tip by adopting the initial condition $C_x = C_0 > 0$ at $t = 0$, with C_0 a constant, and the boundary condition $C_x = C_0 > 0$ as $r \to \infty$ at any time $t > 0$. Equation (2.77) can be solved using the method of characteristics for boundary conditions consistent with the crack and the grain boundary acting as point defect sinks.

This model has been applied to the non-equilibrium segregation of impurity elements such as sulphur in an α-iron matrix and indeed ferritic steels. In α-iron the mechanism has been postulated to account for cracking within the heat affected zone of a weld arising from the post-weld heat treatment, stress relief cracking or the subsequent service temperature cycle, reheat cracking. At a particular time $t(> 0)$ it is possible to obtain a measure of the point defect flow to either the crack surface or the grain boundary ahead of the crack tip. The exact form of the segregation and the distribution between these sinks depends upon the particular loading conditions experienced by the system (Figures 2.19a to c).

2.3.4   Ceramics, Minerals and Ionic Materials

The surfaces of minerals and ceramics have features in common to the extent that the latter are often the result of combining different mineral structures (Myhra *et al.*, 1988, and White and Toor, 1996). Similarities in phase structure and composition, surface structure and surface sites, microstructure and surface reactivity are observed on many occasions. Certainly, in the case of ceramics and minerals the grain boundaries are recognised to have an important role in modifying properties. By contrast, for glasses there is greater interest in reactions that occur at the free surface and less concern with grain boundaries and intergranular films.

Segregation to grain boundaries and surfaces is important to the behaviour of ceramic materials during the fabrication process. Solid state reactions, sintering, grain growth and resulting mechanical and electronic properties are all affected directly. Moreover, in principle, local changes in composition are of similar importance in the study and behaviour of minerals, particularly for multiphase composites or ores. Ceramic materials are mainly non-stoichiometric compounds such as metal oxides, carbides and nitrides. Hence, they contain substantial concentration of point defects, defect complexes and defect clusters. It is this area concerning the relationship between non-stoichiometry and properties of ceramics that assumes a high priority. Defects in ionic solids and related chemical defect reactions are usually addressed via mass action relationships (Smart and Nowotry, 1998) so that, in the context of defect chemistry, of lattice charge neutrality in the bulk phase at any point there is charge compensation

$$\sum z[A^{z-}] - \sum z[D^z] = 0 \qquad (2.78)$$

where $[A^{z-}]$ and $[D^z]$ are the concentration of acceptor and donors for both ionic and electronic charges respectively and z is the valency. This approach has been successful for describing the bulk properties of non-stoichiometric compounds. However, in the case of grain boundaries and interfaces it cannot be readily applied for several reasons. First, the condition in equation (2.78) does not apply since the

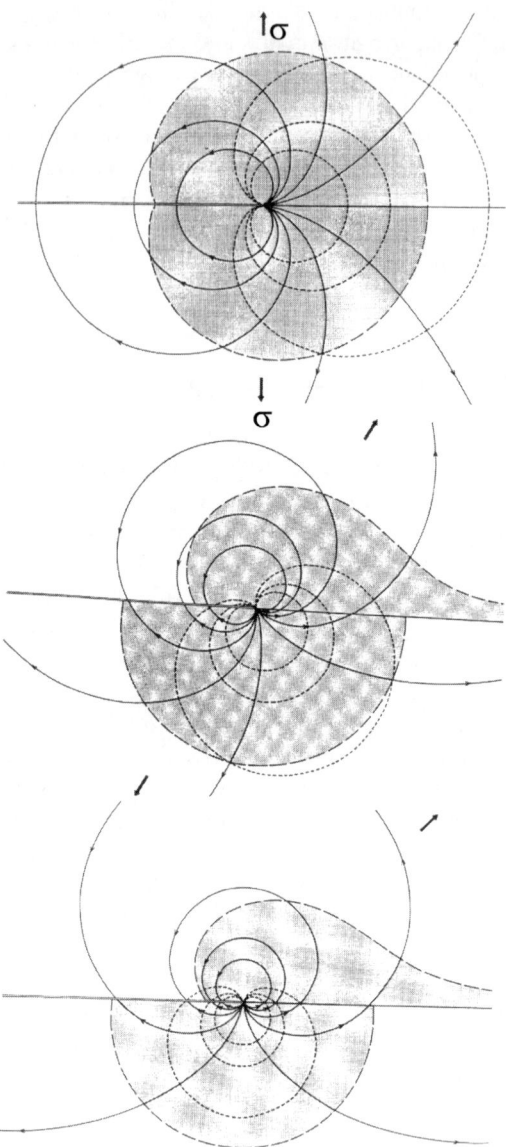

Figure 2.19 Equipotentials (- - - -), flow line (——), and an expanding characteristic (– – –) in the vicinity of a crack-tip when the point defect relaxation volume $\Delta V < 0$. Inside the characteristic (shaded region) the point defect concentration is zero, and outside it retains the initial value C_0: (a) mode I loading ($\alpha = \pi/2$), flow of point defects is entirely to the crack surfaces; (b) mixed-mode loading ($\alpha = \pi/3$), flow of point defects is to both the crack surfaces and the grain boundary; (c) mixed-mode loading ($\alpha = \pi/4$). Flow of point defects is to both the crack surfaces and the grain boundary. Reproduced by permission of the Royal Society from Rauh, H. and Bullough, R. (1985) from *Proc. R. Soc. (Lond.) A*, **397**, 121

charge neutrality requires the charge at the boundary to be compensated by that in the space charge layer:

$$\int\int z[A^{z-}]_s \, ds - \int\int\int z[D^z]_{sc} \, dV = 0 \tag{2.79}$$

where s is the area of the surface layer, V is the volume of the space charge layer and subscripts s and sc denote the surface layer and the space charge layer. A second complication is that the concentration of defects in the boundary or interface layer is larger than for the bulk phase. As a consequence, the ideal defect model does not apply and, therefore, defect activity has to be considered rather than concentration. Moreover, the extent of the defect interactions is a fraction of the distance from the interface. Impurity elements will also have a significant contribution if segregated to these grain boundaries or interfaces, and certainly the concentration enrichment can be several orders of magnitude in the case of, for example, silicon in zirconia (Hughes, 1991).

Figure 2.20 shows schematically a simple $10°$ tilt grain boundary in an ionic ceramic, such as MgO or Al_2O_3, where an excess vacancy at the boundary is created to accommodate the need to conserve charge for the overall system (Kingery, 1974). It is this net charge at interfaces, which arises from creating dipoles due to the discontinuities in the periodic order of the adjacent crystals, that provides the major difference compared with other materials (Lehovec, 1953; Johnson, 1977; Tau, 1975). At thermodynamic equilibrium in an ionic material an excess charge at a grain boundary or interfaces creates a dipole which can function as a vacancy sink if there is a difference between the cation and anion vacancy formation energies. To compensate for the excess charge there is a balancing space charge of vacancies of opposite sign that extends for distances of 2–10 nm from the grain boundary (Lehovec, 1953). It is this balancing space charge vacancy contribution that enhances the segregation of solute atoms under either equilibrium or non-equilibrium conditions, as described previously in this chapter. The magnitude of this additional contribution depends upon the strength of the dipole in the particular ceramic and the solute–vacancy binding energy. Tau (1975) modelled the segregation of divalent impurity elements such as Pb^{++} in AgBr using the contribution of a positive space charge and measured enrichment of Pb at a free surface. Unfortunately, differences in the formation energy of the defects and the solute–vacancy binding energy, which are input parameters to the model, have to be assumed, and this is a difficulty for modelling almost all ceramic oxides (Johnson and Stein, 1974). An alternative to the space charge model is to eliminate this additional contribution and use models appropriate to metallic and other systems described earlier (McCune and Ku, 1984). Certainly, the equilibrium segregation of Ca to grain boundaries in saturated alumina (aluminium oxide) can be modelled using the Langmuir–McLean equation discussed in Section 2.2.1. However, in this case the magnitude of the internal energy change arising from the segregation of Ca is given by considering the high strain energy introduced into the alumina by the substitution of the larger Ca^{++} ion of 0.099 nm radius for the smaller Al^{+++} ion of 0.05 nm radius (Table 2.2) (Tau, 1975).

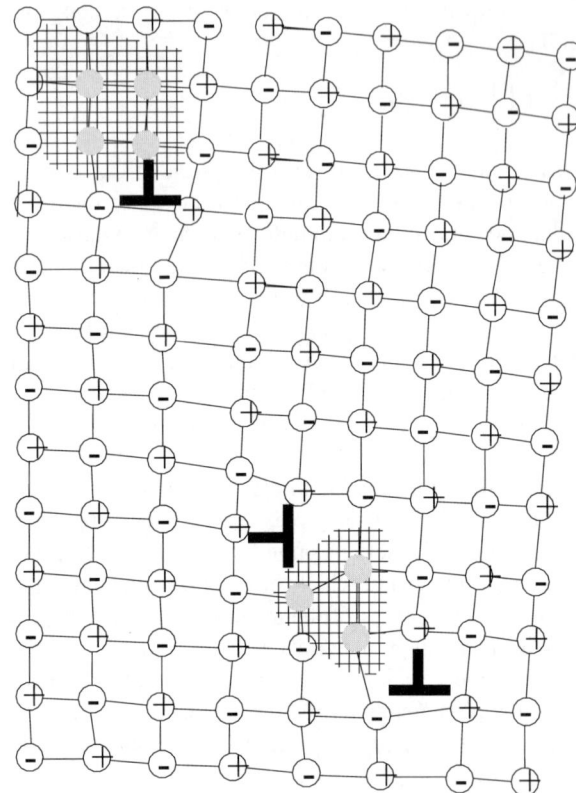

Figure 2.20 Schematic diagram of an ionic 10° tilt boundary in a ceramic material, showing the additional contribution of ionic bonding compared with that in, a metal (after Johnson, 1977)

Table 2.2 Misfits and estimated segregation energies in Al_2O_3

Solute	Ionic radius (nm)	Misfit	ΔG_{seg}
Ca^{2+}	0.099	+0.98	117 kJ/mole (28 kcal/mole)[a]
Y^{3+}	0.093	+0.86	92 kJ/mole (22 kcal/mole)
Ni^{2+}	0.072	+0.44	23 kJ/mole (5.6 kcal/mole)
Mg^{2+}	0.065	+0.30	11 kJ/mole (2.6 kcal/mole)
Si^{4+}	0.041	−0.18	4 kJ/mole (0.9 kcal/mole)

[a]Empirical value

2.4   INTERPHASE BOUNDARIES

There are several types of interphase boundary where solute enrichment has been observed. In solid systems, segregation to a coherent interphase boundary can be described by the models that are applied to grain boundaries in polycrystalline materials. However, during solidification there is the potential for segregation to solid–liquid interfaces and, certainly, if the surface energy is small, which is usually the case, then the segregation is not dissimilar to that occurring at grain boundaries. In this section we will deal specifically with interphase boundaries encountered in metal systems.

Several types of interphase boundary may be classified in a solute system, depending upon the mismatch between the parent and precipitate phase (Aaronson *et al.*, 1970 and 1995), Christian, 1966, and Russel, 1970):

(a) *Coherent*—lattice match between the parent and product phase provides a boundary without disorder and with a structure close to that of the parent phase.
(b) *Semicoherent*—the lattice mismatch between the parent and product phase is small so that dislocations accommodate the strain at the boundary.
(c) *Incoherent*—the lattice mismatch between the parent and product phase is large and the associated mismatch produces a disordered boundary of high dislocation density.

In the case of the coherent boundary there is a lack of excess free volume and as a consequence no segregation is observed at these boundaries. However, as this is relaxed in the semicoherent interface by the presence of misfit dislocations, some enrichment by segregation is encountered. This occurs to an extent where the misfit is sufficiently large for a substantial dislocation density to be present, and the boundary can be effectively classified as incoherent. As a consequence, it is mainly the incoherent boundaries that can accommodate solute or impurity atoms and are subject to equilibrium and non-equilibrium segregation. Certainly, either the reduction in interfacial energy or the accommodation of excess strain energy associated with solute atoms would lead to the possibility of equilibrium segregation and partitioning during growth of one of the phases.

2.4.1 Equilibrium Segregation

The mechanisms of equilibrium segregation to intraphase boundaries can be described by the methodology given in Section 2.3. However, the parameter C_i is defined as the difference between the concentration of component i and the amount that would be present if both phases were homogeneous up to the boundary plane. For a grain boundary this is independent of the choice of surfaces, whereas for an interphase boundary it is a function of the interface plane location (Cahn and Hilliard, 1959). Figures 2.21a and b show a comparison between enrichment at a grain boundary and an interphase boundary between α and β phases. If the boundary is at position (A) rather than at position (B) then C_i will be smaller by an amount equal to the distance between these surfaces multiplied by the difference in

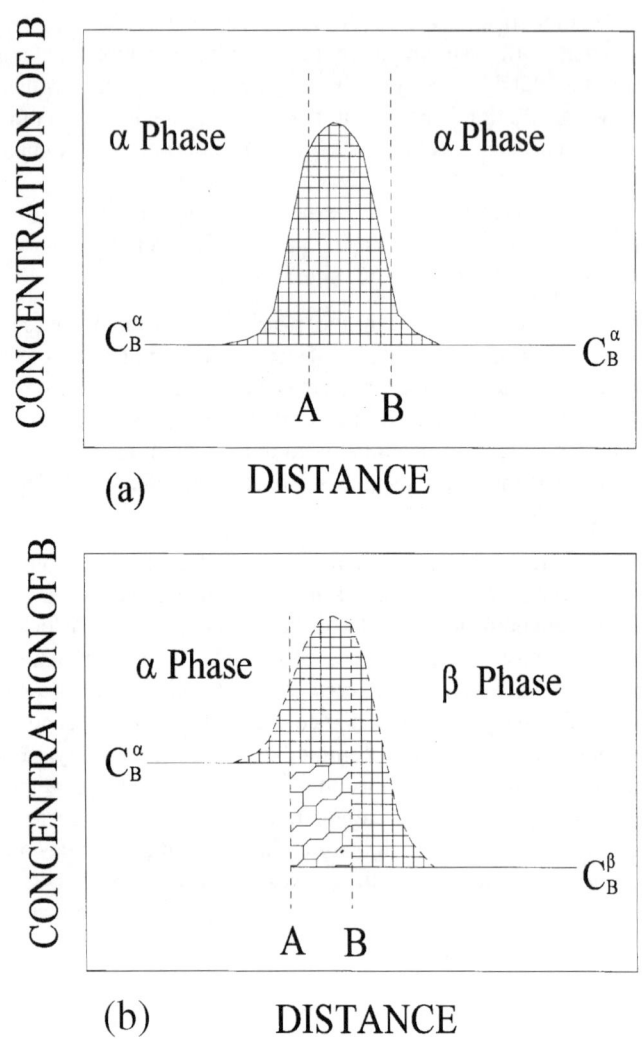

Figure 2.21   Schematic diagram comparing (a) the enrichment at a grain boundary compared (b) with that which occurs at an interphase boundary between an α and β phase (after Johnson, 1971)

composition between the two phases. Hence, for a two-component system the Gibbs–Dunhem equation becomes

$$d\gamma = -\{C_1 - [X_1/(1 - X_1)]C_2\}\, d\mu_1 \qquad (2.80)$$

where X_1 is the mole fraction of the segregating element in either the α or β phase. This equation has limited application because of the sparseness of interfacial energy and chemical potential data available for interphase boundaries. However, as for grain boundaries, a modified Langmuir–McLean isotherm, equation (2.4), can be used to describe segregation of solute atoms for either phase (Figure 2.21b) to the interphase boundary. However, since the segregation can occur from either the α or β phase, the analysis is complicated because the absorption energies have to be related to the separate properties of both phases which are coupled via the chemical potential of the system at equilibrium, and there will also be site competition as the atoms arrive at the interface (Johnson, 1971).

2.4.2 Non-Equilibrium Segregation

Of particular importance to the interphase boundary composition are changes that arise from non-equilibrium partitioning which can develop when one phase grows within another (Hillert, 1967, and Aaronson and Domain, 1966). Hillert (1967) considers the dynamic balance at the interphase boundary, which arises from the

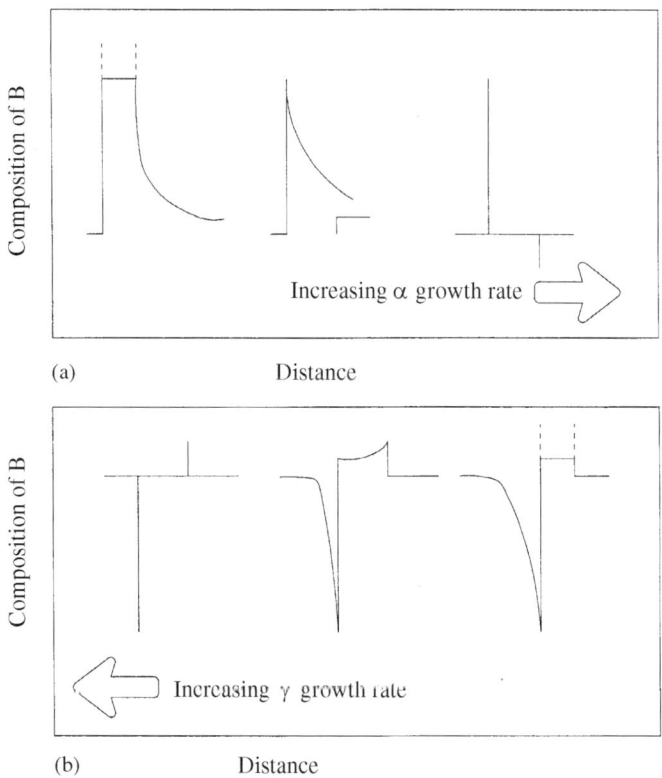

(a)   Distance

(b)   Distance

Figure 2.22 Schematic diagram showing the solute concentration at the interphase boundary for transformation of the α to γ transformation in Fe and the reverse transformation in the Fe – C – X ternary alloy: (a) the transformation of α to γ phase and (b) the transformation of γ to α phase (after Johnson, 1971)

competition between solute rejection and enhancement during growth of a given phase and diffusion from the interface. For the austenite, γ, to ferrite, α, transformation in iron and the reverse transformation in Fe–C–X ternary alloy systems, this leads to a composition dependence at the interface that is a function of growth rate. As shown in Figure 2.22, if the equilibrium alloy composition of the α phase exceeds that of the γ phase there is a significant increase in the concentration at the interphase boundary (Figure 2.22a), whereas growth of γ into α at similar rates leads to a corresponding depletion (Figure 2.22b). Such enrichments or depletions at the interphase boundary will be non-equilibrium and hence may be removed by an appropriate equilibrium heat treatment.

2.5   REFERENCES

Aaronson, H. I. and Domain, H. A. (1966) *Trans. AIME*, **236**, 571.

Aaronson, H. I., Laird, C. and Kinsman, K. R. (1970) *Phase Transformations*, Metals Park, OH: ASM, p. 313.

Aaronson, H. I., Spanos, G., Masamura, R. A., Vardiman, R. G., Moon, D. W., Menon, E. S. K., Hall, M. G. (1995) *Mater. Sci. Eng.*, **B32**, 107.

Adetunji, G. J., Faulkner, R. G. and Little, E. A. (1991) *J. Mater. Sci.*, **26**, 1847.

Allnott, A. R., Barber, A., Franklin, A. D. and Lidiard, A. B. (1983) *Acta Metall.*, **31**, 1307.

Anthony, T. R. (1969) *Acta Metall.*, **17**, 603.

Anthony, T. R. (1975) *Diffusion in Solids*, eds A. S. Norwich and J. E. Burton, Academic Press, p. 353.

Aust, K. T., Armijo, S. J., Kock, E. F. and Westbrook, J. A. (1967) *Trans. Am. Soc. Metals*, **60**, 360.

Aust, K. T., Hanneman, R. E., Niessen, P. and Westbrook, J. H. (1968) *Acta Metall.*, **16**, 291.

Baluffi, R. W. (1979) *Grain Boundary Structure and Segregation in Interfacial Segregation*, ed. W. C. Johnson and J. M. Blakely, Metals Park, OH: ASM, p. 173.

Beeré, W. and Buswell, J. T. (1994) Nuclear Electric Report TIGM/MEM/0039/94.

Bernardini, J., Gas, P., Hondros, E. D. and Seah, M. P. (1982) *Proc. R. Soc. (Lond.) A*, **379**, 455.

Bercovici, J., Hunt, C. E. L. and Niessen, P. (1970) *J. Mater. Sci.*, **5**, 326.

Bozzolo, G., Ferrante, J. and Smith, J. R. (1992) *Phys. Rev.*, **45B**, 493.

Bozzolo, G., Good, B. and Ferrante, I. (1993) *Surf. Sci.*, **255**, 126.

Bozzolo, G., Good, B. and Ferrante, J. (1994) *Surf. Sci.*, **625**, 307.

Briant, C. L. (1990) *Metall. Trans. A*, **21A**, 2339.

Briant, C. L. (1991) *Grain Boundary Segregation in Ordered and Disordered Alloys in Structure and Property Relationships for Interfaces*, eds J. R. Walter, A. H. King and K. Tangri, Metals Park, OH: ASM, p. 43.

Briant, C. L. and Banerji, S. K. (1978) *Int. Metals Rev.*, **222**, 164.

Brunauer, S., Emmett, J. and Teller, E. (1938) *J. Am. Chem. Soc.*, **60**, 309.

Brunauer, S. Deming, L. S., Deming, W. E. and Teller, E. (1940) *J. Am. Chem. Soc.*, **62**, 1723.

Burton, B. (1977) *Vacancies '76*, eds R. E. Smallman and J. E. Harris, London: The Metals Society p. 156.

Cabané, J. and Cabané, F. (1991) *Equilibrium Segregation in Interfaces in Interface Segregation and Related Processes in Materials*, ed. J. Noworthy, Trans Tech Publ. (Switzerland).

Cahn, J. W. (1977) *J. Chem. Phys.*, **66**, 3667.

Cahn, J. W. and Hilliard, J. E. (1959) *Acta Metall.*, **7**, 219.

Carlslaw, H. S. and Jaeger, J. E. (1947) *Conduction of Heat in Solids*, Oxford: Clarendon Press.

Christian, J. W. (1966) *The Theory of Transformation in Metals and Alloys*, Oxford: Pergamon Press.

Coble, R. L. (1963) *J. Appl. Phys.*, **34**, 1679.

Crank, J. (1975) *Mathematics of Diffusion*, Oxford: Oxford University Press.

Damask, A. C. and Dienes, G. H. (1966) *Point Defects on Metals*, New York: Gordon and Breach.

Demangeat, C. (1974) *Acta Metall.*, **22**, 1521.

Doig, P. and Flewitt, P. E. J. (1981) *Acta Metall.*, **29**, 1831.

English, C. A., Murphy, S. M. and Perks, J. M. (1990) *J. Chem. Soc., Faraday Trans.*, **86**, 1263.

Ewing, R. H. (1971) *Acta Metall.*, **19**, 1359.

Eyre, B. L. and Maher, D. M. (1970) AERE Report R6618.

Faulkner, R. G. (1981) *J. Mater. Sci.*, **16**, 373.

Faulkner, R. G. (1985) *Mater. Sci. Technol.*, **1**, 442.

Faulkner, R. G. (1987) *Acta Metall.*, **35**, 2905.

Faulkner, R. G. (1989) *Mater. Sci. Technol.*, **5**, 1095.

Faulkner, R. G., Waite, N. C., Little, E. A. and Morgan, T. S. (1993) *Mater. Sci. Eng.*, **171**, 241.

Faulkner, R. G., Song, S. and Flewitt, P. E. J. (1994) *J. Nucl. Mater.*, **212–215**, 608.

Faulkner, R. G., Song, S. and Flewitt, P. E. J. (1996) *Mater. Sci. Technol.*, **12**, 904.

Floreen, S. and Westbrook, J. H. (1969) *Acta Metall.*, **17**, 1175.

Flynn, C. P. (1972) *Point Defects and Diffusion*, Oxford: Clarendon Press.

Fowler, R. H. and Guggenheim, E. A., (1939) *Statistical Thermodynamics*, Cambridge University Press, p. 429.

Gao, F., Bacon, D. J., Calder, A. F., Flewitt, P. E. J. and Lewis, T. A. (1996) *J. Nucl. Mater.*, **230**, 47.

Gao, F., Bacon, D. J., Flewitt, P. E. J. and Lewis, T. A. (1998) *Mater. Sci. Eng.*, **6**, 543.

Gibbs, J. W. (1957) *Collected Works*, New Haven: Yale University Press, Vol. 1, pp. 219–233

Graham, D. (1953) *J. Phys. Chem.*, **57**, 665.

Guttman, M. (1980) *Residuals, Additives and Materials Properties*, London: Royal Society.

Hanneman, R. E. and Anthony, T. R. (1969) *Acta Metall.*, **17**, 1133.

Hasiguti, R. R. (1967) *Lattice Defects and their Interactions*, ed. R. D. Hasiguti, New York: Gordon and Breach.

Herring, C. (1950) *J. Appl. Phys.*, **21**, 437.

Hillert, M. (1967) *Proc. Int. Conf. on the Mechanism of Phase Transformation in Crystalline Solids*, London: Institute Metals, p. 231.

Hoffmann, M. A. and Wynblatt, P. (1989) *Metall. Trans. A*, **20A**, 215.

Hofmann, S. (1987) *J. Chim. Phys.*, **84**, 141.

Hofmann, S. (1990) *Segregation at Grain Boundaries in Surface Segregation Phenomena*, eds P. A. Dowben and A. Millar, Boca Raton, FL: CRC Press, p. 107.

Hondros, E. D. and Seah, M. P. (1977a) *Metall. Trans. A*, **8A**, 1363.

Hondros, E. D. and Seah, M. P. (1977b) *Int. Metall. Rev.*, **22**, 863.

Hondros, E. D. and Seah, M. P. (1983) *Interfacial and Surface Microchemistry in Physical Metallurgy*, Cahn, R. W. and Haasen, P. (eds) Amsterdam: North-Holland, ch. 3, p. 855.

Howard, R. E. and Lidiard, A. B. (1965) *Phil. Mag.*, **12**, 1176.

Hughes, A. E. (1991) PhD thesis, Royal Melbourne Institute of Technology.

Johnson, W. C. (1971) *Interfacial Segregation*, eds W. C. Johnson and J. M. Blakeley, Metals Park, OH: ASM, p. 351.

Johnson, W. C. (1977) *Metall. Trans. A*, **8A**, 1413.

Johnson, R. A. and Lam, N. Q. (1976) *Phys. Rev.*, **13**, 4364.

Johnson, W. C. and Stein, D. F. (1974) *J. Am. Ceram. Soc.*, **57**, 342.

Joshi, A. (1978) *Interfacial Segregation*, eds W. J. Johnson and J. M. Blakely, Metals Park, OH: ASM.

Karlsson, L. (1988) *Acta Metall.*, **36**, 25.

Kingery, W. K. (1974) *J. Am. Ceram. Soc.*, **57**.

Kumar, V. (1981) *Phys. Rev.*, **23B**, 3756.

Langmuir, I. (1918) *J. Am. Chem. Soc.*, **40**, 1361.

Lehovec, K. (1953) *J. Chem. Phys.*, **21**, 1123.

Lejcek, P. and Hofmann, S. (1995) *Critical Rev. Solid State Mater. Sci.*, **20**, 1.

Lidiard, A. B. (1955) *Phil. Mag.*, **46**, 1218.

Lidiard, A. B. (1999) *Phil. Mag.* **A79**, 1493.

McCune, R. C. and Ku, K. C. (1984) *Advances in Ceramics*, New York: American Ceramic Society, vol. 10.

McLean, D. (1957) *Grain Boundaries in Metals*, Oxford: Clarendon Press.

Medema, A. R. (1978) *Z. Metallkunde*, **69**, 455.

Militzer M., Ivashchenko, Y. N., Knajnikov, A. V., Lejcek, P., Wieting, J. and Firstov, S. A. (1992) *Surf. Sci.*, **261**, 267.

Murphy, S. M. (1989a) *Phil. Mag.*, **59**, 953.

Murphy, S. M. (1989b) *Phil. Mag.*, **59**, 1163.

Myhra, S., Smart, R. S. and Turner, P. S. (1988) *Scanning Microscopy*, **2**, 715.

Nabarro, F. R. N. (1948) *Strength of Solids*, London: London Physical Society, p. 75.

Naundorf, V. Machted, M.-P. and Wollenberger, H. (1992) *J. Nucl. Mater.*, **186**, 227.

Okamoto, P. R. and Wiedersich, H. (1974) *J. Nucl. Mater.*, **53**, 336.

Rauh, H. and Bullough, R. (1985) *Proc. R. Soc. (Lond.) A*, **397**, 121.

Rauh, H., Hippsley, C. A. and Bullough, R. (1989) *Acta Metall.*, **37**, 269.

Rodriguez, A. Bozzolo, G. and Ferrante, I. (1994) *Surf. Sci.*, **625**, 307.

Russel, K. (1970) *Nucleation in Solids, Phase Transformation*, Metals Park, OH: ASM, p. 219.

Schlapink, F. W. (1965) *Phil. Mag.*, **11**, 1055.

Seah, M. P. (1976) *Proc R. Soc. (Lond.) A*, **349**, 535.

Seah, M. P. (1977) *Acta Metall.*, **25**, 345.

Seah, M. P. (1980) *J. Phys. F. Metal Phys.*, **10**, 1043.

Seah, M. P. and Hondros, E. D. (1973) *Proc. R. Soc. (Lond.) A*, **335**, 191.

Seah, M. P. and Hondros, E. D. (1977) *Int. Metals Rev.*, **222**, 262.

Seah, M. P. and Lea, C. (1975) *Phil. Mag.*, **31**, 627.

Shimoda, T. and Nakamura, T. (1981a) *Acta Metall.*, **29**, 1631.

Shimoda, T. and Nakamura, T. (1981b) *Acta Metall.*, **29**, 1637.

Simonen, E. P., Bradley, E. R. and Jones, R. H. (1989) *Proc. ASTM, Meeting on Effects of Irradiation on Materials*, eds N. H. Packam, R. E. Stoller and A. S. Kumar, ASTM, STP 1046, p. 411.

Smart, R. St C. and Nowotry, J. (1998) in *Ceramic Interfaces. Properties and Applications.* eds R. St C. Smart and J. Nowotry, London: Institute of Materials, p. 5.

Song, S. (1995) PhD thesis, Loughborough University.

Song, S., Xu, T. and Yuan, Z. (1989) *Acta Metall.*, **37**, 319.

Sundaraman, J. and Wynblatt, P. (1975) *Surf. Sci.*, **52**, 569.

Tau, Y. T. (1975) *Prog. Solid State Chem.*, **10**, 193.

Viefhaus, H. and Rusenberg, M. (1978) *Surf. Sci.*, **1** 159.

Westbrook, J. H. (1969) *Interfaces*, ed. R. C. Gifkins, London: Butterworths, p. 283.

White, C. L. and Coghlan, W. A. (1977) *Metall.* Trans. A **8A**, 1977.

White, T. J. and Toor, A. I. (1996) *J. Mater.*, **48**, 54.

Williams, F. L. and Nason, D. (1974) *Surf. Sci.*, **45**, 377.

Wynblatt, P. and Ku, R. C. (1979) *Interfacial Segregation*, eds W. C. Johnson and J. M. Blakely, Metals Park, OH: ASM, 115.
Wynblatt, P. and Ku, R. C. (1977) *Surf. Sci.*, **65** 511.
Xu, T. and Song, S. (1989) *Acta Metall.*, **37**, 2499.

Chapter 3
Composition Changes in Materials

3.1 INTRODUCTION

In the preceding chapters we have considered the reasons and driving forces that determine the composition of grain boundaries, interfaces and interphase boundaries following a range of equilibrium and non-equilibrium processes. Moreover, we will describe in Chapter 4 how it is now possible to select from a range of techniques to measure these changes in composition to both identify and quantify the amount of a given atom species at surfaces, boundaries and interfaces. Prior to considering the effect of such local changes in alloying or impurity element composition on the physical and mechanical properties of a material, it is appropriate to look in more detail at the composition changes that have been observed in materials following a range of equilibrium and non-equilibrium heat treatment cycles and/or exposure to environmental conditions such as neutron irradiation. Certainly this is one component of the total microstructure which determines the structure dependent properties of solids, including the engineering mechanical properties addressed in Chapter 5.

The simple visualisation of the segregation to interfaces, in general, shown schematically in Figure 3.1, is based upon that proposed by Hondros and Seah (1977). Here, a polycrystalline solid is held at equilibrium at a defined high temperature within an isothermal volume. In addition, the chemical potential for each atom is considered to be constant for the total system so that all the interfaces shown become enriched with the surface active atoms. As a consequence, enrichment is a result of the system parameters at equilibrium and not dependent upon the total history, and, indeed, the adsorption will be reversible. For a simple system, such as a binary iron–nitrogen alloy, the level of grain boundary enrichment can arise from either the bulk nitrogen content of the material or the partial pressure of nitrogen in the vapour phase which, for equilibrium conditions, is given by the gas solubility relationship (Hondros, 1967).

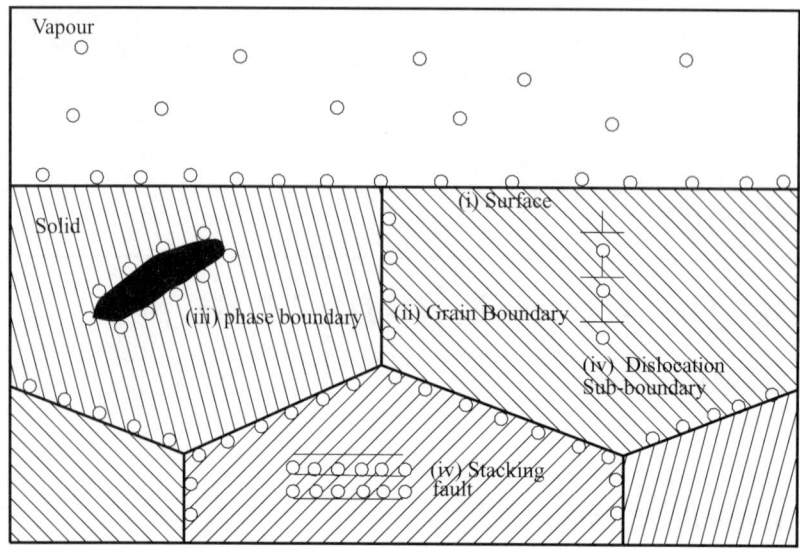

Figure 3.1 Adsorption at different interfaces: (i) surface (ii) grain boundary, (iii) interphase and (iv) defects such as stacking faults or dislocations in a crystalline material (constant chemical potential and temperature)

The need to understand the role of alloying and impurity elements in materials arises from the requirements of a wide range of engineering and scientific disciplines that demand consistently high and quantifiable performance from metallic and non-metallic materials. These extend from the more traditional requirements of mechanical and electrical engineering to the microelectronic, biochemical and nuclear industries. Certainly, in nuclear applications the safety requirements have led to stringent limits for materials ranging from the conventional metals and alloys, usually steels, to those producing long-lasting isotopes and materials such as boron that affect the neutron economy. Moreover, there are a range of issues that affect neutron radiation damage and hence degradation of material properties over the lifetime of the plant. Enhanced compositional control for materials can be achieved by the use of a range of specialist processes, such as vacuum melting, ion exchange purification, ion beam deposition, zone refining and sputter coating supported by composition control and appropriate analytical techniques to provide materials of the specified and, indeed, required compositions. However, even with such control it is necessary to understand which elements, particularly residual impurity elements, are important with respect to their potential for subsequent segregation to surfaces, grain boundaries, interfaces and interphase boundaries. Moreover, deliberate alloy or impurity element additions are often made to achieve the required physical and mechanical properties. Historically, this led Westbrook (1980), when reviewing the preparation of a new material from recycled materials, to focus attention on residual elements (Figure 3.2a). As a consequence, it is essential to understand the elemental

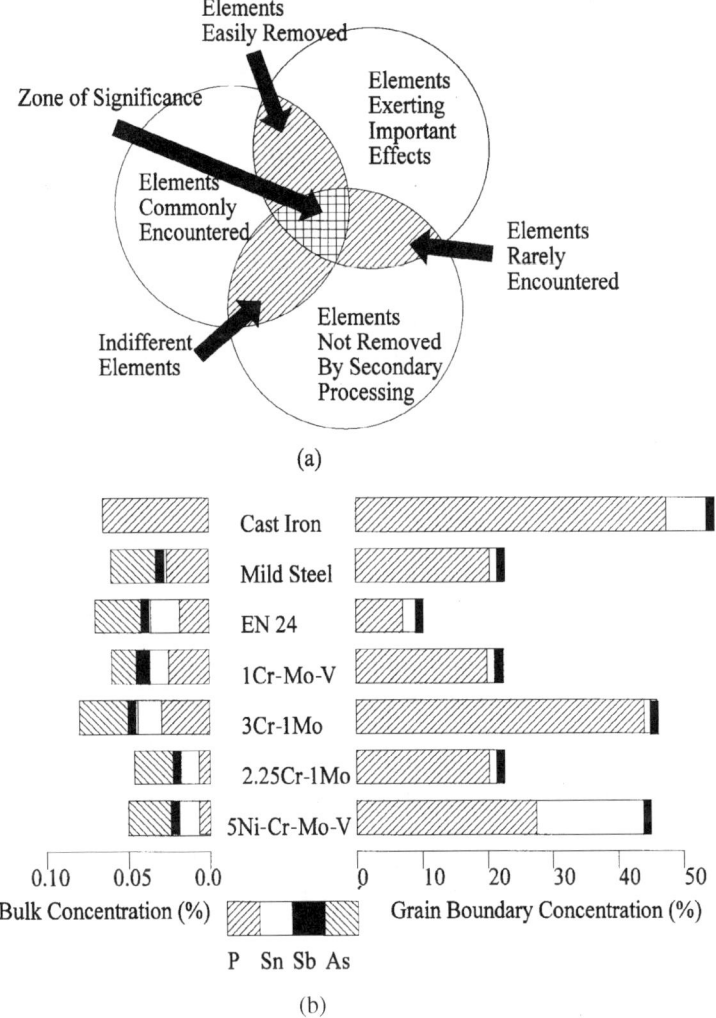

(a)

(b)

Figure 3.2 Residual elements in materials: (a) identification of residual elements should be considered for specification control (Westbrook, 1980); (b) Concentration of the four principal impurity elements at the grain boundaries of common commercial steels, compared with the bulk contents (Hondros *et al.*, 1976)

redistributions that can take place in different materials and the conditions that lead to these. In the case of commercial steels, Figure 3.2b shows a simple, but helpful, comparison of the segregation in several of those more commonly encountered, where the concentration of impurity elements at the grain boundaries has been measured and compared with the bulk composition (Hondros *et al.*, 1976). Although phosphorus is a well recognised and dominant segregating species in both ferritic and austenitic steels, it is clear that other impurities such as Sn, Sb and As as well as substitutional alloying elements may also segregate substantially to the grain boundaries when subject to either heat treatment or an external environment. In all of these cases, grain boundary composition can be addressed by appropriate equilibrium or non-equilibrium segregation theory.

Although segregation is most severe at lower temperatures within a range typical of the service conditions of many major plant components and structures, where the relevant atoms remain mobile, it has to be recognised that both equilibrium and non-equilibrium processes, which may be used prior to introduction of the material into service, also effect segregation. An example is the non-equilibrium thermal cycle associated with welding a steel. In addition there is the segregation of key elements that occurs in structural materials during service. Therefore, it is possible for the composition of surfaces, grain boundaries, interfaces or interphase boundaries to change continually throughout the service life of major structures and components, and this has to be accommodated in the initial design and monitored through subsequent operation of the plant.

3.2   ROLE OF CRYSTALLOGRAPHY ON COMPOSITION

The types of solute or impurity atom segregation that can occur at different surfaces, grain boundaries, interfaces or interphase boundaries in metallic and non-metallic systems under various conditions cover a spectrum of possibilities. At the extreme is the dilute limit where a single low bulk concentration of atoms may segregate weakly to concentrate at special sites within the boundary or interface but there is no change associated in the atomic structure. The other limit is where one or more of the strongly interacting atoms of higher bulk concentration segregate to produce a new multilayer structure at the boundary or interface: this produces changes in crystal structure and, indeed, can result in new phases being formed. Further factors have to be accommodated mainly if the segregated atoms are in equilibrium with those in the adjacent bulk material. Non-equilibrium segregation arises from thermal cycles including cooling and annealing, the application of a stress and neutron irradiation.

It is recognised that considerable variability is observed in the composition measured at grain boundaries and interfaces in a polycrystal. This variability can arise from one or a combination of different sources. The main ones are

(a)   a result of the method by which the local composition is measured,
(b)   differences arising from inhomogeneity in the bulk composition,
(c)   the orientation of the particular grain boundary or interface and
(d)   non-equilibrium rather than a true equilibrium being achieved (Briant, 1985).

The effect of the detailed atomic arrangement within a grain boundary and thereby the role of orientation on the final composition has been examined by several workers. It is this feature that means that within a given polycrystal there will be a change in grain boundary composition unique to each specific grain boundary. Watanabe *et al.* (1980) considered segregation of tin and silicon to grain boundaries in bicrystals of α-iron, demonstrating that for a tilt angle of about 15° the amount of segregation is large (Figure 3.3), the individual values lying within ±20% of the mean. Mulford (1985) investigated sulphur segregation in nickel base alloys and found variations of the order of ±30% which were similar to those reported by Briant (1985) for phosphorus segregation in low-alloy ferritic steels. Similar variations for the segregation of iron to the grain boundaries in the ceramic MgO have been observed by Hall *et al.* (1981) and Mitamura *et al.* (1979). The orientation dependence of grain boundary composition in silicon:iron shown in Figure 3.3 was originally fitted by a monotonic relationship, but this had to be revised when no silicon segregation was detected at tilt boundaries of less than about 20° misorientation. It is only at misorientations above this value that the composition slowly increases with angle up to about 60°. However, in the case of tin, the enrichment is more than for the silicon (Figure 3.3) and, indeed, increases with misorientation angle even within the lower angular range. Lejček and Hofmann (1995) considered these data with respect to the misorientation angle for the specific adjoining grains used to define the grain boundaries, without specifying the rotation axes and the degree of symmetry for these particular boundaries. The fit as shown in Figure 3.3 by the 'solid circle' line indicates that it is possible to detect small concentrations of tin segregation at low index grain boundaries. By contrast, phosphorus segregation

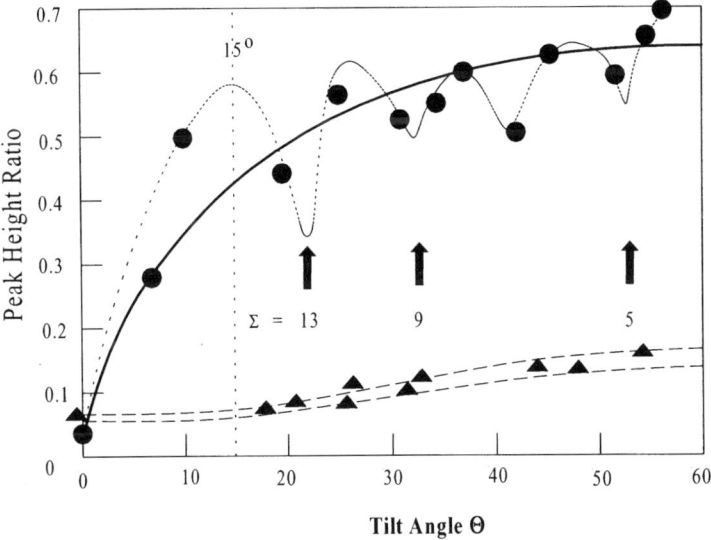

Figure 3.3 Variation in the Auger electron spectroscopy peak-to-peak height ratios, *r*, for (▲)Si/Fe and (●) Sn/Fe with the grain boundary tilt angle after electron spectroscopy (Watanabe *et al.*, 1980). The dashed line is an alternative fit

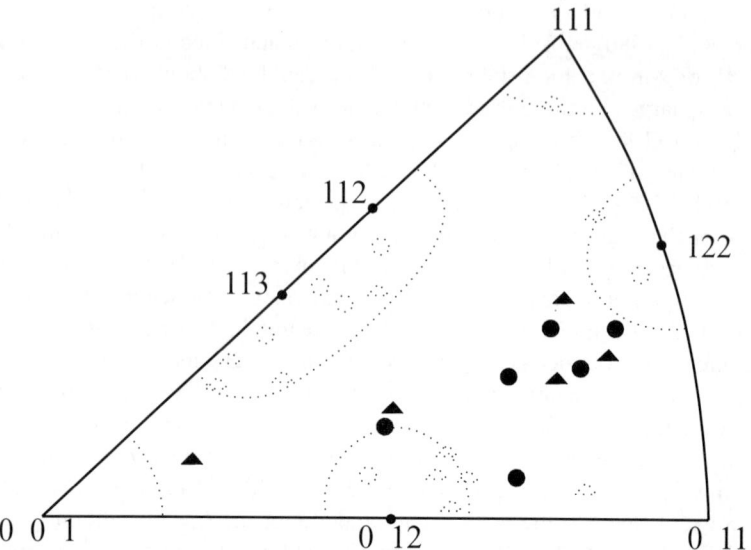

Figure 3.4 Auger electron spectroscopy peak-to-peak height ratio for phosphorus and the orientation of grain boundary planes. The ratios are: ◌ < 0.4; △ 0.4–0.5; ● 0.5–0.6; ▲ > 0.6 (Suzuki *et al.*, 1980)

to known grain boundaries in polycrystalline α-iron is proposed to depend upon the crystallographic orientation of the boundary plane rather than the degree of misorientation. As shown on the stereographic projection Figure 3.4, the amount of phosphorus segregation is larger for high index grain boundary planes and smaller on low index planes (Suzuki *et al.*, 1980). This latter model is certainly consistent with the measurements undertaken for a range of metallic and non-metallic systems (Bernardini *et al.*, 1985; Tatsumi *et al.* 1986; Ogura *et al.*, 1987; El' Rabat and Priestner, 1988; Romarro *et al.*, 1989). As a consequence, when considering the measurements presented in this chapter, the inherent variations in grain boundary composition together with other sources of scatter have to be recognised.

3.3 RELATIONSHIP BETWEEN DIFFERENT SURFACES, INTERFACES AND BOUNDARIES

The similarity between solute and impurity segregation to free surfaces, grain boundaries and interfaces in both metallic and non-metallic systems has been used to justify the comparison of measurements made on free surfaces to provide a guide to the composition to be anticipated at grain boundaries. Certainly, it is well established that segregation to a free surface varies with temperature (Polak, 1990; de Rugg and Vielfhaus, 1986) and crystallographic orientation in a way that is similar to that for grain boundaries (Johnson *et al.*, 1978; Zhou *et al.*, 1981; Bark and White, 1987). Measurements of solute and impurity segregation to both grain boundaries and free surfaces in polycrystalline Fe–Sn alloys containing different amounts of bulk tin was undertaken by Lea and Seah, (1975a) and Seah and Lea (1975) and correlated with predictions of the Langmuir–McLean equation (2.4) (Figure 3.5). Here, the variation in the amount of segregant, X_{sn}^{ϕ}, at a grain boundary and a corresponding free surface with the bulk concentration of tin, X_{Sn}^{B} was established by a thermodynamic argument using an approximation for the vibrational, anharmonic and multiplicity entropy terms. Based upon the similarity of these dependences for the two interfaces, a method was proposed to evaluate grain boundary composition for dilute alloys measured from the free surface composition where the concentration at the grain boundaries is insufficient to promote intergranular fracture so that it is not possible to make measurements by techniques such as Auger electron spectroscopy.

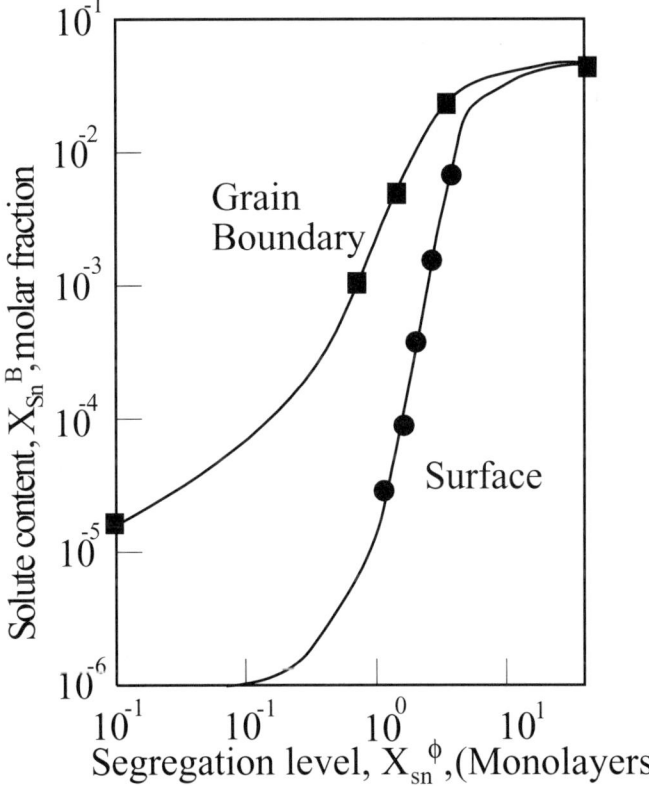

Figure 3.5 Relationship between the equilibrium surface and grain boundary segregation levels, X_{sn}^{ϕ}, as a function of bulk content, X_{sn}^{B}, at a temperature of 823 K (Seah and Lea, 1975)

Similar behaviour for elemental segregation for both free surfaces and grain boundaries has been observed in several solid state systems (Fukushima and Birnbaum, 1984), thereby providing support to this method for characterising the chemical composition of grain boundaries in polycrystalline materials (Tvrdý *et al.*, 1985a, b). However, it has to be recognised that, even if this correlation is valid for simple binary alloys, the presence of a third element can result in a breakdown (Bernadini *et al.*, 1985). One reason for the differences is the strong site competition between dissimilar atom species at free surfaces and at grain boundaries. An example is nickel, where repulsive interactions between sulphur and carbon atoms and attractive interactions between sulphur and calcium can occur at free surfaces but these are not observed at grain boundaries.

3.4   IRON ALLOYS AND STEELS

In this section we will consider both ferritic and austenitic alloys and steels, where there is considerable evidence that the amount and rate of segregation to grain boundaries of both impurity and alloying elements depend upon the total bulk composition of the particular material. For example, in commercial steels, which contain both many major alloying elements such as Cr, Ni, Ti, Al, V, Mo and C and minor impurity elements such as H, B, As, Sb, Sn and P, there are known to be complex interactions that influence the final composition of the surfaces, interfaces, grain boundaries and interphase boundaries, although, at present, few of these alloys and steels are well characterised by reference to the Periodic Table of elements (Figure 3.6). However, alloying and impurity elements can be divided into five broad categories based upon the changes in local grain boundary composition which are known to have an influence on the fracture and mechanical properties of ferritic and austenitic alloys and steels. The five categories are:

(a) Impurity elements that lie within Groups IV to VI that generally affect intergranular embrittlement, although carbon and boron appear to increase grain boundary cohesion. In addition, hydrogen is recognised to have grain boundary embrittling characteristics. These will be discussed more fully in Chapter 5.

(b) Elements that promote segregation of a specific atom species, usually an impurity, by cosegregation.

(c) Elements such as carbon and boron that segregate preferentially to grain boundaries and preclude further segregation of other impurity elements as a result of competition at the grain boundary sites.

(d) Elements that assist diffusion and thereby promote segregation of other, specific elements to the grain boundaries but do not themselves segregate.

(e) Elements that scavenge the matrix and interact with the impurity elements so that they become fixed and are unable to redistribute.

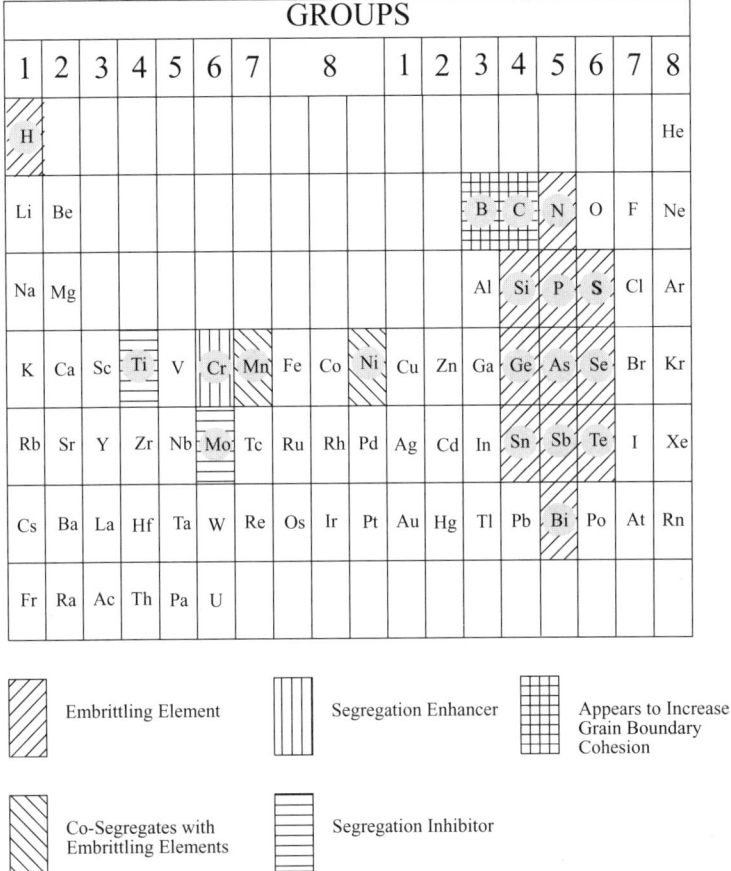

Figure 3.6 Periodic Table of elements summarising the role of various elements that influence segregation to grain boundaries together with their effect on embrittlement of steels

It is difficult to predict, *a priori*, the composition of surfaces, grain boundaries, interfaces or interphase boundaries in either ferritic or austenitic steels that have been subject to equilibrium or non-equilibrium heat treatment or neutron irradiation. As a consequence, there is substantial literature describing the measured local compositions in ferritic and austenitic alloys and steels. Table A.1 in the Appendix provides data compiled by Hondros and Seah (1978) for iron base alloys, but this has obviously been developed extensively since.

3.4.1 Iron-base Alloys

Of the widely recognised impurity elements in iron base alloys and steels that segregate to grain boundaries, three elements, namely, P, Sn and Sb, are proposed to behave similarly. However, they are all recognised to be sensitive to synergistic interactions with alloying elements present in the material. Phosphorus has the greater propensity to segregate and this is achieved at higher temperatures than for Sb and Sn in both alloyed and unalloyed ferritic materials, whereas in austenite steel this difference is less pronounced (Guttman, 1975; Guttman *et al.*, 1974; McMahon, 1976; Doig and Flewitt 1983). The differences are a simple reflection of the segregation energy for P which is much larger than for either Sn or Sb.

Although there is much literature available that purports to consider the segregation of impurity additions in isolation, this is indeed rare. More usually, the alloy is so complex that synergistic interactions and site competition at the grain boundaries or interfaces cannot be discounted. Even in the case of α-iron containing a known element addition, there are usually either further impurity or solute atoms present in very small concentrations, even down to the parts per billion level. As a consequence, this section addresses data relating to 'simple' austenitic and ferritic alloys and steels where there is evidence of known single-element segregation, usually the metalloid elements contained within Groups IV to IV of the periodic table (Figure 3.6), together with carbon and boron. Erhart and Grabke (1981) considered Fe–P alloys containing 0.003–0.3 wt % phosphorous, subject to a series of isothermal heat treatments in the temperature range 673–1073 K. The composition of the grain boundaries was measured using Auger spectroscopy undertaken on intergranularly fractured surfaces. Figure 3.7 shows the change in the measured grain boundary concentration in polycrystalline α-iron as a function of the bulk concentration of phosphorus after a 873 K heat treatment. For bulk concentrations of phosphorus of > 0.1 wt % there is monolayer coverage of the grain boundaries which become almost saturated. The temperature dependence of the measured phosphorus segregation in the α-iron (Figure 3.8) reveals a relationship between temperature, bulk concentration and degree of coverage at the grain boundaries that obeys the Langmuir–McLean equation (2.4). Moreover, sputtering experiments undertaken in the Auger electron spectrometer show that the phosphorus is concentrated in one to a few atom layers at the grain boundary (Figure 3.9), and this width is independent of time and temperature. Certainly, the spatial extent of similar grain boundary concentration profiles is supported by field ion microscopy measurements (Sakurai *et al.*, 1980).

In the case of antimony it has been demonstrated that segregation to grain boundaries occurs in the presence of nickel. This was originally explained as cosegregation (Gas *et al.* 1981; Ohtami *et al.*, 1979) but was later attributed to the fact that nickel lowers the solubility of antimony in α-iron and steels (Briant, 1987), leading eventually to precipitation of NiSb compounds at grain boundaries (Wirth *et al.*, 1986). More recently, Möller *et al.* (1986) and Jenko and Lucas (1996) examined the equilibrium segregation of antimony in iron base alloys, a Fe–C–Sb alloy and a silicon steel. Figure 3.10a illustrates that the coverage of Sb increases with the bulk Sb concentration and decreasing heat treatment temperature. The Langmuir–Mclean relationship plot based on equation (2.4) (Figure 3.10b) provides the segregation enthalpy $\Delta H_{Sb} = 18 \times 10^3 \, \mathrm{J\,mol^{-1}}$ and the entropy

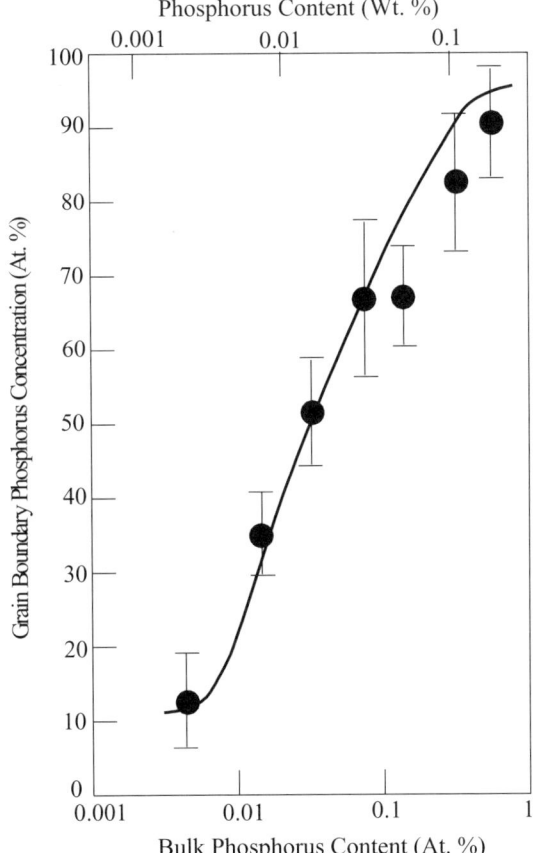

Figure 3.7 Equilibrium concentration of phosphorus at a grain boundary as a function of the bulk phosphorus content for binary Fe–P alloys heat treated at 873 K (Erhart and Grabke, 1981)

$\Delta S_{Sb} = 29.1 \, \mathrm{J \, mol^{-1} \, K^{-1}}$ and hence a measure of the free enthalpy for segregation in α-iron. This rather low value for the segregation enthalpy, when compared, for example, with that for phosphorus, explains why antimony is less likely to segregate to interfaces and grain boundaries in α-iron.

The behaviour of the more reactive chalcogen elements O, S, Se and Te is relatively simple since they are surface active and segregate with a large enrichment ratio provided their solubility in the parent phase is not sufficiently low that they are not retained in solution. Sulphur segregates more strongly than phosphorus at higher temperatures in pure α-iron. However, owing to strong interactions with major alloying elements such as Mn, this element can be precipitated within the grain interior and hence this interaction limits segregation in steels. Both Se and Te behave similarly to sulphur but are not usually encountered in commercial steels.

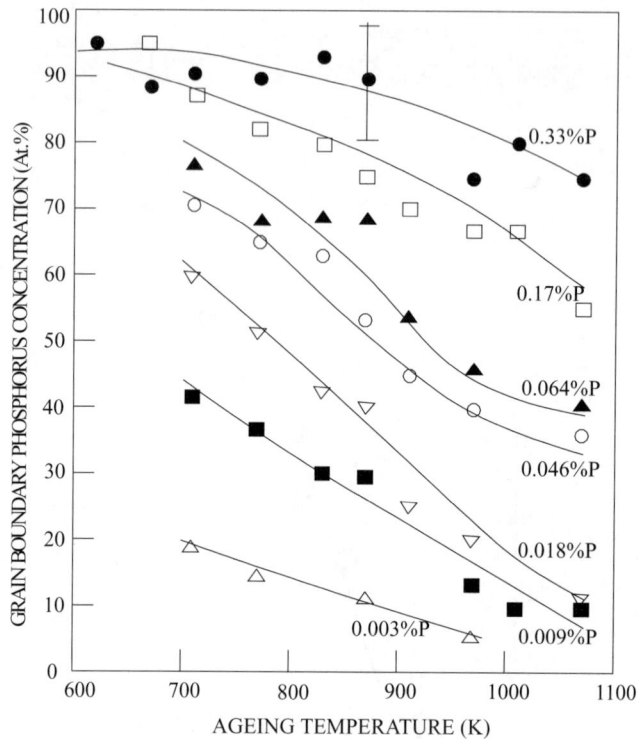

Figure 3.8 Variation in grain boundary phosphorus concentration with equilibrium temperature for different Fe–P alloys (Erhart and Grabke, 1981). Error bar applies to all measurements

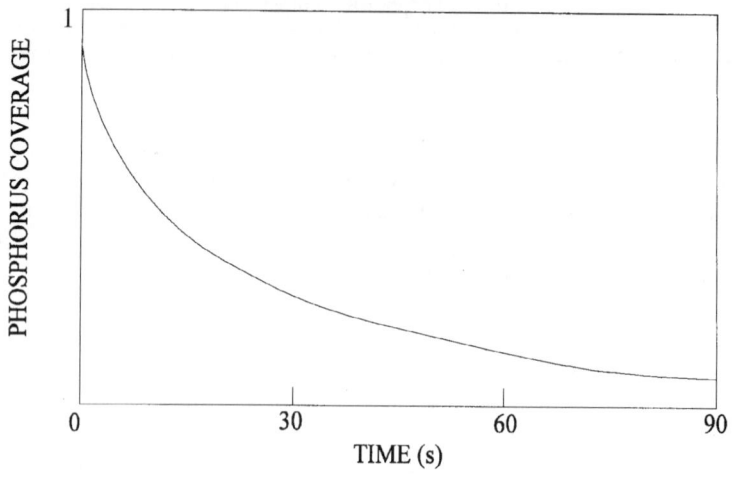

Figure 3.9 Typical sputter depth profile for phosphorus on an intergranular fracture surface of an Fe–P alloy (60 s sputtering corresponds to removal of two atomic layers)

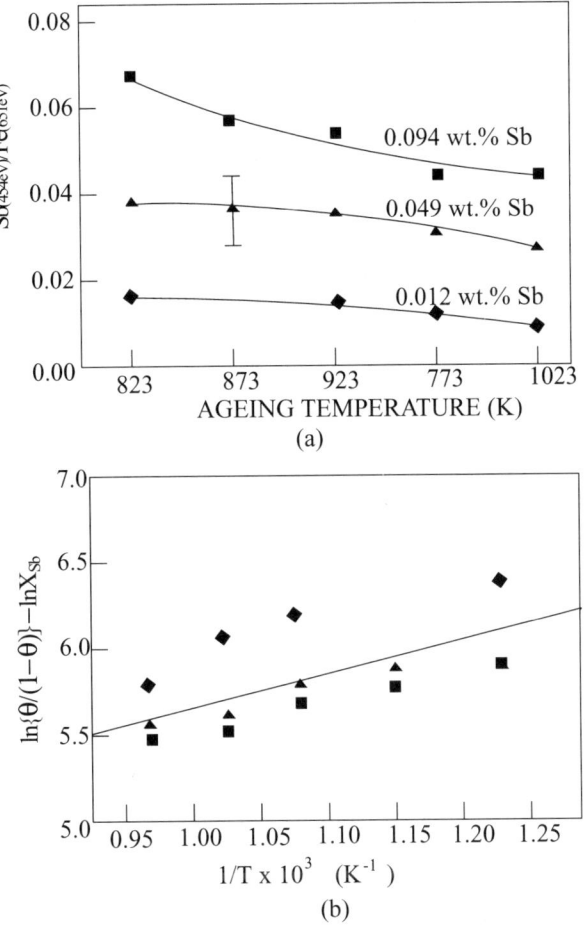

Figure 3.10 (a) Grain boundary concentration of antimony ratio of Auger peak heights plotted as a function of equilibration temperature for different Fe–Sb alloys. (b) The corresponding Langmuir–McLean plot (Jenko and Lucas, 1996)

Nitrogen is surface active in both the ferrite and austenite phase (Hondros, 1967) but, generally, in steels it is partially scavenged by Al and Si additives. However, it is recognised that free nitrogen is often present and segregation of this element to grain boundaries occurs. In the case of the other major interstitial alloying elements in α-iron and ferritic steels, it is carbon that segregates strongly to grain boundaries (Papazin and Besher, 1971) and, indeed, interacts with many metallic alloying elements. In steels the behaviour is complicated because the concentration of carbon is usually well in excess of the solubility limit and both soluble and carbon rich precipitates are involved in controlling the segregation of free carbon to the grain boundaries.

3.4.2   Cosegregation

The extent of impurity element segregation in ferritic and austenitic iron-base alloys and steels can be exacerbated by cosegregation with minor alloying elements (Kearns and Burstein, 1985). These processes usually occur when the segregating elements interact with each other but the interaction energy is insufficient for precipitation to occur within the matrix. If the latter were to occur it would effectively scavenge these elements (Guttmann, 1980). For example, tin and antimony are known to interact strongly with alloying additions of fcc nickel, but less so with bcc chromium; the reverse is the case for phosphorus. On the other hand, the affinity of manganese for both tin and antimony is intermediate between that for chromium and nickel. The segregation of phosphorus and antimony is enhanced by Cr, Mn and N (Mulford *et al.*, 1976, Dumoulin and Guttmann, 1980, Krahe and Guttmann, 1973, and Clayton and Knott, 1982). There is also enhancement of both Ni and Cr segregation by Mn and Si and surface interaction between segregating pairs of elements Ni–Sb and Ni–Sn in steels (Ciancelli *et al.*, 1978).

Nitrogen has been identified by Auger electron spectroscopy to segregate to grain boundaries in Si–Mn steels (Balasubramaniam and Stein, 1973). Cosegregation of phosphorus with chromium and molybdenum has been observed in 12 % chromium steels (Wild and Hickey, 1997). Here a large number specimens containing 0.2 wt % carbon, 0.02 wt. % phosphorus, 11 wt. % chromium and 0.7 wt. % molybdenum were exposed at a temperature of 763 K for up to 40 000 h. The samples were then fractured in ultrahigh vacuum and the grain boundary composition was determined by Auger electron spectroscopy. The phosphorus concentration was then plotted against the chromium and molybdenum concentrations (Figure 3.11a and b). Typically the phosphorus level increased with both chromium and molybdenum such that, at the maximum levels, 10 at. % phosphorus, 25 at. % chromium and 3 at. % molybdenum was observed at the grain boundary. Chromium and molybdenum are known to form Cr–Mo rich carbides at the grain boundaries in these steels and it was initially considered that the interfaces between such carbides and the matrix provided a more energetically favourable site for the phosphorus. However, a plot of phosphorus as a function of carbon composition shows no correlation (Figure 3.11c) and there is similarly no correlation between carbon and chromium or molybdenum. It is therefore presumed that the enhanced levels are the result of cosegregation between phosphorus, chromium and molybdenum. Consequently, it is clearly evident that the exact role of these interactions between alloying elements and impurity elements is far from well established and relies upon observation without a formal basis for rationalisation.

Several workers have attempted to rationalise these various observations. One approach, the application of grain boundary segregation isotherms to intergranular segregation, has been summarised by Misra and Rao (1993) (Figure 3.12). They show that the following can be considered using grain boundary isotherms:

(a) the separation and the kinetics of different stages of segregation;
(b) the interaction processes by using time–temperature space interaction maps and the role of alloying elements in the interaction processes;

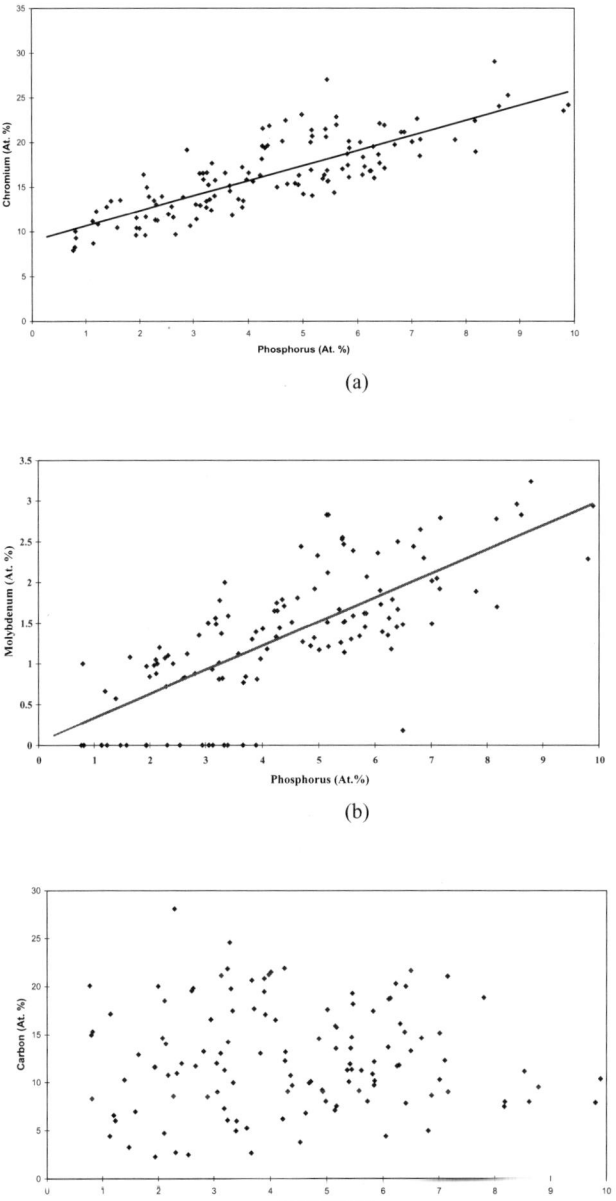

Figure 3.11 Grain boundary compositions of (a) chromium, (b) molybdenum and (c) carbon plotted against the phosphorous composition for grain boundaries a nominal 12 % Cr steel (Wild and Hickey, 1998)

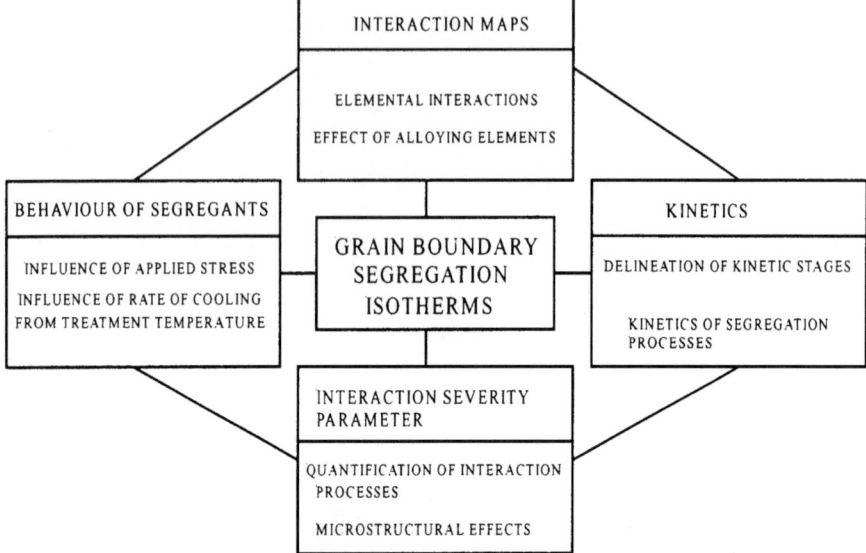

Figure 3.12 Applicability of grain boundary segregation isotherms in understanding intergranular segregation (after Misra and Rao, 1997) showing the interrelationship between behaviour of segregants, interaction severity parameter, kinetic parameters and interaction maps.

(c) the effect of microstructure or alloying elements on the segregation to grain boundaries, quantified with respect to the severity of the interaction;

(d) the effect of cooling rate after solution annealing or ageing;

(e) the effect of applied tensile stress on grain boundary segregation.

This approach originally adopted by Misra *et al.* (1987) considers in some detail the kinetics of segregation at constant temperature in the temperature range 773–853 K (Figures 3.13a to c) for a commercial ferritic steel with a nominal composition 2.5 % Ni, 0.4 % Cr, 0.3 % Mo and 0.1 % V (wt %) that has been quenched and tempered to produce a martensitic microstructure. These data are the normalised peak-to-peak height ratio, H_x, as a function of heat treatment time for the Auger electron spectrometry measurements covering the elements Fe, Cr, N, Sn, Sb, S and C. There are two distinct kinetic regimes revealed in the segregation isotherms in Figure 3.13. The first stage shows rapid enrichment of the grain boundaries in N and Cr and this is accompanied with desegregation of carbon. The simultaneous increase in Cr and N is attributed to the interaction between this pair of atoms and their cosegregation. This stage is followed by an increase in the concentration of P at the grain boundaries with little change in Sn and Sb. By comparison, at temperatures of 823 and 853 K there is a second stage that leads to increases in both P and V at the grain boundaries, the difference in the kinetics being a reflection of the different heat treatment temperatures. A third stage follows which produces a continuous increase

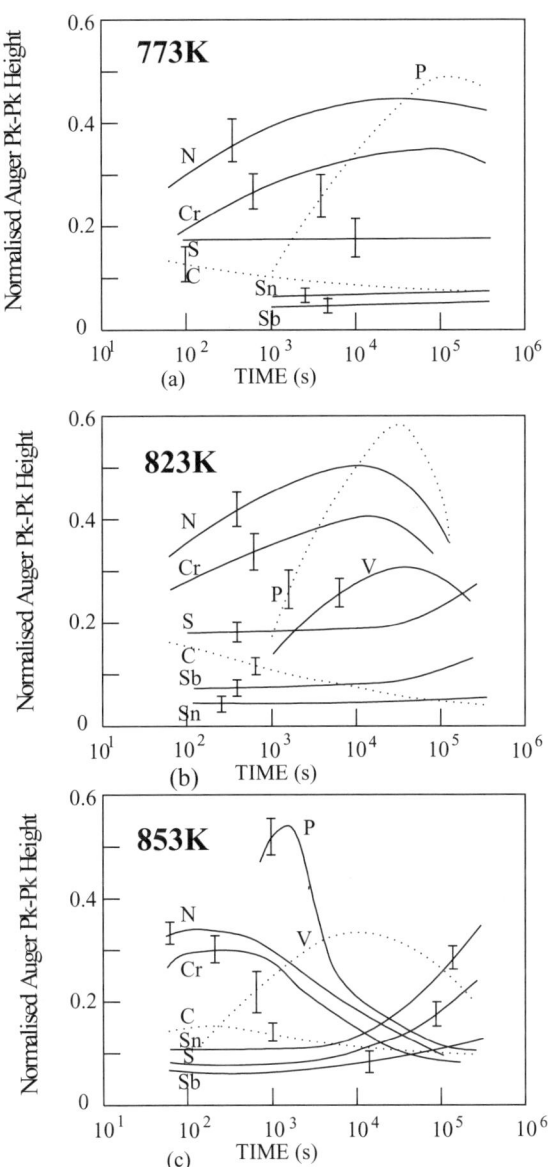

Figure 3.13   Segregation isotherms recorded at temperatures of (a) 773 K, (b) 823 K and (c) 853 K. Bars indicate the range over which values were measured (Misra *et al.*, 1987)

in grain boundary coverage with S and Sb. Manganese certainly enhances the segregation of the free nitrogen in these steels owing to the relatively strong interactions between these two elements (Guttman, 1976). In a high purity ternary Fe–2 % Mn–0.12 % Sb alloy containing less than 15 ppm N and 10 ppm C, segregation to grain boundaries is always accompanied by Mn and Sb in the temperature range for equilibrium segregation. Such observations are supported by those of Edwards *et al.* (1976) who correlated rapid segregation of Mn and N to grain boundaries in an Fe–8 % Mn alloy.

Misra and Balasubramanian (1989) observed quantitatively similar cooperative segregate and site competition at grain boundaries when these low-alloy steels were heat treated to produce bainite and ferrite–pearlite microstructures by comparison with those shown in Figure 3.13 for the martensite microstructure. However, there are differences in the kinetics and the overall amount of solute and impurity element segregation. For a constant heat treatment time, at a given temperature, the amount of grain boundary segregation for all elements is generally greater for the martensitic than for the ferrite–pearlite microstructures. Segregation of V occurs only in the martensitic and bainitic microstructures, whereas enrichment of the grain boundaries in Cr develops in all three microstructures. The differences in segregation of V and Cr arises because the interaction of Cr–N is less than for V–N. The lack of segregation of V in the ferrite–pearlite microstructure is consistent with differences in both the activity coefficients for Cr and V and with arguments that segregation of V occurs after dissolution of carbide precipitates has started. Moreover, when this carbide-forming element is present in solid solution, grain boundary segregation of P and N is small since the interaction energy between V and these elements is high. As a consequence, V in solid solution suppresses the diffusion of these elements to grain boundaries (Ustino and Shchikov, 1983), whereas any depletion by V_4C_3-type carbide precipitation enhances P and N segregation.

To provide a rationale to these complex interactions in this particular low-alloy ferritic steel, Misra and Balasubramanian (1990) have developed interaction maps. This approach enables the kinetic stages and the interaction processes for grain boundary segregation isotherms to be presented in the concise form of time–temperature space diagrams—interaction maps described in Figure 3.12. Figure 3.14a presents such maps for the martensitic and bainitic microstructures respectively, where, although the maps are similar, the interaction severity, ϕ (Figure 3.14b and c), differs for each microstructure. These are to be compared with Figure 3.15 for the corresponding ferrite–pearlite microstructure where both the map and the severity differ. The maps in Figures 3.14a and 3.15a are each constructed using those kinetic regimes of isothermal grain boundary segregation profiles that give significant cooperative interaction and site competition where the interaction process is active. To map these experimental boundaries for the interaction process, the time interval over which the particular interaction occurs, at the different temperatures, is obtained from the segregation isotherms. The time interval range for the interaction process is specified on the ordinate for each temperature, and the mapped regions define regimes for experimentally observed start and final temperatures of the interaction process. The broken lines show that measurement of the interaction between segregating elements has been interpolated beyond the time–temperature limits mapped.

These interaction maps provide an overview of possible interactions that may occur during isothermal heat treatment and offer one approach that may lead

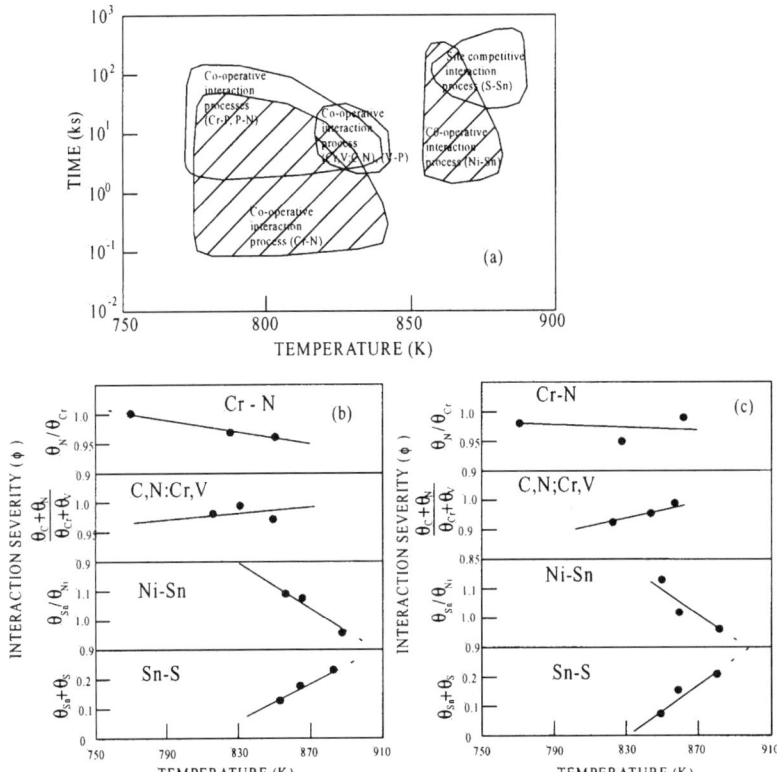

Figure 3.14 (a) Misra and Balasubramanian (1990) grain boundary interaction map for impurity and alloying elements in an Ni–Cr–Mo–V steel heat treated to give either a martensitic or bainitic microstructure, and the corresponding interaction severity ϕ as a function of temperature for (b) the martensitic and (c) the bainitic microstructure. Broken lines indicate that experimental evaluation of the interaction process was not undertaken beyond the time–temperature limits indicated in (a), (b) and (c)

ultimately to relating the grain boundary composition to the mechanical behaviour of engineering steels. Moreover, it is evident that there is a particular time–temperature regime where several cooperative processes between elements such as Cr–N, Cr–P, P–N, Cr–V : C–N and V–P may take place simultaneously. Comparing the maps for the martensite and bainite microstructures (Figure 3.14), and that for the ferrite–pearlite microstructure (Figure 3.15) reveals an interesting similarity, except for the absence of V : C–N and V–P interactions in the latter case.

Unfortunately, these Misra, Balasubramanian interaction maps fail to accommodate the strength of coupling between the interacting elements. Hence, they have to be used in conjunction with a knowledge of the amount of grain boundary coverage attained by particular elements. To accommodate this, the degree of interaction or

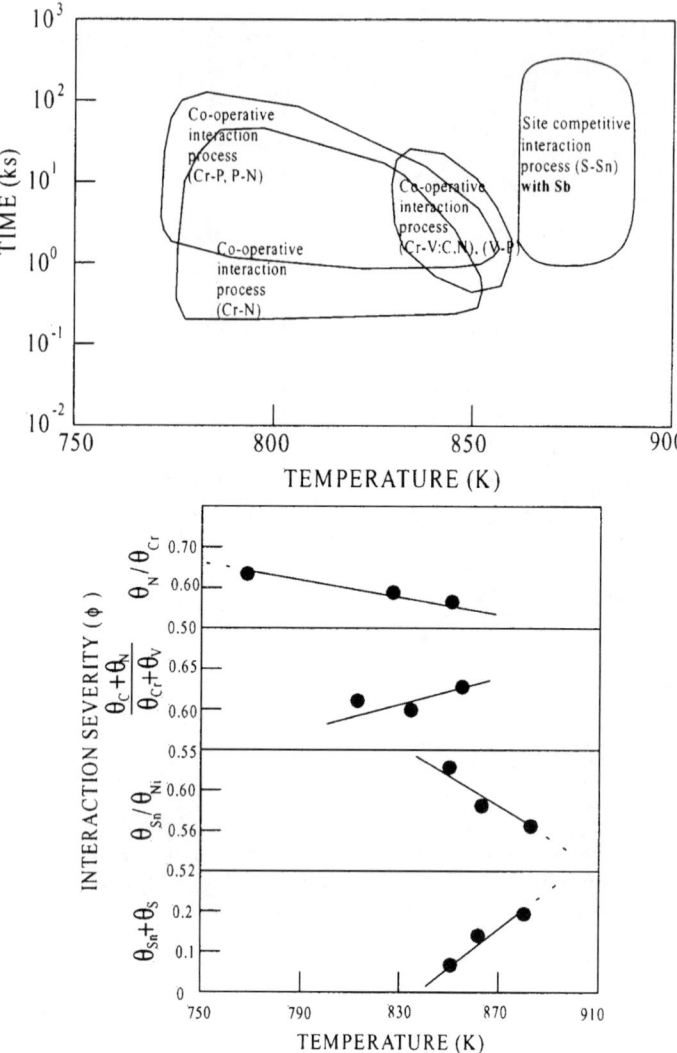

Figure 3.15 (a) Misra and Balasubramanian (1990) grain boundary interaction map for impurity and alloying elements in an Ni–Cr–Mo–V steel heat treated to give a ferrite–pearlite microstructure. (b) The corresponding interaction severity ϕ as a function of temperature. Broken lines indicate that experimental evaluation of the interaction processes was not carried out beyond the time–temperature limits indicated in (a) and (b)

interaction severity is shown as a function of time in Figures 3.14b and 3.15b for the three microstructures mapped. Here, the interaction severity, ϕ, is defined as the ratio of grain boundary coverage, θ, produced by the interacting elements X and Y, such that $\phi = \theta_y/\theta_x$. For site competition it is defined as the sum of the coverage of these interacting elements: $\phi = \theta_y + \theta_x$. This leads to a value of ϕ equal to unity for strong interaction between the segregating elements, whereas for a cooperative process ϕ will depend upon the stoichiometry of the compound formed. For this particular low-alloy ferritic steel, the degree of interaction follows the microstructural sequence martensite>bainite>ferrite plus pearlite, which is consistent with differences both in the driving force and the kinetics for segregation. These factors are controlled by the grain boundary crystal structure, the presence of interphase boundaries such as carbide–ferrite interfaces in the bainite and the density of dislocations (Ohtami and McMahon, 1994, Viswanathan and Toshi, 1975; Toshi, 1975). A further feature of the plots (Figures 3.14b and c and 3.15b) is that they reveal the contribution of emerging higher temperature interactions compared to those that occur at lower temperatures together with their relative magnitude. For example, in this particular case, site competition between S–Sn occurs when the Ni–Sn cooperative interactions are completed, and the severity of interaction between S–Sn is at a maximum when that for Ni–Sn is at a minimum.

3.4.3 Competition

Although the arguments in Section 3.4.2 have referred to site competition, it is appropriate to address this more specifically for ferritic and austenitic steels in view of the overall importance that this interaction has in defining the composition of grain boundaries and interfaces. Mulford *et al.* (1976) demonstrated that Ni and Cr both enhance the enrichment of P at the grain boundaries in Ni–Cr steels. However, the cosegregation of Cr and P and N and P was deduced from more general observations in low-alloy ferritic steels. Studies in Fe–C–S and Fe–N–S alloy steels by Tauber *et al.* (1978) show that both carbon and nitrogen can displace sulphur from grain boundaries in α-iron and thereby reduce their local concentration. The sulphur is progressively expelled from the grain boundaries as the bulk concentration nitrogen is increased (Figure 3.16a), for concentrations that are more than an order of magnitude greater than for sulphur. In the case of grain boundaries, these results can be rationalised on the basis of differences in the sizes of the two atoms, whereas for free surfaces different considerations have to be invoked (Lea and Seah, 1975b). Interstitial elements are able to occupy relatively small sites within the lattice at the grain boundaries and, as such, do not compete with substitutional elements such as Sn, Si and Te. However, although sulphur is a substitutional atom in the matrix, it is small compared with many corresponding substitutional atoms and, therefore, may be accommodated within the grain boundary region by distortion of the sites occupied by interstitial atoms.

In α-iron base alloys such as Fe–C–P, Fe–C–P and Fe–Cr–C–P, Erhart and Grabke (1981) used Auger electron spectroscopy to measure the effects of the alloying elements on the impurity segregation of P to the grain boundaries. Figure 3.16b shows the change in grain boundary composition for both C and P compared with the bulk carbon content for the particular alloy. Clearly, with increasing bulk carbon content, the concentration of P at the grain boundary decreases with an associated

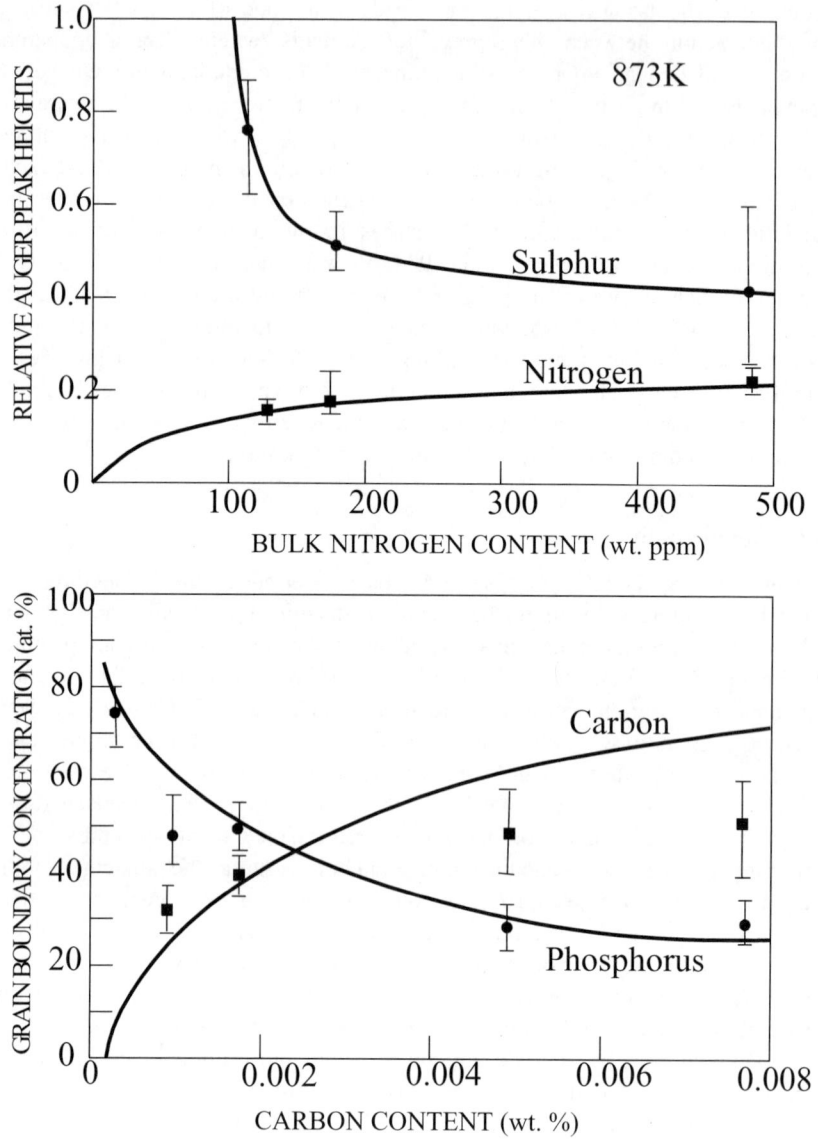

Figure 3.16   Effect of competition between elements at grain boundaries. (a) Influence of the bulk N content on the segregation of N and S in α-iron; bulk S iron. Bulk S content = 17 wt ppm (Tauber *et al.*, 1978). (b) Variation in the intergranular concentration of phosphorus and carbon with the bulk carbon concentration in Fe–0.17P annealed at 873 K (Erhart and Grabke, 1981)

increase in carbon level. This mutual displacement of phosphorus and carbon on the grain boundaries is widely observed and the coverage of boundaries by these elements, θ_p and θ_c, respectively can be described by

$$\theta_p = (1 - \theta_p - \theta_c)(x_p \exp -\Delta G_p^0/RT) \tag{3.1}$$

$$\theta_c = (1 - \theta_p - \theta_c)(x_c \exp -\Delta G_c^0/RT) \tag{3.2}$$

where x_m is the composition and ΔG_m^0 is the free enthalpy for the segregation of species m to a grain boundary. These equations do not accommodate energetic interactions, which are generally small compared with the main contribution from the competition of phosphorus and carbon for sites at the grain boundaries. Clearly, the carbon activity and concentration are each sufficient for the carbon to displace phosphorus atoms when reaching a grain boundary, even when the bulk phosphorus composition is large.

Morrisey (1997) demonstrated the dramatic influence that site competition can have on grain boundary segregation levels of phosphorus in iron. Samples of iron containing 1200 ppm phosphorus and 30 ppm carbon were solution annealed at 1473 K and then aged for 20 h at 923 K. Equilibrium segregation for iron containing this level of phosphorus predicts a grain boundary phosphorus level of 19.5 at. %, with 90 % of equilibrium segregation being achieved in less than 1 h. Measured levels indicated a grain boundary phosphorus composition of only 10 at. % and a carbon level of 30 at. %. However, when the carbon was removed during annealing by exposure to an oxygen atmosphere, then on aging the phosphorus segregation increased to 20 at. % and the carbon decreased to 1 at. %.

In a more complex alloy system such as Fe–Cr–C–P, which may be considered to be representative of a commercial ferritic steel, Auger electron spectroscopy has been used to identify enrichment of the grain boundaries in phosphorus, chromium and carbon. Clearly, the grain boundary concentration of phosphorus (Figure 3.17) at a given heat treatment temperature is influenced by the presence of Cr as a result of the interaction between Cr and C atoms. The carbon activity and the concentration of both free carbon and segregated carbon at the grain boundaries are decreased by the presence of the Cr; for a solid solution, carbon activity is decreased by interaction between the Cr and C atoms. However, at higher carbon compositions, carbide precipitates such as $(Fe, Cr)_3C$ and $(Cr, Fe)_7C_3$ are formed (Kellar et al., 1971). Since carbon is no longer able to displace this element from the grain boundary as a result of the decrease in carbon activity, there is further enrichment of the grain boundary with phosphorus. In general, alloying elements such as Cr, V and Mn, which form stable carbides, will decrease carbon activity and thereby promote P segregation to interfaces and grain boundaries, whereas Mo and Ti have a different role since they form transient compounds such as Mo–P (Yu et al., 1974).

Boron is a light element and, unlike many other elements that segregate to grain boundaries, does not induce embrittlement in steels (Martrepierre et al., 1980; Thomas and Henry, 1980; Hashimoto et al., 1984). However, it does present special problems when attempting to measure the magnitude and spatial extent of the

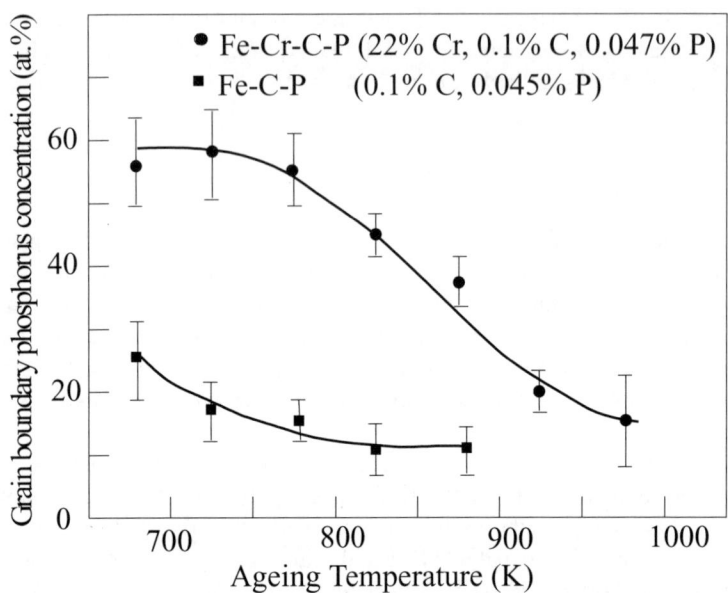

Figure 3.17 Variation in the grain boundary concentration of phosphorus with annealing temperature for Fe–0.1C–0.045P and Fe–2.2Cr–0.1C–0.047P alloys (Erhart and Grabke, 1981)

concentration of this element at grain boundaries (Karlsson *et al.*, 1988). The most fruitful way to obtain a measure of the coarse- and fine-scale segregation to interfaces and grain boundaries is to use a combination of the techniques described in Chapter 4. Boron is added to ferritic steels to increase the hardenability and it is well established that this element segregates to prior austenite grain boundaries and thereby increases the hardenability by suppressing the nucleation of ferrite (Taylor, 1993; Hway and Morris, 1980). Equally, in austenitic stainless steels, such as the Type 300 series, boron can be present either as a trace impurity or as an alloying addition, albeit in low concentrations (< 100 ppm by weight). In each of these cases, the segregation of the boron to the grain boundaries can occur either by equilibrium or by non-equilibrium processes. Karlsson *et al.* (1988), following work of Williams *et al.* (1976), successfully modelled the shape and variation of the segregation profiles developed under different start temperature and cooling rates for austenitic and ferritic steels. Figure 3.18a compares the calculated segregation profile for an austenitic stainless steel containing 130 at. ppm boron with values measured by He *et al.* (1982). Certainly, in Type 316 austenitic stainless steel the segregation of boron is mainly non-equilibrium, and Figure 3.18b shows the distribution of boron at grain boundaries in a high-boron, molybdenum-free steel, imaged using field ion microscopy with $^{11}B^{16}O_2^-$ ions. Segregation of this type is observed for a range of cooling

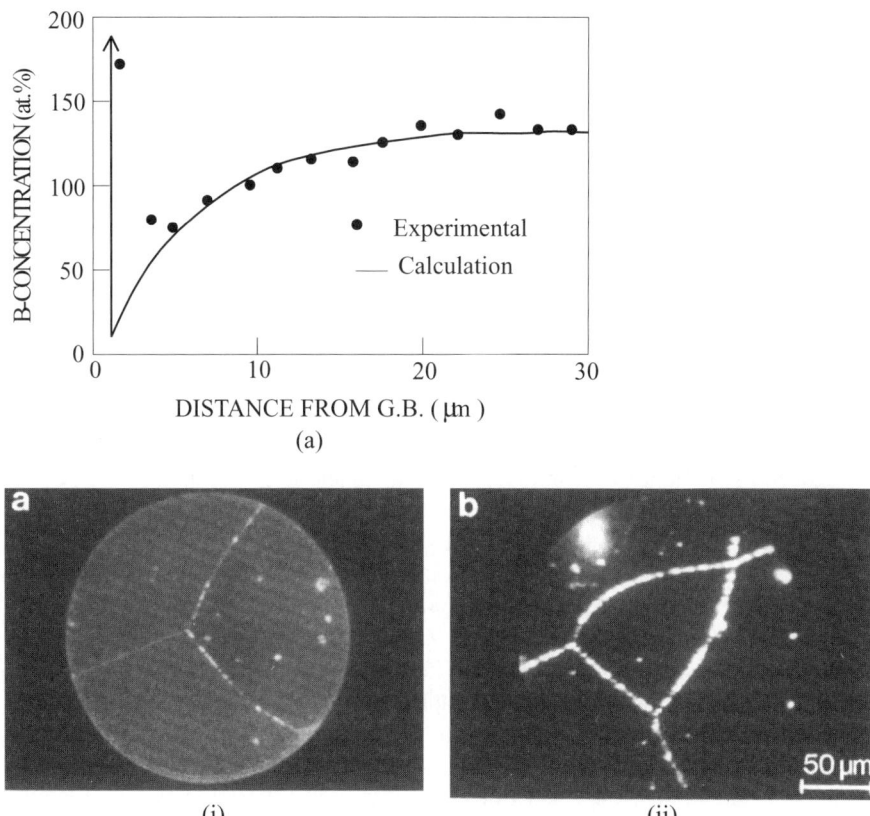

Figure 3.18 Segregation of boron to grain boundaries in austenitic stainless steel. (a) Calculated segregation profile and experimental segregation profile in austenitic with 130 at. ppm boron zero is position of grain boundary. In this calculation, the boundary was assumed to be a perfect sink for boron atoms. Starting temperature 1423 K, cooling rate 50 K s^{-1}, grain size 150 μm. (Reproduced from Karlsson *et al.* 1988). *Acta Metall.*, **36**, 1. (b) Ion micrographs showing the distribution of B in (i) the 'high-B' steel and (ii) the 'Mo-free' steel after holding at 1073 K for 3 h. The micrographs were obtained with $^{11}B^{16}O_2$ ions

rates and start temperatures: 0.3–500 K s^{-1} and 1073–1523 K. The amount of segregated boron increases with increasing start temperature and is greater at the intermediate cooling rates. The width of the boron-enriched zone increases with higher start temperatures and lower cooling rates. Certainly, a narrow boron-rich grain boundary layer, < 2 nm, forms in high-boron Type 316 austenitic stainless steel, whereas monolayer enrichment is found in lower boron steels.

Three features, in particular, are associated with the segregation of boron in austenitic steels:

(a) the interactive effect of B and Mo on grain boundary composition and precipitation,
(b) the formation of boride phases other than those predicted by the equilibrium phase diagram and
(c) the marked difference between grain boundary and matrix composition.

The presence of Mo in Type 316 austenitic stainless steel, for example, has a pronounced effect on the subsequent precipitation, with the Mo-containing steel giving Mo-rich borides of the M_3B_2 and M_5B_3 types and tetragonal M_2B, whereas for Mo-free steels generally, the chromium-rich M_2B boride forms with an orthorhombic structure (Goldschmidt, 1971; Padilha *et al.*, 1982; Finlan *et al.* 1986). Certainly, the enrichment of boron at grain boundaries affects precipitation behaviour, particularly for higher-boron steels where the precipitation of the three types of boride occurs. This leads to compositions at the boundaries that are significantly different from the bulk. For example, in high-boron steels the matrix adjacent to the boundary becomes depleted in Cr and Mo and enriched in Fe, Ni, Mn, B and C, and in boron-free steels Fe, Cr, and Mn are depleted at the boundaries. However, in the latter case there is also Mo, B and C enrichment at the boundary; B and Mo have a synergistic effect on the boundary composition and precipitation. Similar effects have been observed with B and Nb, and the M_3B_2 phase precipitates in high-boron steels and also in Mo plus V and Nb, containing boron steels. Under these circumstances, then, Mo, V and Nb have a similar contribution to the changes leading to final grain boundary boron composition (Karlsson and Norden, 1984, and Martrepierre *et al.*, 1980).

Certainly, the concentration of boron, at a grain boundary increases as the heat treatment temperature is lowered (Morral and Cameron, 1980), following a relationship of the form given by the Langmuir–Mclean equation (2.4):

$$C_B = C_B(1 - f)e^{Q/RT - \Delta S/R} \tag{3.3}$$

where Q is the activation energy, ΔS is the binding entropy of boron to the grain boundary and f is the fraction of relevant grain boundary sites occupied by boron or another atom species. For the case of a system containing carbon atoms, f is $(C_B + C_C)$, where C_C is the fraction of relevant grain boundary sites occupied by carbon atoms; this does not accommodate either elastic or chemical boron–carbon interactions in the grain boundary. Moreover, the beneficial effect of boron in suppressing intergranular embrittlement of both ferritic and austenitic steels (Taylor, 1993, and Hway and Morris, 1980) is a direct result of site competition, whereby the selective occupancy of the grain boundary by boron and carbon suppresses the ability of elements such as phosphorus and others in Groups V to IV of the periodic table (Figure 3.6) from segregating to a boundary.

Seah *et al.* (1979), Lea (1980) and Yishi and Huiliang (1983) have shown that P segregation to grain boundaries in low-alloy ferritic steels can be reduced substantially by additions of lanthanum, since this element promotes the formation of phosphorus-rich precipitates by selective competition. Cerium which is also a rare-earth element segregating readily to grain boundaries during isothermal or non-equilibrium heat treatments. Figures 3.19a and b show the grain boundary composition for C–Mn ferritic steel both with and without a cerium addition measured by secondary ion mass spectroscopy after ageing for different times at a temperature of 723 K (Yuan *et al.*, 1994). The segregation of Ce increases with time; the P generally reaches equilibrium at the grain boundary, but Ce or Mn do not achieve equilibrium. Since the Ce competes very effectively when reaching the grain boundaries compared with either Mn or P, then, with increasing heat treatment time, such that it is enriched at grain boundaries which leads to desegregation of Mn and P. Moreover, since Mn combines with M_3C-type carbide precipitates the segregation of Mn is further reduced. As a consequence, the segregation of the Ce to grain boundaries results in a decrease in the grain boundary composition for these two important elements, P and Mn.

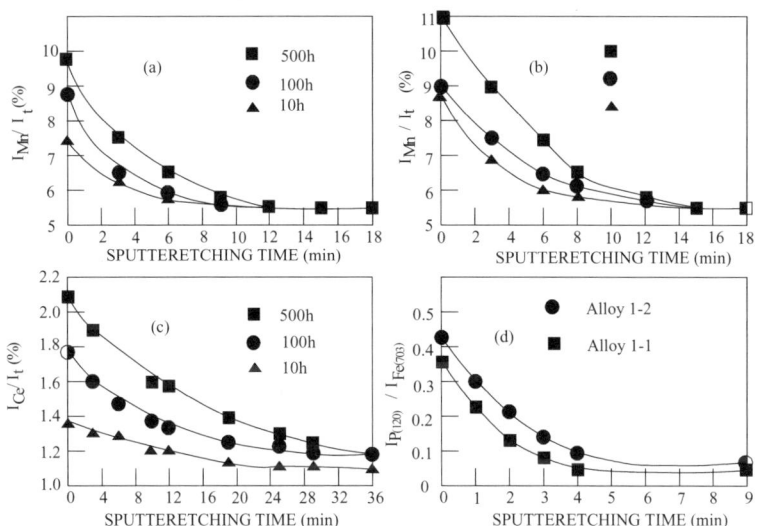

Figure 3.19 Ratio of the SIMS secondary ion peak height of Mn to the total secondary ion peak height of all elements, I_{Mn}/I_t, as a function of the sputter-etching time of an argon ion beam on the fracture surfaces of carbon manganese steel (a) with Ce (alloy 1-1) and (b) without Ce (alloy 1-2) aged for different times at 773 K. (c) Ratio of the SIMS secondary ion peak height of Ce to the total secondary ion peak height of all elements, I_{Ce}/I_t, as a function of the sputter-etching time of an argon ion beam on the fracture surface of alloy 1-1 aged for different times at 773 K. (d) Ratio of P 120 eV Auger peak height to Fe 703 eV Auger peak height, $I_{P120/I_{Fe703}}$, as a function of the sputter-etching time of an argon ion beam on the fracture surfaces of (■) alloy 1-1 and (●) alloy 1-2 aged for 100 h at 500 °C (after Yuan *et al.*, 1994)

3.4.4   Scavenging

Titanium is typically added to commercial steels, in particular low-alloy ferritic steels, for selective scavenging of impurity elements such as P and Sb. As a consequence, the segregation of both P and Sb to the prior austenite grain boundaries is suppressed during equilibrium and non-equilibrium heat treatments. Similarly, Mo can also limit segregation of P to grain boundaries and interfaces in ferritic steels (Schultz and McMahon, 1972; McMahon *et al.*, 1977), but the effectiveness of this element depends upon the concentration of both the phosphorus and carbon as well as the thermal history. Kaneko *et al.* (1965) have shown that both Ti and Mo readily form phosphide precipitates in the matrix, thereby precluding the availability of phosphorus for segregation to the grain boundaries or interfaces. Carbon can decrease significantly the amount of soluble alloying metals in a low-alloy ferritic steel by precipitating these reactive metals as carbide-type precipitates. Thus, in turn, it will decrease the amount of scavenger available and, thereby, release the impurity elements into solution to effect grain boundary segregation. Conversely, in the case of Cr, which again can readily form carbides, the smaller amount of metallic cosegregant available reduces grain boundary enrichment with impurity elements.

3.4.5   Effect of Stress

The effect of an applied tensile or compressive stress at temperature to ferritic and austenitic steels on the segregation of impurity and solute atoms to surfaces, interfaces and grain boundaries is complex. In particular, it is the role of stress on the diffusion of elements such as sulphur and phosphorus to grain boundaries that has been considered in some detail. For example, Misra (1996) examined segregation of sulphur in an Ni–Cr–Mo–V ferritic steel containing 0.01 wt. % S using tensile geometry test specimens. The grain boundary isotherms for the various constituent elements of this steel aged at a temperature of 883 K (Figure 3.20a), reveal that sulphur segregates significantly in the unstressed condition. However, when a tensile stress is applied (Figure 3.20b), there is an increase in the grain boundary sulphur coverage, as measured by Auger electron spectroscopy, over the initial heat treatment period of 3 h, but this is followed by a decrease to equilibrium coverage after 25 h ageing under stress. The activation energy for sulphur diffusion under a tensile loading is $106.3 \, \text{kJ} \, \text{mol}^{-1}$ (Misra and Rao, 1993) and the initial sulphur enrichment of the grain boundary is attributed to the rapid diffusion of this element from one side of the grain boundary to the other under the associated potential gradient. However, since this follows predominantly grain boundary paths, the subsequent decrease in sulphur as the system undergoes further ageing with the stress applied is consistent with stress induced inhomogeneous distributions of sulphur which disappear as more slowly diffusing atom species migrate, producing Coble creep, and reduce the transient sulphur concentration gradient. Hence, thermodynamically the tensile stress facilitates the transfer of sulphur atoms from those on grain boundaries parallel to the stress axis to those perpendicular to it.

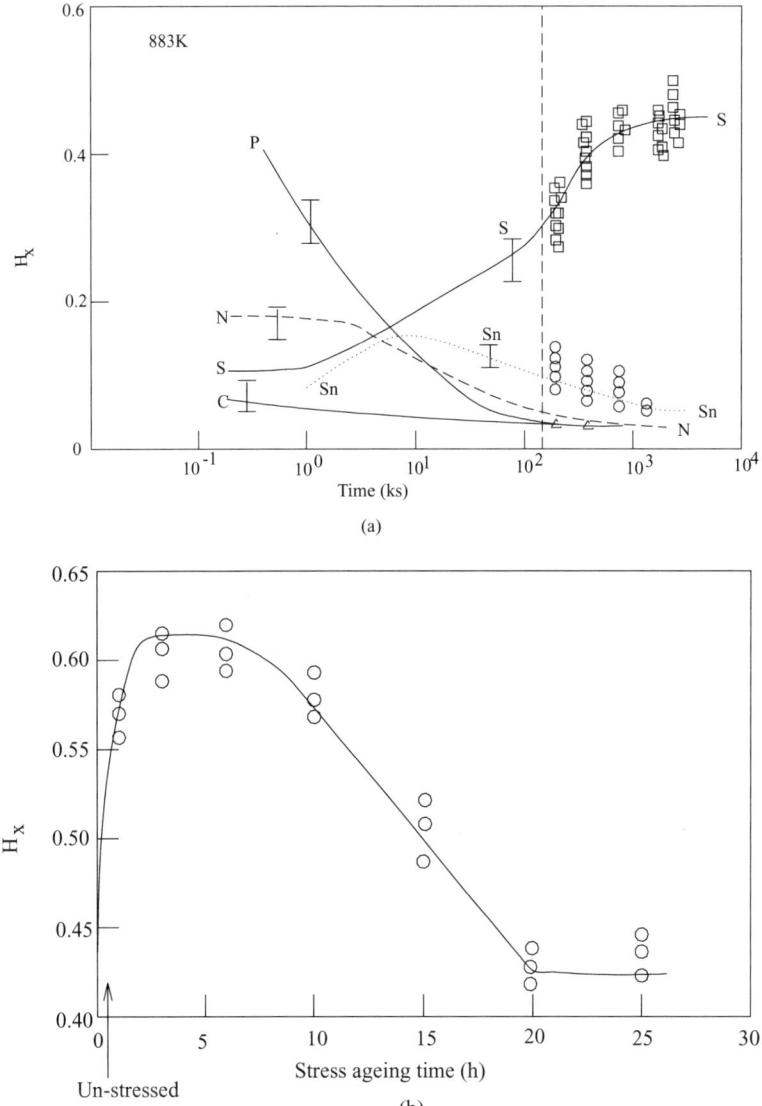

Figure 3.20 Grain boundary segregation isotherms (a) recorded at 883 K on unstressed low-alloy steel and (b) recorded on steel in the stressed condition after prior ageing this in the unstressed condition for 2160 ks at 883 K (after Misra, 1996)

In addition, there have been a range of observations where sulphur has concentrated on grain boundaries in regions just ahead of cracks in low-alloy ferritic steels that have been subject to a tensile stress at a given temperature over an extended period of time (Hippsley *et al.*, 1980; 1982, 1984; Bowen *et al.*, 1984; Lewandowski *et al.*, 1987; Hippsley, 1987). This is consistent with the predictions of the Rauh and Bullough model described in Chapter 2. It is addressed further in Chapter 6 where the effects of grain boundary segregation are considered with respect to stress relief and reheat cracking in weldments and creep for both ferritic and austenitic steels.

3.4.6 Precipitation

One of the most well-documented examples of the role of grain boundary precipitation is the sensitivity of austenitic stainless steels to intergranular corrosion, and in particular stress corrosion cracking brought about by the precipitation of Cr-rich carbides of the $M_{23}C_6$ type at grain boundaries (Cowan and Tedmon 1973; Bain *et al.* 1933). Here, the formation of the carbide requires the diffusion of chromium towards the grain boundaries, and this causes a zone that is depleted in chromium to form in the immediate vicinity of the boundary. This Cr depletion model for sensitisation to stress corrosion cracking was originally proposed by Bain *et al.* (1933), and since then refined thermodynamic and kinetic models have been developed on the basis of these proposals by Stawstrom and Hillert (1969) and Tedmon *et al.* (1971). These approaches are based upon models developed by Aaron and Aaronson (1968) who assume that the whole grain boundary area acts as a collector plate to supply the growing $M_{23}C_6$ precipitates with chromium (Figure 3.21a) (Thorvaldsson and Dunlop, 1983). Here, the valid assumption is that the rate of diffusion of chromium at the grain boundary is much faster than within the adjacent bulk grain. These models predict that the width of the Cr-depleted regions adjacent to grain boundaries in these austenitic stainless steels can be up to 500 nm wide [full width at half-height (fwhh)] when heat treated to produce the 'sensitized' condition.

High spatial resolution microanalysis undertaken on a range of austenitic stainless steels, including the Type 316 series and aged niobium- and titanium-stabilized stainless steels, usually on thin foil specimens and evaluated in the scanning transmission electron microscope using energy dispersive X-ray spectrometry, supports these predictions both in magnitude and spatial extent (Figure 3.21b). This figure shows specific chromium profiles (i), (ii) and (iii) measured about grain boundary segments between $M_{23}C_6$ carbide precipitates in a Ti-stabilized austenitic stainless steel containing ~ 17 % Cr, for increasing ageing times at a temperature of 1023 K. This follows a water quench to suppress carbide precipitation where initially a pronounced, but relatively narrow, depletion profile of ~ 14 wt % Cr develops. As ageing time increases, the maximum in the depletion is progressively modified and reduced and the profile width increases, until, after extended ageing, it is removed. In terms of stress corrosion cracking, the latter is referred to as the healing stage, since the susceptibility to increased corrosion is removed. Recently, Laws and Goodhew (1991) have shown that the specific grain boundary orientation has a significant influence on the magnitude of the depletions developed in these steels. This is consistent with the observations of Thorvaldsson and Dunlop (1983) and Chastell and Flewitt (1979), where Cr depletion zones do not form at coherent twin

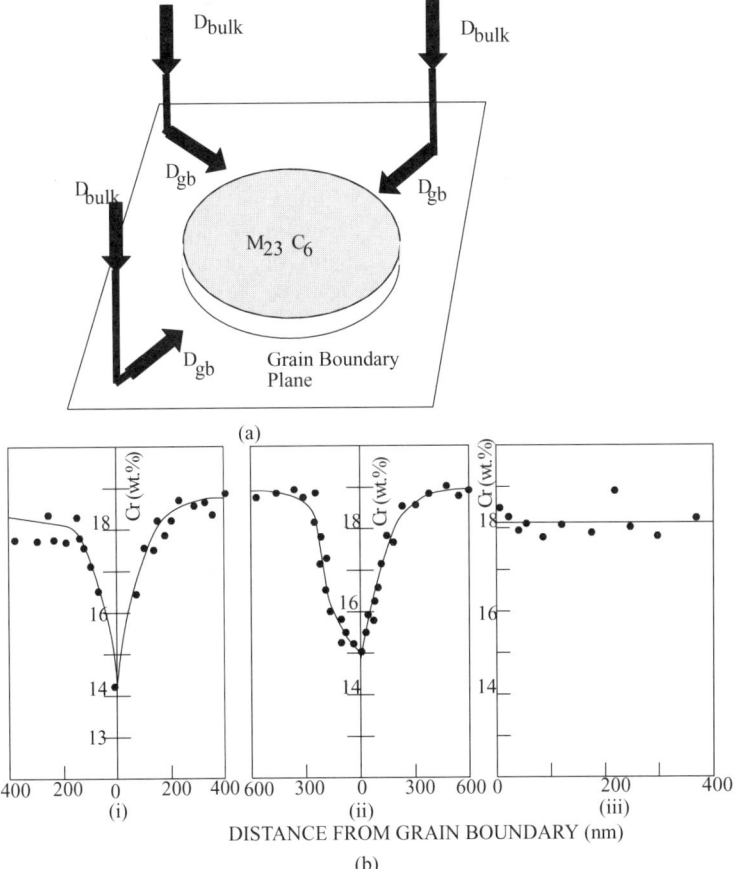

Figure 3.21 (a) Schematic diagram of the collector plate model. The grain boundary area acts as a collector plate to supply the growing $M_{23}C_6$ carbide with chromium (after Aaron and Aaronson, 1968). (b) Variation in the grain boundary chromium composition profile in a Ti-stabilised austenitic stainless steel heat treated at 1023 K with increasing ageing times: (i) initial quench, (ii) intermediate time, (iii) heated condition

boundaries, whereas the depleted zones that form at incoherent twin boundaries are similar to those that form at grain boundaries.

3.4.7 Irradiation of Steels

It is well established that, for metals and alloys subject to irradiation by neutrons, ions and electrons, the diffusion of point defects to sinks such as grain boundaries, free surfaces or interphase boundaries can lead to a change in the local composition in the vicinity of these sinks (Anthony, 1971; Marwick, 1978). This irradiation-induced segregation is attributed to a preferential coupling to point defects, usually interstitial atoms, with minor alloying elements or impurities. Diffusion of the complex to the sinks is driven by either the point defect flux or the differential diffusion rates of the various elements, arising from the species jumping at relatively different rates into adjacent vacancy sites: both lead to enrichment of grain boundaries and interfaces.

As we have shown in the preceding Section 3.4.6, sensitisation in austenitic stainless steels is generally related to the depletion of chromium in regions at and adjacent to the grain boundaries. This is mainly the result of a heat treatment that leads to the precipitation of a Cr-rich, $M_{23}C_6$-type carbide precipitate at the grain boundaries with a concomitant reduction in the chromium concentration in adjacent regions. Neutron irradiation provides an alternative driving force, so that irradiation-induced segregation can produce similar chromium depletions but in the absence of precipitation. Such changes in the composition of grain boundaries have been measured in for example 20 wt. % Cr–25 wt. % Nb-stabilised stainless steels after neutron irradiation (Norris *et al.*, 1992). These steels contain a range of other minor alloying and impurity elements, but those of particular interest are Fe and Si, the latter being typically 0.45–0.75 wt. %. Figure 3.22a, (Norris *et al.*, 1986) shows that after neutron irradiation, at a temperature of 691 K, the grain boundaries become decorated with intermittently distributed elongated carbide precipitates and there is an increase in the density of dislocations, loops and precipitates within the grains. The composition of the grain boundary regions between these precipitates, measured using high spatial resolution STEM-EDS X-ray microanalysis on thin foil specimens (Figure 3.22b), shows chromium depletion and nickel enrichment near to the grain boundaries. Generally, these composition profiles become wider in extent but smaller in magnitude as the neutron irradiation temperature is raised. Moreover, silicon segregation accompanies these composition changes, but again the effect is more pronounced at the lower irradiation temperatures. Figure 3.23 shows corresponding composition profiles measured on a similar material, but in this case it has been thermally sensitised by ageing at 823 K for 500 h. It is clear that the chromium depletion and nickel enrichment associated with the neutron irradiation is similar to that developed by thermal ageing, although in the latter case there is no evidence for Si enrichment at the grain boundaries. Similar irradiation-induced segregation for impurities such as Si and P have been measured in a range of austenitic stainless steels, particularly those in the Type 300 series, including Types 316, 304 and 348 (Garjanolli *et al.*, 1988; Rehn *et al.*, 1979; Brimhall *et al.*, 1984, 1983. Titchmarsh and Dumbill, 1996). Where chromium profiles in austenitic stainless steels have been examined in detail for both the irradiated and non-irradiated conditions, these reveal a Cr depleted region adjacent to a grain boundary, but an increase in concentration at the position of the boundary (Figure 3.24). However, there is no carbide precipitate found at the grain boundary for the irradiated material. These measurements, made using high-resolution $\sim$ 1 nm diameter electron probe, STEM-EDS X-ray microanalysis, suggest that depletion is not readily derived by this microanalysis technique; the true profile is limited by the spatial resolution of the

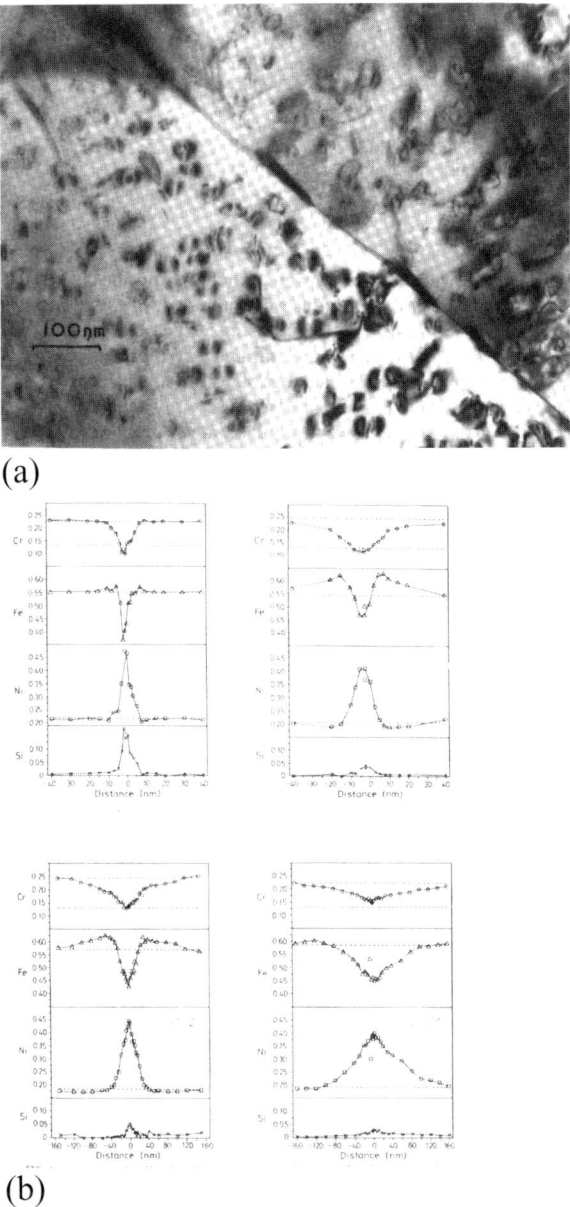

Figure 3.22 (a) Transmission electron micrograph of a foil from 20 wt % Cr–25 wt % Nb stainless steel, neutron irradiated at a temperature 691 K. Precipitates are visible in the grains and in the boundary. (b) Compositional profiles (Cr, Fe, Ni and Si = I) across boundaries from four positions along the grain boundary in (a) (Norris *et al.*, 1986) shows depletion of Fe and Cr but enrichment of Ni and Si

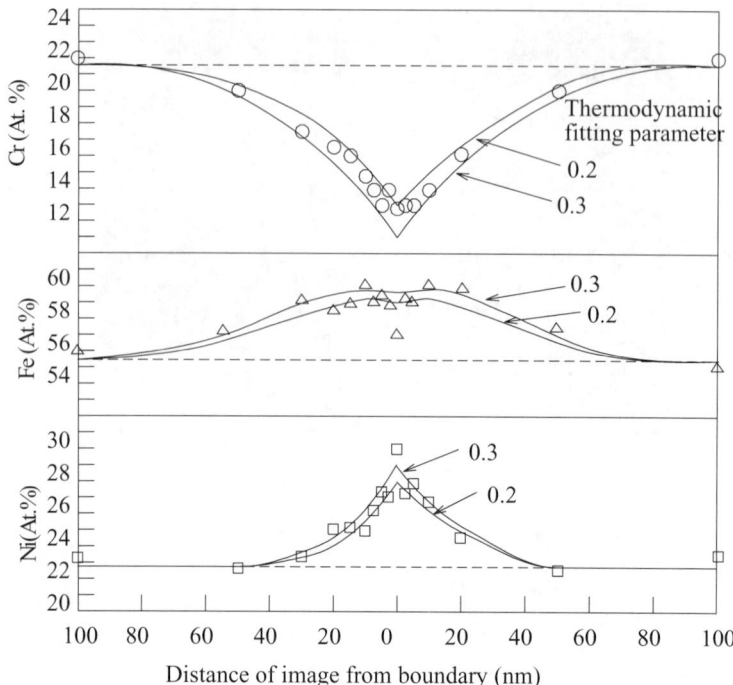

Figure 3.23 Compositional profile across a boundary in the thermally aged stainless steel of composition given in Figure 3.22. The solid lines are model fits compared with the experimental data (Norris *et al.*, 1986)

analysis even of these high-resolution electron probes. The exact form of these grain boundary segregation profiles has to be derived using appropriate deconvolution procedures of the type described by Doig *et al.* (1981) and Carter *et al.* (1994).

Segregation to grain boundaries after irradiation using 1 MeV electrons and 52 MeV Cr^{++} ions at temperatures in the range 623–923 K has been measured in ferritic stainless steels of the type containing Fe–12 wt. %Cr (McMahon *et al.*, 1986). The microstructure of these steels, in general, is martensitic, consisting of packets of lath martensite contained within prior austenitic grains. Table A.2 (Appendix) summarises the composition changes at the prior austenitic grain boundaries produced after irradiation at a temperature of 823 K. These results are complicated by two separate, but small, effects:

(a) irradiation-induced precipitation and
(b) irradiation-induced segregation.

Generally, there is depletion at grain boundaries of elements such as chromium, molybdenum and vanadium and enrichment in nickel, silicon and phosphorus; undersize solute atoms segregate to the grain boundary sinks, whereas the oversize solute atoms diffuse away. Table A.3 (Appendix) shows the size of the atoms in

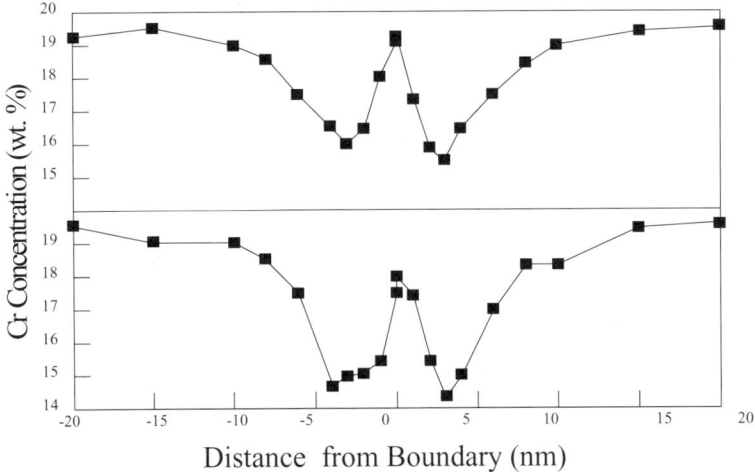

Distance from Boundary (nm)

Figure 3.24 Experimental measurements of the Cr concentration at two boundaries in neutron-irradiated stainless steels. Each profile clearly demonstrates the presence of a depletion adjacent to but not at the boundary, in addition to segregation at the boundary (Titchmarsh and Dumbill, 1996)

relation to the host lattice atoms. As a consequence, solute enrichment at grain boundaries in irradiated ferritic steels can be explained for interstitial–solute boundary segregation since all the solute atoms described diffuse more rapidly than iron atoms, with the exception of the nickel. However, elements such as phosphorus and silicon are related to the interstitial flux, whereas the depletion of elements such as Cr, Mo and B can be described by either mechanism. Certainly, in the case of heavy ion and electron irradiation for these particular steels, the sense of the segregation is given by the size factor: oversize leads to depletion and undersize to enrichment. Moreover, precipitation of $M_{23}C_6$-type carbide is enhanced by electron and ion irradiation if appropriate concentrations of either vanadium or molybdenum are also present.

The complexity of the martensite plus carbide precipitate microstructures in these types of steel can lead to misinterpretation of the role of irradiation. At prior austenite grain boundaries, Takahashi *et al.* (1981) observed only chromium depletions, whereas Ohnuki *et al.* (1982) found chromium enrichment and, in the presence of silicon, Muroga *et al.* (1989) observed chromium depletion, silicon enrichment and a dependence of nickel segregation on the type of solute. Clausing *et al.* (1986) detected enrichment of these prior austenitic grain boundaries in chromium, nickel and silicon. However, in a dedicated fast neutron irradiation investigation of martensite lath boundary composition, Morgan *et al.* (1992) measured both nickel and silicon enrichment. The chromium showed a more complex behaviour, with localised enrichment at lath boundaries and associated depletion in the adjacent matrix. This provides clear evidence of the important role of martensite lath boundaries as major point defect sinks during neutron irradiation.

3.5   Non-Ferrous Metals and Alloys

In the case of non-ferrous metals and alloys, the distribution of solute and impurity elements within the overall microstructure is just as important as for steels in controlling the physical and mechanical properties. Indeed, some of the earlier observations of segregation to grain boundaries in polycrystalline materials were made in non-ferrous metals and alloys. In the case of segregation to the surface of non-ferrous solid solutions, Burton and Machlin (1976) reviewed published data and proposed an empirical method for identifying the segregating components in a binary terminal solid solution based upon the characteristics of the binary equilibrium phase diagram (Figure 3.25). These data, given in Table A.4 (Appendix), show in the predicted column that, if the distribution coefficient is less than unity, namely $X_{solid}/X_{eliquid} < 1$ in Figure 3.25, then solute will segregate to the surface, whereas, if $X_{solid}/C_{eliquid} > 1$, solvent atoms segregate. This approach is rationalised on the basis that liquid in equilibrium with a solid is characteristic of a surface in equilibrium with the same solid, since the liquid and the surface have a lower coordination and are more disordered than the bulk. Predictions of this correlation are given in Table A.4 and compared with the experimental observations, and there is good agreement apart from platinum-base alloys. Certainly, as an empirical approach it offers a simple predictor which agrees with most observations. We now consider some more important and commonly encountered polycrystalline non-ferrous metals and alloys, including nickel, copper, silver, gold and aluminium.

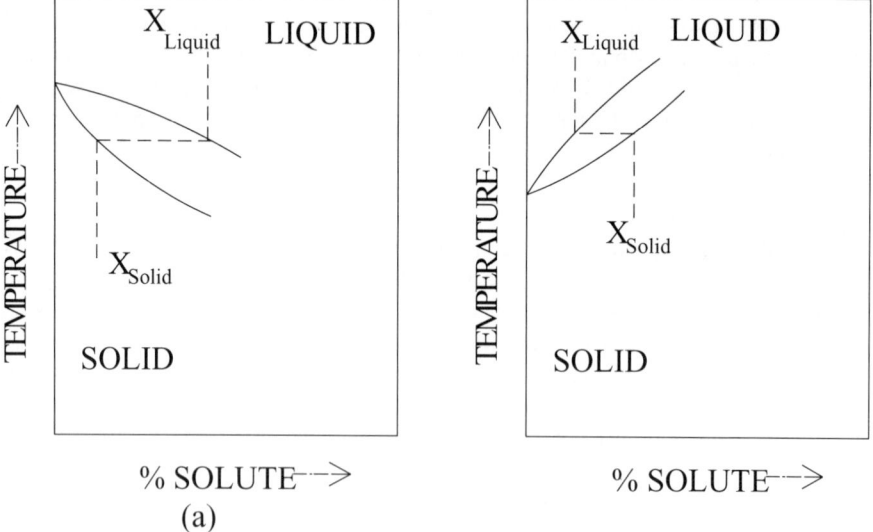

Figure 3.25   Equilibrium diagrams for the terminal solid solution region leading to (a) segregation of the solute and (b) segregation of the solvent (Burton and Machlin, 1976)

3.5.1 Nickel-base Alloys

One of the major applications of nickel is in the nickel base superalloys which are used for their superior high-temperature creep and fatigue mechanical properties. Examples of these are resistant Nimonic and Inconel series, and the high-tempera-ture corrosion and stress corrosion alloy 600 and alloy 690. These alloys are based upon an fcc microstructure, the γ phase, which contains various proportions (up to 100 %) of the order phase Ni_3Al, (γ^1 phase), with an Li_2-type ordered intermetallic structure (Stickler, 1974). However, the properties of these alloys are sensitive to the presence of small amounts of minor alloying and impurity elements (Kent, 1994; Bieber and Dekker, 1961; Holt and Wallace, 1976). For example, Bi is a recognised element that may segregate to grain boundaries together with a range of other impurity elements such as Te, Pb, S, Sb, Sn, As, Ag and Cr. Walsh and Andersen (1976) measured the segregation of Bi, Te, Pb and S to grain boundaries in B-1900 and Mar M-200 alloys and showed larger segregation ratios than in corresponding commercial purity nickel. Moreover, Hf, Zr, Ti and La are recognised scavenging elements effecting precipitation with both sulphur and bismuth. Certainly, in alloy 600 and alloy 690, Stillen *et al.* (1996) considered the detailed microchemistry of the grain boundaries following a series of equilibrium and non-equilibrium heat treatments, in the temperature range of 1297 K down to 638 K for different cooling rates and isothermal heat treatments. Measurement techniques such as secondary ion mass spectroscopy, where O_2^+, Ar^+ and Ga^+ ions are used as primary ions, can determine segregation of elements such as boron to grain boundaries (Figure 3.26).

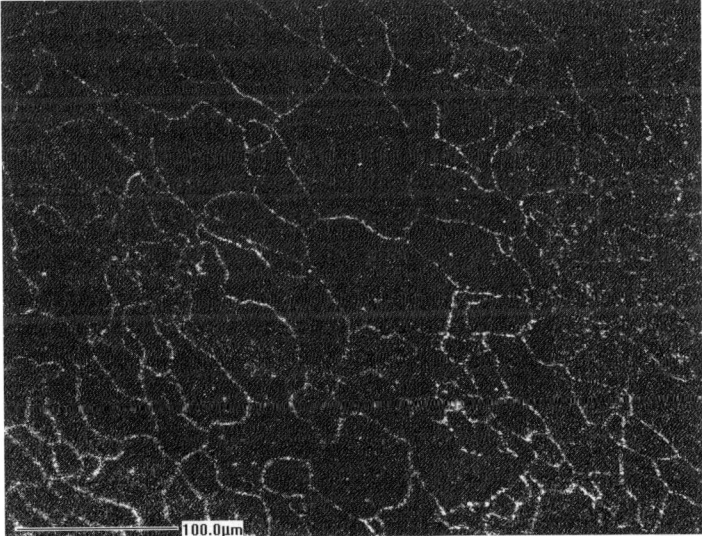

Figure 3.26 Secondary ion mass spectroscopy map showing the boron distributed at grain boundaries in alloy 690

Associated with the grain boundary enrichment were precipitates of Cr-rich carbides leading to a corresponding depletion of chromium at the grain boundaries. This observation is similar to that arising in austenitic stainless steels both in relative magnitude and spatial extent, (Figure 3.27) (Section 3.4.6). It is noteworthy that, for these alloys, enrichment in carbon and boron and depletions in chromium were also observed at special boundaries, in this case twin boundaries.

The segregation of boron in Ni_3Al has been studied in some detail in recent years because of the potential high-temperature application of this ordered intermetallic compound. Indeed, a variety of Auger electron spectroscopy measurements on the grain boundaries show enrichment in nickel both in the absence and presence of boron (Liu *et al.*, 1985; George *et al.*, 1989). Moreover, boron segregation to grain boundaries has been observed with field ion microscopy, which also shows a variation in the boron concentration along individual boundaries (Bremer and Ming-Jian, 1991; and Krzanowski, 1989). In general, segregation of boron is confined to within a distance of 1–2 atom planes of the grain boundary in cast and homogenised material (Mills, 1989), which is consistent with Monte Carlo simulations of the grain boundaries based upon an embedded atom model. These latter predictions made by Foiles (1987), suggest that nickel cosegregates with interstitial boron atoms to grain boundaries.

More recently, Muller *et al.* (1996) and Muller and Mills (1999) have considered the segregation of boron and nickel to the grain boundaries in this compound, both with and without boron additions of up to 1000 ppm. Energy dispersive X-ray microanalysis undertaken to a high spatial resolution on thin foil specimens shows enrichment of nickel at grain boundaries (Figure 3.27a). In this case, a known grain boundary with a normal direction $\sim [121]$ and a $15 \pm 5°$ rotation about the $[312]$

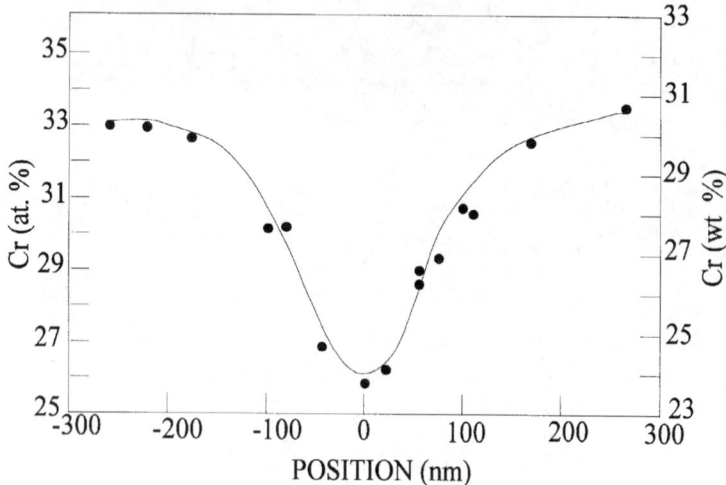

Figure 3.27   Chromium grain boundary composition profiles in the nickel base alloy 690; note the correspondence with that in austenitic stainless steel shown in Figure 3.21

direction was investigated and the width of the nickel peak, ~ 4 nm (fwhh), was measured using a 1 nm diameter incident electron beam. The independent detection of nickel in the annular dark field image (Figure 3.28b) supports this change in composition. However, boron is also present in specific parts of the boundary, as shown by the strain contrast. Figure 3.28c shows the corresponding parallel electron energy loss spectrometry scan across the boundary, supporting segregation of boron. Figure 3.28d compares the electron energy loss spectra taken at positions A and B in Figure 3.28b, which reveals a non-uniform distribution along the boundary. Boron is

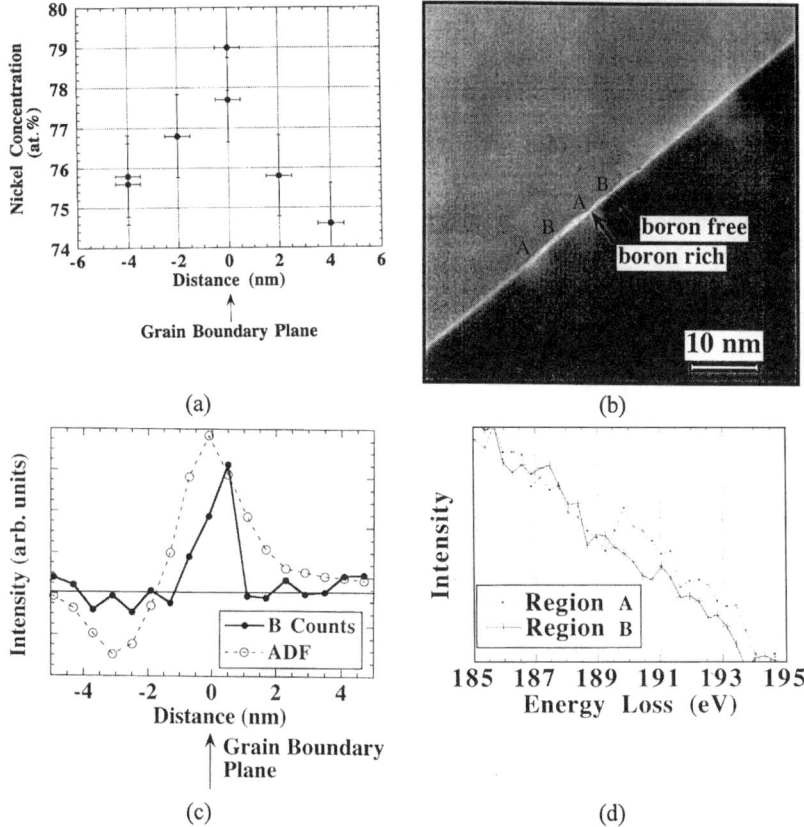

(a)

(b)

(c)

(d)

Figure 3.28 Energy dispersive microanalysis across the boundary in Ni-rich Ni$_3$Al doped with 1000 ppm B. (b) Annular dark field image of the large angle boundary in (a). The boundary appears bright because of Ni enrichment. Boron is present at parts of the boundary (A) that show strain contrast (light). Unstrained (dark) regions have no detectable boron. (c) Electron energy loss line scan in the B–K edge across the boundary as in (b) showing boron segregation to the boundary. The peak of the ADF signal marks the boundary. (d) EELS spectra from regions A and B of (b). The boron K edge is present only at parts of the boundary (A) that show strain (light). Reproduced by permission of Elsevier Science Ltd from Muller, D. A. and Mills, M. J. (1999) *Mater. Sci. Eng.*, **A260**, 12

present at parts of the boundary, A, where there is white contrast, while regions B are not similarly enriched. However, there is no corresponding fluctuation in the nickel enrichment which occurs irrespective of the presence of boron atoms. Figure 3.29 compares the NiL_2 edge spectrum recorded from bulk Ni_3Al with those recorded for regions A and B in Figure 3.28b. In the presence of boron, region A (Figure 3.28b), the NiL_2 edge is similar to that for the grain boundary, whereas in the absence of boron, region B, the edge at the grain boundary is clearly quite different. From this, Muller *et al.* (1996) deduced that the nickel d band is less filled at the grain boundary in the absence of boron; it compares with that for pure nickel. These high spatial resolution results suggest that for the segregated boron rich regions the binding at the grain boundary is similar to that in Ni_3Al, whereas in the absence of boron the binding is similar to pure nickel. From a tight binding model, insufficient filling of the nickel d band would lead to weaker Ni–Al bonds at the grain boundaries. The B–Ni hybridisation replaces lost Ni–Al bonds with Ni–B, bonds filling the Ni d band and restoring bulk-like bonding to the grain boundaries. Thus, boron is ineffective in controlling the segregation and properties of an Al-rich boundary since it will not restore bulk-like Ni–Al bonding for the boundary Al atoms, because the smaller Al atom has the same number of valence electrons as boron, thus splitting the s–p bands. Since Al–B interactions are less favoured than Al–Al, it is unlikely that boron will segregate to Al-rich boundaries in these nickel base alloys; this supports the observations of Lui *et al.* (1985).

As for the case of austenitic stainless steels, the nickel-base alloys, when subject to neutron irradiation, particularly fast neutrons, induce segregation of certain

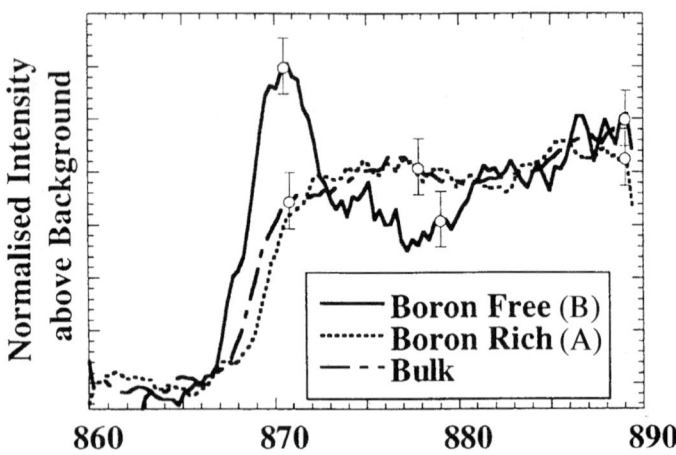

Figure 3.29 NiL_2 edge recorded at regions A and B of 3.28b. A spectrum taken well away from the grain boundary is also shown. Nickel atoms at the boron-doped regions of the grain boundary (A) and in the bulk both have nearly filled $d_{3/2}$ bands. Nickel atoms at regions of the grain boundary without boron (B) have less filled $d_{3/2}$ bands. Reproduced by permission of Elsevier Science Ltd from Muller *et al.* (1996) *Acta Mater.*, **44**, 1637

elements to the grain boundaries. When the superalloy PE16, which is an alloy in the Nimonic series, is subject to fast neutron irradiation, enrichment of the alloying element nickel and associated depletion of chromium and iron at the grain boundaries occurs (Nettleship and Wild, 1990). However, in addition to these composition changes, segregation of the minor impurity elements phosphorus and silicon occurs, producing non-equilibrium segregations that can be reversed by a simple heat treatment. These measurements made by Auger electron spectroscopy are supported by high-resolution STEM-EDS X-ray microanalysis of grain boundaries in thin foil specimens.

3.5.2 Copper, Gold, Silver and Platinum

In the case of copper, gold and silver and their alloys, the main information on segregation to the grain boundaries centres on the corresponding binary metal–metal solid solutions and metal–impurity systems. The observations are various and disparate and as a result we will consider only a few examples. Bismuth was recognised as embrittling copper over a century ago (Hampe, 1874), and more recently Auger electron spectroscopy has shown that the concentration of Bi at grain boundaries can be as high as 30 at. %, but its distribution is limited to a few atom distances about the grain boundary (Joshi and Stein, 1991; Powell and Mydura, 1973; Ranasubramaman and Stein, 1975). A number of localised measurements have been made on the fractured surfaces of polycrystalline copper to detect differences in the amount of segregation at individual boundaries.

For copper containing impurity additions of Bi, Powell and Woodruff (1976) found quite significant differences in the segregation of Bi to the various boundaries without establishing a detailed correlation. Donald (1976) observed pronounced faceting of grain boundaries in copper containing ~ 0.01 at. % Bi, whereas none developed in pure copper, but, unfortunately, no crystallographic detail was described. In considering the kinetics of segregation in this system, Johnson *et al.* (1976) and Fraczkiewicz and Biscondi (1985) observed differences in the times to achieve significant concentrations of bismuth at grain boundaries. More recently, Chang *et al.* (1999) found that the kinetics of grain boundary segregation at higher temperatures can be slower than at lower temperatures. Figure 3.30 shows a temperature–concentration–time diagram for the copper–bismuth system. Here segregation kinetics are regulated by volume diffusion in the single-phase region of the diagram, while enhanced transport occurs in the two-phase region. This difference is explained by invoking dislocation diffusion and, within these two-phase regions, the stability of the bismuth-rich liquid phase is significant because pipe-like precipitation of a bismuth-rich liquid phase forms along the dislocation cores.

Table A.5 (Appendix) was compiled by Cabané and Cabané (1991) for copper (Fe, S) and copper (Ni, S) solid solutions (Pineau *et al.* (1983); Pierantoni *et al.* (1985). These are contrasting systems since bcc iron has limited solubility in copper, whereas fcc nickel is fully soluble. Clearly, the addition of the iron to the copper significantly increases both the amount of sulphur segregation and the segregation

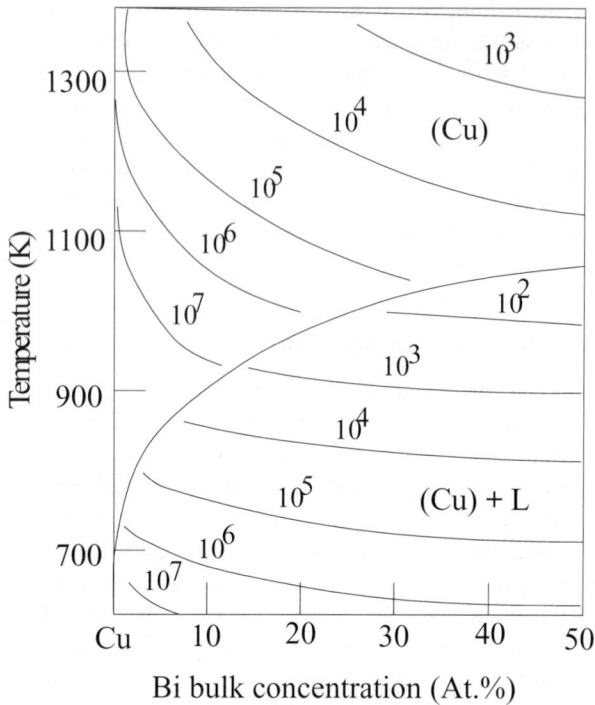

Figure 3.30 Temperature–concentration–time diagram for the Cu–Bi system. The numbers on the lines indicate the required annealing time (s) to achieve saturation of grain boundaries (Chang *et al.*, 1999)

coefficient at a given sulphur chemical potential. A similar result is obtained for segregation to free surfaces. In the case of the polycrystalline Cu (Ni, S) alloys, there is not such a large difference compared with pure copper even when the nickel content is increased to 7 at. %. From these results of surface and grain boundary segregation in Cu (Ni, S) alloys it is concluded that the addition of nickel results in only a small increase in S segregation owing to Ni–S cosegregation. Moreover, by comparison, it is established that segregation of S to the surface of the ternary Ag (Ni, S) solid solution results from Ni and S cosegregation. Here, the maximum amount of segregation in the binary Ag–S and Ag–Ni systems is similar to that measured in the ternary alloy (Aufray *et al.*, 1986). However, it is noteworthy that for the Ni–S system the segregation of nickel is confirmed in the outermost atomic layer on a fracture surface analysed by Auger electron spectroscopy, which is not the case for the Ag–S alloy. As a result, the cosegregated layer is made up of two regions:

(a) an outer atomic layer similar to a two-dimensional compound of the type Ni_2P and
(b) an underlying layer of atoms enriched in nickel.

This is supported by an experimental observation where a nickel layer was vapour deposited, up to two monolayer thickness, on to a Ag–S solid solution substrate. Following heat treatment, the Auger electron spectroscopy peak signals change with time as shown in Figure 3.31 (Cabané and Cabané, 1991). Here, the nickel becomes covered with silver following Ni segregation in pure Ag, but as time increases this is followed by Ni and S cosegregation to the surface. More recently, Way *et al.* (1993) have shown that, for a Cu–Ni alloy containing grain boundaries with a range of misorientations, segregation of copper is observed. Under these circumstances, the concentration of the first layer adjacent to a boundary increases monotonically with the misorientation, and regions with grain boundaries subject to a tensile stress show greater copper segregation than compressive regions; a large shear stress reduces segregation compared with regions subject to a small shear stress.

Platinum and its alloys are widely used as construction materials for operating in aggressive environments. One example is platinum-base thermocouples used to measure temperature in oxidising atmospheres. In Pt–10 wt. % Rh alloys it has been shown that oxygen, aluminium, silicon, carbon and manganese can segregate to the grain boundaries, leading to decreased ductility (Denisov *et al.*, 1987). More recently, Adamek *et al.* (1999) have shown that, in a Pt–17.4 % Rh thermocouple alloy, boundary enrichment of carbon and depletion of rhodium occurs after heat

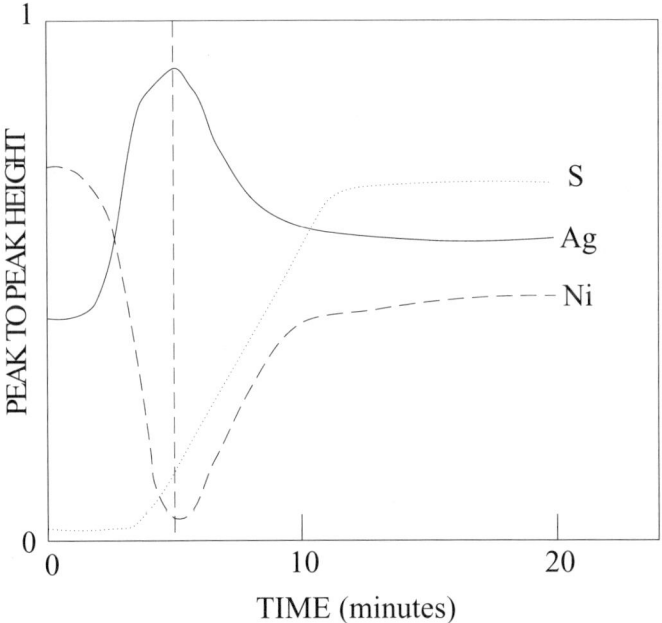

Figure 3.31 Surface composition changes Auger electron spectroscopy (peak height ratio) with heat treatment time for nickel vapour deposited on AgS, showing Ni–S cosegregation (Cabané and Cabané, 1991)

treatment at 1100 K for 2 h followed by air cooling. Indeed, Auger electron spectroscopy showed that the carbon that segregated to the grain boundaries formed as graphite.

3.5.3 Aluminium

Aluminium alloys such as the 7000 series alloys are important for their strength–weight ratio for industries such as aerospace. The strength is achieved by intra-granular precipitation (precipitates such as $MgZn_2$). However, the commercial application of these alloys is often restricted by poor properties such as susceptibility to stress corrosion cracking which depends upon the size and distribution of these grain boundary precipitates of $MgZn_2$ and the widths of the associated precipitate-free zones and grain boundary composition arising from localised segregation. As a consequence, there have been numerous investigations to measure the solute concentration of grain boundaries in aluminium alloys, mainly directed towards understanding the contribution to the stress corrosion behaviour (Cornish and Day, 1971; Taylor and Edgar, 1971; Doig and Edington, 1975a). Taylor and Edgar (1971) examined barrier layer anodic films and concluded that both magnesium and zinc segregate to grain boundaries in Al–Zn–Mg alloys. This supported earlier measurements of Clark (1964) and agreed with later measurements of Shastry and Judd (1972) who both used a relatively low-resolution electron microprobe analysis technique to demonstrate that significant amounts of zinc and magnesium segregate to the grain boundaries in quenched and aged high-purity Al–Zn–Mg alloys.

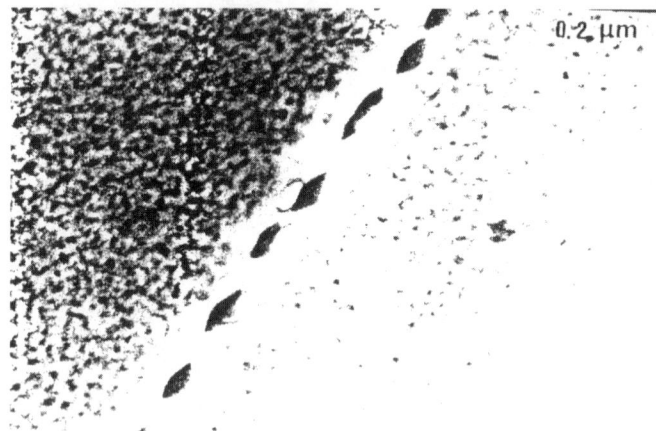

Figure 3.32 Microstructure developed in an Al–6.42Zn–2.0Cu–2.35Mg alloy, showing the typical precipitate size, inter-particle spacing and the width of the precipitate-free zone (specimen aged 2 h at 435 K after solution treatment). Reproduced by permission of Elsevier Science Ltd, from Jiang, H. and Faulkner, R. (1996b) *Acta Mater.*, **44**, 1857

Figure 3.32 shows a transmission electron micrograph of a typical microstructure produced in an Al–6.42Zn–2.0Cu–2.35Mg (wt %) alloy after 2 h ageing at a temperature of 453 K following a solution heat treatment (Jiang and Faulkner, 1996). Clearly, the main features of the microstructure are the intragranular precipitates, the grain boundary precipitates and the precipitate-free zone adjacent to the grain boundary. The composition changes occurring in the vicinity of the grain boundaries owing to differences in heat treatment are generally quite complex. Conventional analytical techniques, including electron microprobe analysis, have really only sufficient spatial resolution to distinguish the main features of the precipitate-free zones and the grain boundary precipitates. However, high spatial resolution electron beam techniques using probes down to about 1 nm have provided further information. Using electron energy analysis, Cundy *et al.* (1968) and Doig and Edington (1975a and b) showed that, under certain conditions, solute segregation occurs in rapidly cooled Al–Mg, Al–Cr and Al–Zn–Mg alloys. Figure 3.33 (Doig and Edington (1975a) maps energy loss profiles across grain boundaries and

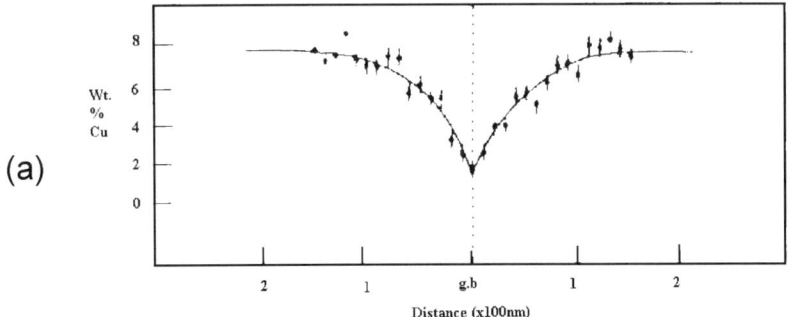

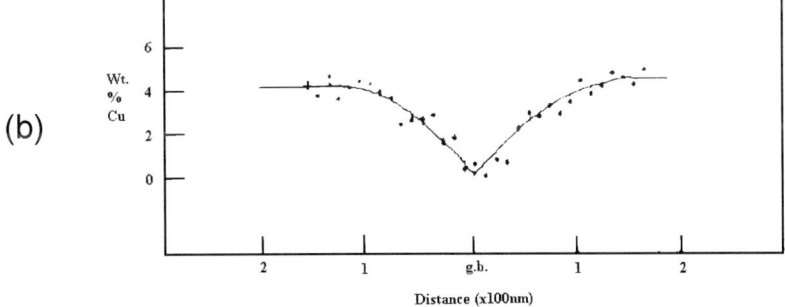

Figure 3.33 Energy loss profile across a grain boundary in (a) aged Al–7.2 % Mg alloy and (b) Al–4.4 wt % Cu alloy, showing depletion of Mg and Cu at the grain boundaries respectively (reproduced by permission of The Minerals, Metals and Materials Society from Doig and Edington, 1975a)

relates the measured changes in energy loss to composition. Although these studies provide interesting and useful compositional information near to the boundary, they are not of sufficient resolution to describe segregation of solute which would be within a region of less than 1 nm. In this respect, workers such as Doig *et al.* (1977), Joshi *et al.* (1981) and Paine *et al.* (1986) examined the effects of heat treatment of a 7075 aluminium alloy on solute redistribution at the grain boundaries using Auger electron spectroscopy and STEM-EDS X-ray microanalysis. Figure 3.34 (Joshi *et al.*, 1981) shows the measured solute concentrations of copper, zinc and magnesium at the grain boundaries as a function of the solute heat treatment temperature. As the temperature is increased from 666 to 800 K, the solute concentration at the boundaries decreases and passes through a minimum at 711 K for all these elements and then increases. These observations can be explained by equilibrium and non-

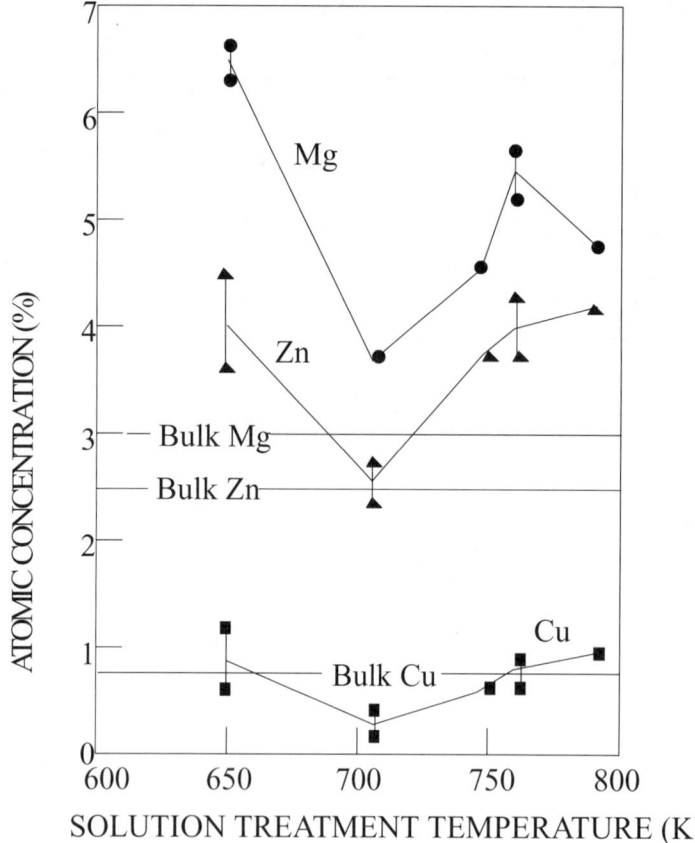

Figure 3.34 Variation in the copper, zinc and magnesium atomic concentrations on the fracture surface of aluminium 7075 alloy with the temperature of solution heat treatment (Joshi *et al.*, 1981)

equilibrium segregation models. The non-equilibrium enriched regions can be described by a vacancy–solute drag mechanism and the subsequent formation of grain boundary precipitates which cause local depletion of the enriched zone. Faulkner and Jiang (1993) and later Jiang and Faulkner (1996a and b) provided a model that takes account of the combined effects of solute segregation and precipitate growth kinetics by linking the flux expected at the grain boundaries of these alloys from both segregation and precipitation mechanisms.

3.6 NON-METAL SYSTEMS

3.6.1 Introduction

The term non-metallic solid is used to define a very wide field in materials and as a result it is not surprising that the behaviour of these materials is very different in a variety of situations. However, the general physical principles used to describe segregation to interfaces, grain boundaries and interphase boundaries in metallic solids can be used to describe segregation to similar features in non-metals. However, to achieve this, appropriate modifications have to be included, as described in Chapter 2, to accommodate the presence of charged defects. For example, in the case of ceramic materials the complicated techniques that usually have to be adopted for their fabrication result in complex microstructures (Rice, 1966). The solid state mass transport that is responsible for sintering and hot pressing often results in second-phase precipitation and pores at grain boundaries. In these circumstances, as shown by Kanlea and Schneider (1993) and Schneider *et al.*, (1994), the micro-structures in mullite, a ceramic containing high Al_2O_3 compositions formed from coprecipitated precursors calcined at 1373 K, develop an equiaxed mullite micro-structure together with some a α-Al_2O_3 grains. High-resolution electron microscopy of a typical grain boundary between these mullite crystals reveals a relatively simple microstructure with no evidence of a glassy phase at the grain boundary (Figure 3.35) (Schneider *et al.*, 1994).

Generally, equilibrium segregation experiments are more difficult to undertake in non-metals compared with metals because it is not so easy (a) to control the composition of pure materials and residual impurities and (b) to measure the low concentration of impurities. Moreover, the defect structure of these ionic materials tends to be complex because of charge compensation, and therefore the relationship between properties and intrinsic vacancy content can be complex. Cutler (1970) has, for example, estimated the magnitude of this problem in MgO by calculating the effect of Ti^+ ions on the defect structure. By assuming that this impurity ion substitutes for Mg^{+2} sites and that charge compensation is a function of Schottky defects, then additions as low as 1 at. ppm Ti^{+4} will give an extrinsic vacancy concentration that is essentially constant at temperatures below 1723 K. This makes it difficult to study vacancy-controlled segregation since any impurities will effect extrinsic behaviour. One approach adopted to overcome this limitation is to alter the

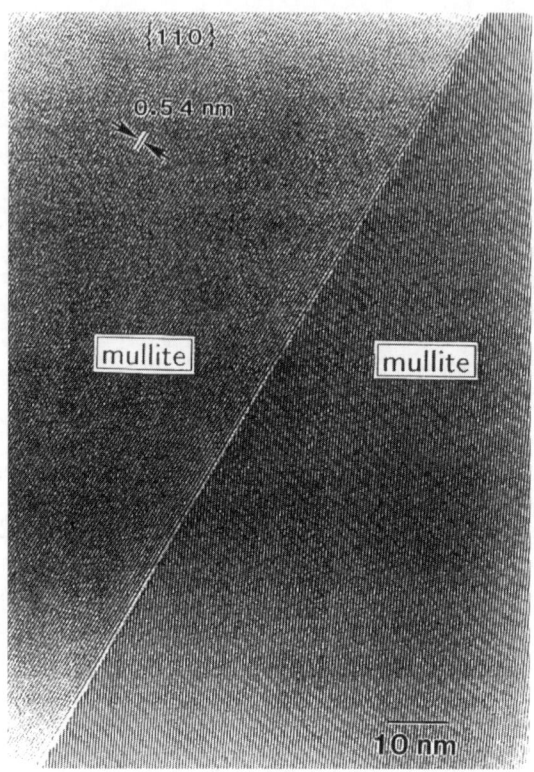

Figure 3.35 High-resolution electron micrograph of a typical grain boundary between mullite crystals in a mullite produced from coprecipitated precursors (bulk composition 78 wt % Al_2O_3 and 22 wt % SiO_2). Reproduced by permission of John Wiley & Sons from Schneider *et al.* (1994), *Mullite and Mulite Ceramics*, p. 164

extrinsic vacancy level by doping the material. Despite these difficulties, similar techniques to those adopted for metals can be used to measure the composition of surfaces, interfaces, grain boundaries and interphase boundaries in non-metallic systems. In this section we will concentrate on the segregations in ceramic and semiconductor materials.

3.6.2 Ceramics

The term ceramics describes those products that are made from inorganic crystalline materials and that have non-metallic properties (Chapter 2). Simple examples are ionically bonded magnesia, MgO, and covalently bonded silicon carbide, SiC. These have crystal structures of the sodium chloride and diamond type respectively. Apart from their known high-temperature applications, some polycrystalline electronic ceramics are used extensively for communications, electronic and appliance industries. Among the best known of the ceramics for these applications are the ZnO varistors, boundary layer capacitors, ferrites and positive temperature coefficient devices. They owe their unusual electrical properties to the presence and character of the grain boundaries since single crystals of these materials do not exhibit the same phenomena as the polycrystals.

Ceramic oxides have been considered extensively and there are several overviews to which the reader is referred (Johnson, 1977; Wynblatt and McCure, 1981; Kingery, 1984; Nowotny and Sloma, 1988, Nowotny, 1988. Segregation in ceramic materials is often inferred from property measurements arising from particular grain growth data changes (Coble, 1962, and Jorgensen, 1957), but difficulties can be encountered in obtaining sufficient purity to prevent macrocomposition variations masking solute segregation to grain boundaries. Hence, many of these observations and data must be subject to question. Some of the oxides such as MgO and Al_2O_3 and their derivatives, mullite $Al_2(Al_{2+2x}Si_{2-2x}O_{10-x})$ (Cameron, 1977), have been considered in greater detail. One of the early experiments that positively identified the segregation of impurities to grain boundaries was for a ceramic oxide in the form of MgO bicrystals where autoradiography recorded impurity segregation. Ionic interfaces, but particularly grain boundaries, can become charged owing to a break in the periodic structure so that in MgO this has to be accommodated by excess vacancies in these boundaries. In this respect it is interesting to consider the segregation to grain boundaries of chemical species of different ionic states. For different ions such as Ca^{++} and Sn^{++} there is pronounced segregation to the grain boundaries and, indeed, the free surface of MgO, accompanied by strain relaxation. As a result, the segregated ions become distributed at the grain boundaries as monolayers (Masri and Tasker, 1985). In the case of trivalent ions, such as Fe^{+++}, Cr^{+++} and Sc^{+++}, the segregation measured at both grain boundaries and free surfaces extends for distances of up to about $\sim 2\,nm$ to allow the strain energy for these trivalent ions to be minimised (Chiang *et al.*, 1981).

Calcium enrichment has been observed at the grain boundaries in MgO using Auger electron spectroscopy. Figure 3.36a (Johnson, 1977) shows a typical grain boundary composition for hot pressed MgO containing 150 at. ppm Ca as an impurity where it has been possible, using ion sputtering to demonstrate that this enrichment is localised to the grain boundaries (Figure 3.36b). When a sample is heat treated at 1723 K for 5 h and then quenched, the concentration of Ca at the grain boundaries is only 0.15 that attained after furnace cooling from the same temperature. The segregation of the Ca^{2+} can be rationalised with the size misfit for the Mg^{2+} sites. (Marcus *et al.*, 1971; Johnson *et al.*, 1974; Jupp *et al.*, 1980). Subsequently, there have been a number of measurements of boundary segregation in thin foils of MgO undertaken using STEM-EDS X-ray microanalysis. (Kingery *et al.*, 1979a, b; Mitamura *et al.*, 1979). This approach has identified grain boundary segregants such as Ca, Ti, Si, Fe^{++} (ferric ions), La, Na and F. A broader profile at the grain boundary is usually observed, $\sim 20\,nm$ [half-width at half-height (hwhh)]

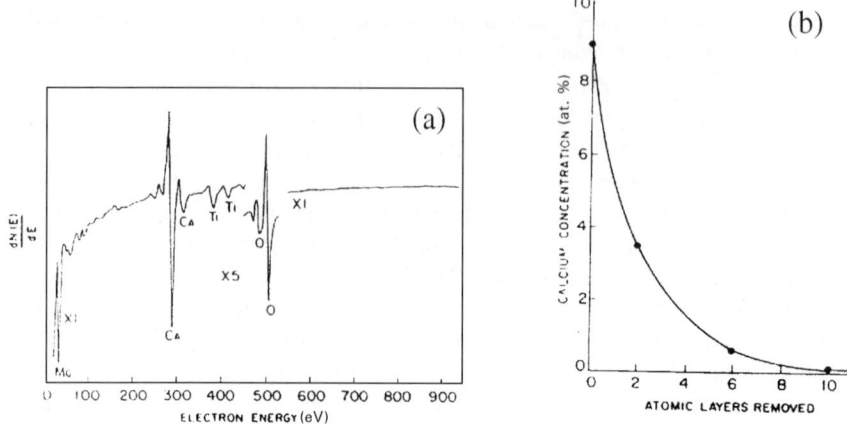

Figure 3.36 Composition at grain boundaries in hot pressed MgO: (a) Auger spectrum from grain boundary fracture of hot pressed MgO; (b) Ca sputter etch profile from MgO grain boundary (Johnson, 1977)

compared with 6–10 atom layers measured by Auger electron spectroscopy. However, overall the observations are consistent and the differences accounted for by the need to deconvolute the true concentration from the measured concentration, taking account of various components of the electron beam spreading the thin foil specimens. Similar enrichment of the grain boundaries occurs in sintered Al_2O_3, and the observations of Johnson and Stein (1975) may be described by an equilibrium segregation mechanism, where the uniform mono- or multilayer coverage of the grain boundaries occurs. Similar studies have identified Ca, Y and Ni enrichment at grain boundaries in Al_2O_3.

The results of these grain boundary segregations may be interpreted using the Langmuir–Mclean model, equation (2.4), and it is possible to fit the temperature dependence of the Ca concentration in Al_2O_3 to obtain a heat of segregation of $-26\,kcal\,mol^{-1}$ (Johnson *et al.*, 1975) and similar data for a range of other cations in Al_2O_3. This approach provides a basis for ranking the relative propensity for segregation of the cations Ca, Y, Ni and Mg in this ceramic. More recently it has been shown that Mg segregates to both grain boundaries and the free surface in Al_2O_3 (Baik *et al.*, 1985, and Mukhopadhyay *et al.*, 1988). Chemical etching techniques showed grain boundary enrichments in Ca and Y for a manganese–zinc spinel ferrite $[(Fe_2O_3)_{53}(MnO)_{28}(ZnO)_{19}]$ which were justified on the basis of a size misfit model where by the solute atoms are incorporated into the spinel lattice.

In the case of zirconia, ZrO_2, containing, in solid solution, 15 at. % Y, the grain boundaries and surfaces can become enriched with Y by segregation. The enrichment is in the form of a thin layer $\sim 2\,nm$ wide which leads to a maximum concentration of about twice that in solid solution. This follows isothermal heat treatment in the temperature range 670–1220 K and implies that the fractional concentration of Y at the grain boundaries relative to the cation lattice $(Y + Zr)$ amounts to 0.33 which is consistent, in general, with the values given by Hondros (1985) for metal alloys. This segregation is a result of strain relaxation since the ionic radius of Y is greater than that of Zr and there is an associated electrostatic potential with a compensating space charge layer.

Nickel oxide, NiO, has been investigated because of its practical use as a catalyst. It is often considered as a model oxide both since the effect of oxygen pressure in equilibrium and the nature of the defect structure are well characterised. This has allowed the charge distribution near to grain boundaries to be calculated for both doped and undoped NiO (Duffy and Tasker, 1984). Of the various systems, it is that doped within Cr_2O_3 that has been studied extensively and SIMS analyses of single crystal surfaces reveal significant enrichment in chromium (Hirshwald *et al.*, 1981). As shown in Figure 3.37, at a temperature of 723 K this Cr-enriched layer is about two monolayers (fwhh) and the surface concentration corresponds to about 17 at. % Cr^{3+} which compares with a bulk value of 0.56 at. %. In the undoped condition, the interface for NiO may be described by a negative local charge with excess cation vacancies compensated by a positive space charge formed by electron poles. However, when doped with Cr_2O_3, since the ionic radii of both Cr^{3+} and Ni^{2+} are similar, elastic strain energy will not drive the segregation. Rather, there is segregation of Cr^{3+} cations and cation vacancies to the surface with a charge separation; when Cr^{3+} cations are in excess at the surface it is positively charged, and

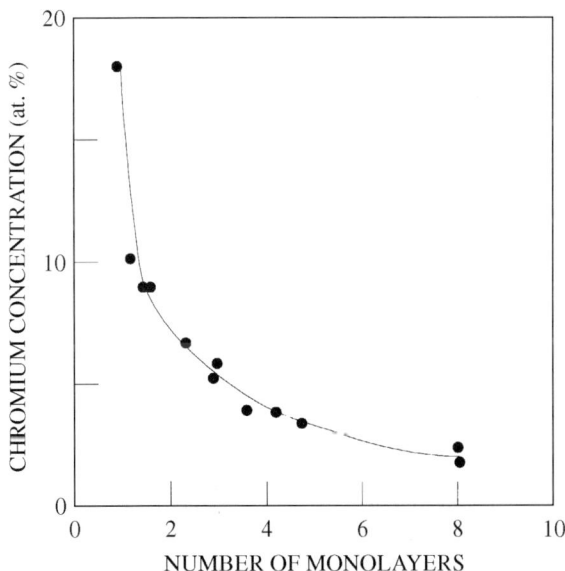

Figure 3.37 Concentration profile of Cr measured by secondary ion mass spectroscopy at the surface of a single crystal of Cr-doped NiO (Hirshwald *et al.*, 1981)

where cation vacancies are in excess it is negatively charged (Nowotry and Wagner, 1981). It is interesting to note that this can be compared with cosegregation phenomena observed in metals (Section 3.2.4) since Cr^{3+} and cation vacancies cosegregate to the surface to form a new two dimensional enriched layer. Within the space charge region it is possible for complexing to occur between trivalent cations and cation vacancies. Hirschwald *et al.* (1985) have shown that, in NiO doped with Cr_2O_3, segregation is also dependent upon the oxygen partial pressure; low oxygen pressures promote chromium segregation.

Thin films of lead titanate, $PbTiO_3$, are used for devices because they are ferroelectric at room temperature. Above the Curie temperature of 673 K, this material has a perovskite crystal structure, but, on cooling below this temperature, the symmetry reduces to a tetragonal arrangement as a result of movement of the O and Ti ions relative to the Pb ions. Typically, these films are grown on {001} oriented MgO substrates to produce polycrystalline films consisting of domains where [001] directions are aligned into one of the six possible variants. As such, these domains meet in special grain boundaries (Ernst *et al.*, 1999). Figure 3.38 shows a high-resolution electron image of a 90° domain boundary in [010] projection. Using Fourier filtering of the image, the positions of the Pb columns were extrapolated by cross correlating the image with the pattern for each Pb column in the simulated images of undisturbed $PbTiO_3$.

The electrical properties of strontium titanate, $SrTiO_3$, ceramics are strongly influenced by grain boundary segregation of charged point defects such as dopant atoms, impurity elements, vacancies or self interstitials. Grain boundary segregation of charged point defects causes the formation of space charges distributed into the adjoining crystals (Maier, 1995; Chaing and Takagi, 1990; Desu and Payne, 1990). InSrTiO$_3$ polycrystalline ceramics doped with iron to a level corresponding to Fe/Ti of 0.4 at. %, a significant number of $\Sigma 3$ grain boundaries were found by Ernst *et al.* (1999). In Figure 3.39 the mean orientation of the grain boundary planes deviate from {111} planes in the two grains so that the boundary decomposes into two facets of types A and B. Facets of type A correspond to {111} planes in both crystals and have features of a $\Sigma 3$ boundary, and there is no evidence of Fe segregation. By contrast, type B facets are of random orientation with associated Fe segregation up to ~ 0.46 atoms nm^{-2}. In the case of completely randomly oriented grain boundaries in these ceramics, the existence of an amorphous film of up to ~ 0.8 nm thickness has been revealed, again with an associated enrichment in iron (Figure 3.40). Table 3.1 summarises the results obtained for a range of different grain boundaries in $SrTiO_3$ doped with iron.

Silicon nitride Si_3N_4 is a ceramic used for high-temperature applications, but owing to the covalent bonding, the sintering requires the addition of aids that oxidise the Ai_3N_4 grains. As a consequence, SiO_2 covers the grains as a liquid phase before densification, producing an amorphous film in all randomly oriented grain boundaries. Tanaka *et al.* (1994) have shown that this amorphous film has a thickness that depends upon the sintering additives. In pure Si_3N_4 the SiO_2 wetting film has a thickness of ~ 1 nm, whereas with further additions of CaO the film thickness

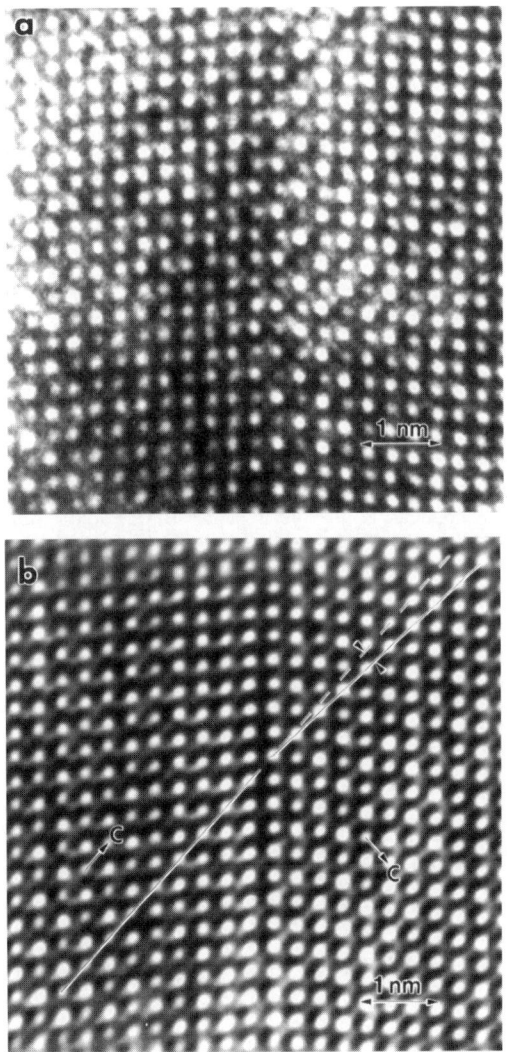

Figure 3.38 (a) Experimental HRTEM image of a 90° domain boundary in PbTiO₃. The viewing direction coincides with the common *a* direction of the two grains. The boundary plane lies vertical in this image and corresponds to {101} on both sides. (b) The image after adaptive Fourier filtering. The angle between *c* on one side and *a* (or *b*) on the other side reflects the tetragonality of the low-temperature structure: $c/a = 1.09$. Reproduced by permission of Elsevier Science Ltd from Ernst *et al.* (1990) *J. Eur. Ceram. Soc.*, **19**, 665

initially reduces but then results in an increase in thickness (Figure 3.41). Here, the equilibrium thickness is considered to be a balance of attractive Van der Waals forces, repulsive forces arising from steric effects and electronic double layer forces

Table 3.1   Properties of different grain boundaries in $SrTiO_3$

Sample	Fe doping C_{Fe}/C_{Ti}	Fe excess θ_{Fe}
	Bicrystals	
$\Sigma = 3$, (111) (Verneuil method)	0.04 at. %	<detection limit
$\Sigma = 5$, (310) (Diffusion bonded)	0.01 at. %	< detection limit
Near $\Sigma = 13$, (510) (Diffusion bonded)	0.47 at. %	$(0.46 \pm 0.13)\,\text{atoms nm}^{-2}$
	Polycrystal	
Near $\Sigma = 3$, (111)	0.40 at. %	
(111) facets		< detection limit
'Random facets'		$(0.46 \pm 0.13)\,\text{atoms nm}^{-2}$
'Random' boundary	0.40 at. %	$(4.0 \pm 0.26)\,\text{atoms nm}^{-2}$

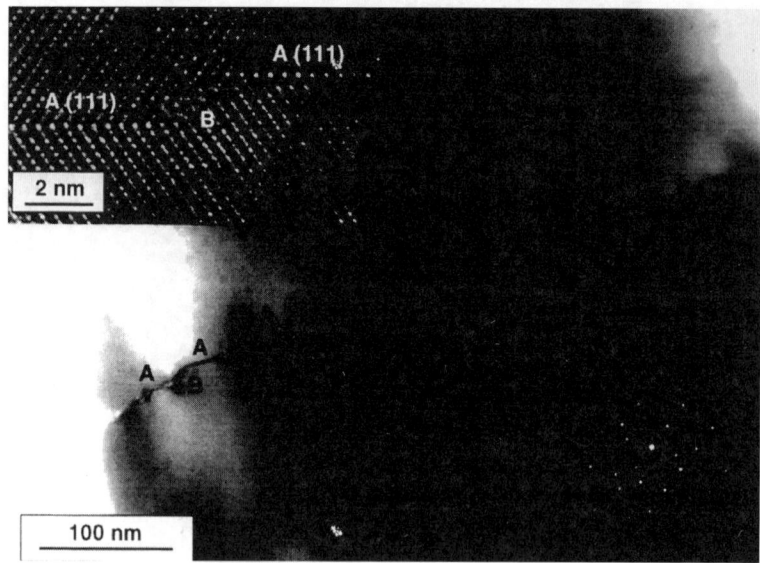

Figure 3.39   A $\Sigma = 3$ grain boundary in polycrystalline $SrTiO_3$. On a macroscopic scale, the mean orientation of the boundary plane deviates by $4°$ from the orientation $\{111\}$ planes of the two crystals. On the microscopic scale, however, the grain boundary decomposes into two types of facets, A and B. Facets of type A correspond to $\{111\}$ planes in both crystals, while facets of type B follow another, presumably 'random' plane. Reproduced by permission of Elsevier Science Ltd from Ernst *et al.* (1999) *J. Eur. Ceram. Soc.*, **19**, 665

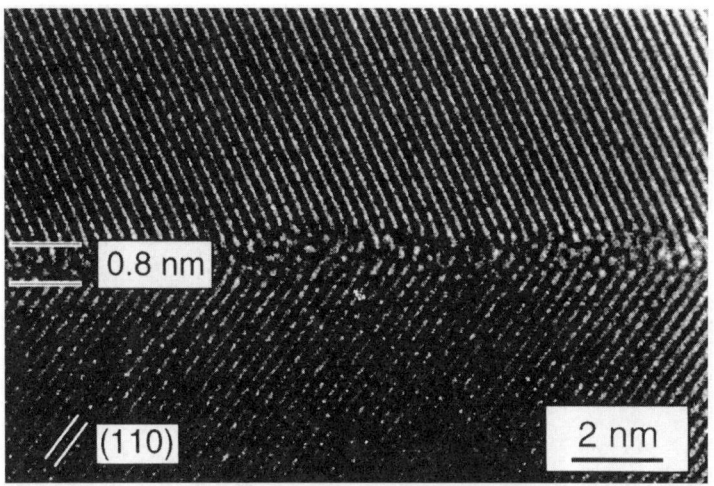

Figure 3.40 A high resoltion transmission electron micrograph of a 'random' grain boundary in polycrystalline SrTiO$_3$ (same material as in Figure 3.39), revealing an amorphous film between the two crystals. Reproduced by permission of Elsevier Science Ltd from Ernst *et al.* (1999) *J. Eur. Ceram. Soc.*, **19**, 665

(Ernst *et al.*, 1999). Various impurity additions produce different equilibrium thickness amorphous layers, but different elemental species may also produce the same equilibrium thickness film. Increasing the content simply increases the volume of triple junctions. Similar observations have been made for Si$_3$N$_4$ densified with lanthanide oxides, La$_2$O$_3$, Nd$_2$O$_3$, Gd$_2$O$_3$ and Yb$_2$O$_3$ as well as yttria, Y$_2$O$_3$ (Wang *et al.*, 1996).

Turan and Knowles (1995) investigated the effect of pretreatment on the resulting composite microstructure of silicon carbide, SiC, reinforced with different proportions of silicon nitride Si$_3$N$_4$ in the range 10–20 wt %. Intergranular films were found to develop between SiC–SiC and Si$_3$N$_4$–Si$_3$N$_4$ grains and various combinations of these (Figure 3.42a). Although it is difficult to analyse these films using conventional STEM-EDS X-ray microanalysis, high-resolution PEELS analysis undertaken at triple point junctions of three SiC grains (Figure 3.42b), shows enrichment in amorphous silica. Other examples have been found where there is evidence of an enrichment in carbon.

It is important to distinguish between electronic ceramics, such as the ZnO used for varistors (Matsuoka, 1971), and the polycrystalline silicon and gallium arsenide materials that we will consider in the next Section 3.6.3. Although the electrical properties are similar in a number of respects, the grain boundaries in these electronic ceramics are expected to be different because of the various methods used to introduce dopants and then move them to the grain boundaries (Figure 3.43a). A segment of a grain boundary in a commercial ZnO–Bi$_2$O$_3$ varistor (Clarke, 1988), when analysed using high-resolution X-ray microanalysis, shows the bound-

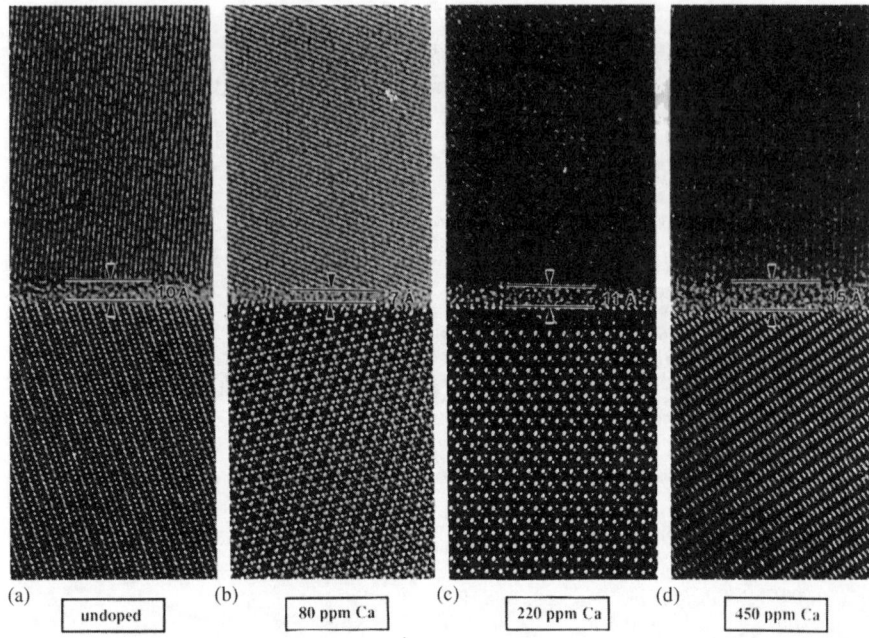

(a) undoped (b) 80 ppm Ca (c) 220 ppm Ca (d) 450 ppm Ca

Figure 3.41 High resolution transmission electron micrographs revealing the dependence of the SiO_2 wetting layer thickness in Si_3N_4 grain boundaries on the amount of Ca sintering aid: (a) 0 ppm Ca, (b) 80 ppm Ca, (c) 220 ppm Ca and (d) 450 ppm Ca. Reproduced by permission of Elsevier Science Ltd from Ernst *et al.* (1999) *J. Eur. Ceram. Soc.*, **19**, 665

aries to be enriched in bismuth with an attendant concentration profile extending from the plane of the boundary (Clarke, 1978; Kingery *et al.*, 1979a, b). One method for producing boundary layer capacitors uses a two-step process where powders are sintered and then cooled. The dopant compound is then painted on to the surface and the ceramic heated to high temperature so that it diffuses along the grain boundaries to change their local chemistry. Figure 3.43b (Clarke, 1978) shows a transmission electron micrograph where the contrast reveals the presence of a grain boundary Bi precipitate layer and an associated enrichment in Bi within a boundary layer capacitor produced by diffusion of Bi into $SrTiO_3$. This is clearly a two-layer boundary structure comprising (a) a discrete Bi-rich intergranular phase and (b) a segregation layer within the $SrTiO_3$ grains on either side. These segregation layers contain about 1 at. % Bi and are revealed because of the larger atomic scattering factor of the Bi compared with that of either Sr or Ti atoms (Bougers and Franken, 1981).

An extension to the investigations of the polycrystalline ceramic boundaries has been the heterogeneous interface between metals and ceramics because of their use in a range of applications including structural and electronic materials (Rühle *et al.*, 1990). In particular, $Nb-Al_2O_3$ and $Cu-Al_2O_3$ interfaces prepared by either molecular beam epitaxy or diffusion bonding have been studied (Bruley *et al.*,

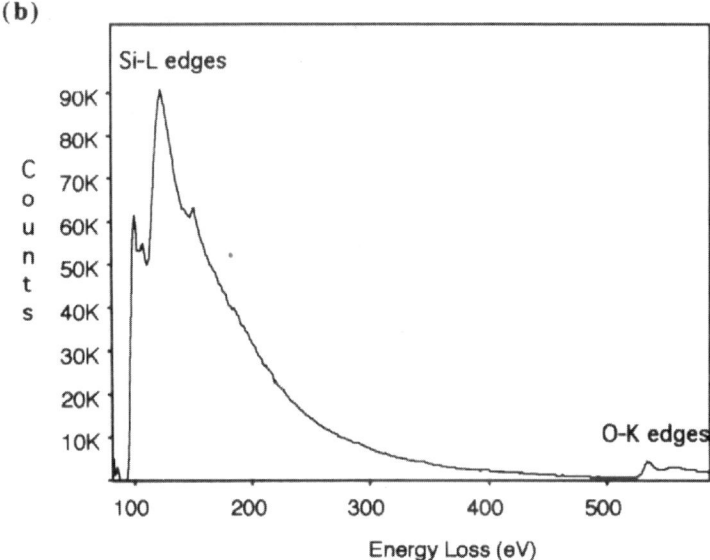

Figure 3.42 Silicon carbide reinforced with silicon nitride: (a) an example of a large triple-junction phase between three SiC grains; (b) an electron energy loss spectrum showing stripped silicon L-edges and unstripped oxygen K-edges of the triple junction phase, showing that it is rich in amorphous silica Reproduced by permission of Blackwell Science Publishers from Turan, S. and Knowles, K. (1995) *J. Microscopy*, **177**, 287

1994, and Scheu *et al.*, 1998). In the case of a Nb–α–Al$_2$O$_3$ interface, the diffusion-bonded material has an interface terminated by aluminium and this plane is coordinated tetrahedrally to three oxygen anions on the ceramic side and a Nb atom on the metal side. By contrast, the molecular beam epitaxy-formed material has

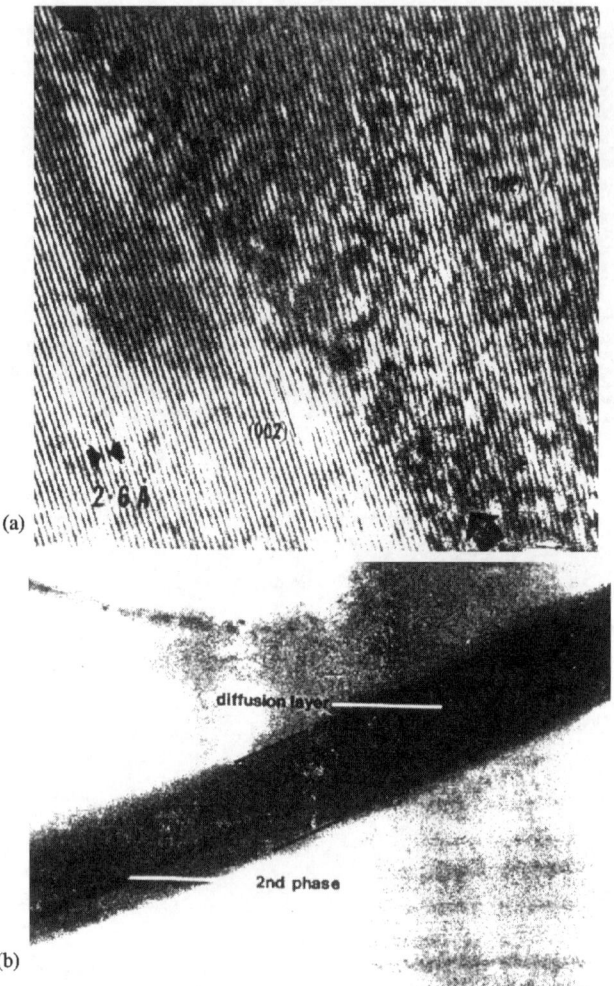

Figure 3.43 (a) Commercial ZnO varistor showing a lattice fringe image of a grain boundary segment; the boundary plane is edge-on (solid arrows). The (002) lattice fringes in the adjacent grains are continuous right up to the boundary, demonstrating that the boundary is devoid of an intergranular film. (b) Bright-field electron micrograph of a grain boundary region in a boundary layer produced by inward diffusion of Bi into a $SrTiO_3$ ceramic capacitor. A discrete, bismuth-rich intergranular second phase is together with a bismuth diffusion layer extending into the adjacent $SrTiO_3$ grains. A segregation or diffusion layer, not usually present but is seen on account of the particularly strong absorption contrast due to the presence of bismuth atoms (reproduced by permission of the American Institute of Physics from Clarke, (1978))

an interface terminated with oxygen, with the Nb donating some charge to the ceramic. For the molecular epitaxy-bonded Cu to $\alpha-Al_2O_3$ the oxygen is directly bonded to the copper at the interface, forming a monolayer of copper nominally in the Cu^+ oxidation state. The basal plane of the ceramic is oxygen terminated so that the aluminium does not participate in the bonding at the interface.

3.6.3 Semiconductors

Many of the traditional semiconductor materials have crystal structures that are related to the simple diamond cubic crystal lattice where each atom is tetrahedrally coordinated, but the local atomic environment is not identical for all atoms. Most semiconductor compounds and alloys are designed according to the principle of keeping the average electron to an atom ratio of four. The simplest illustration of this is given by the range of AB-type semiconductor materials found between Group III and Group V elements, the so-called III to V semiconductors, which include GaAs and have a sphalerite superlattice structure, whereas the types II to VI and IV to VI compounds usually have structures of sphalerite and rock salt respectively (Table A.6) (Grovenor, 1989). Owing to the complexity of the physical chemistry of semiconductor materials, the conditions required to enable a systematic study of equilibrium segregation to grain boundaries are not easily achieved. However, high-resolution Auger electron spectroscopy and secondary ion mass spectroscopy have been used to examine the composition of grain boundaries of silicon following intergranular fracture under ultrahigh vacuum conditions. Figure 3.44 (Pollock *et al.*, 1982) shows a series of secondary ion images obtained for a polycrystalline silicon wafer coated with a 30 nm thick copper film following heat treatment at a temperature of 823 K for 1 h. Clearly, the impurity elements H, C and O have accumulated at the grain boundaries (Pollock *et al.*, 1982). Moreover, similar distributions of P at the grain boundaries were observed in other wafers heat treated for a total time of 3 h (Figure 3.45). Intergranular segregation of hydrogen has also been established by high-resolution autoradiography (Aucoututier *et al.*, 1982).

Oxygen is usually present in silicon as an impurity element derived from the various methods of preparation and heat treatment history. This element is known to precipitate slowly even when the system is supersaturated simply because the concentration is low and the rate of diffusion is small. However, oxygen has been detected at grain boundaries using a range of high spatial resolution microanalysis techniques (Bourret and Colliex, 1983; Russell *et al.*, 1982) where it occupies interstitial positions between two silicon atoms with an associated relaxation of the strain energy. This increase in the local grain boundary composition is often accompanied with cosegregation and even coprecipitation with other impurity elements such as titanium and aluminium (Kazmerski, 1982). Here, the presence of Ti segregated at the grain boundaries has been linked with an increase in the barrier height, and certainly several other impurity or dopant elements that also segregated to grain boundaries have a similar effect; when segregated to the grain boundaries, hydrogen, by contrast, reduces the barrier height (Grovenor, 1989). As a

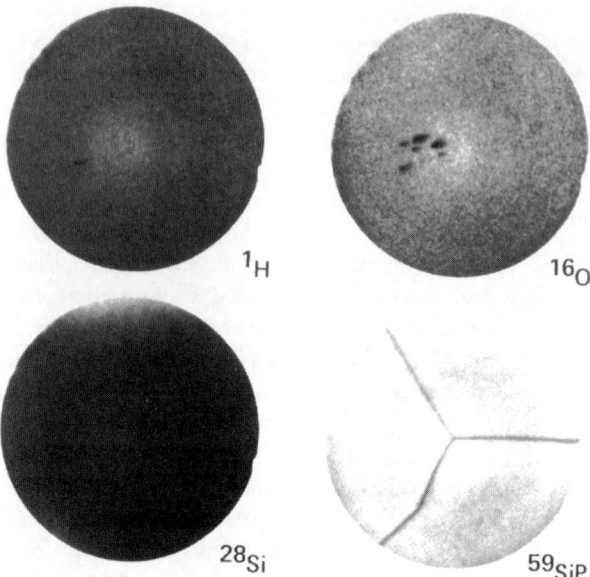

Figure 3.44   Polycrystalline silicon wafer coated with copper following heat treatment at 823 K for 2 h. Bombardment with cesium ion source produced H, O, Si and P images (Pollock *et al.*, 1982)

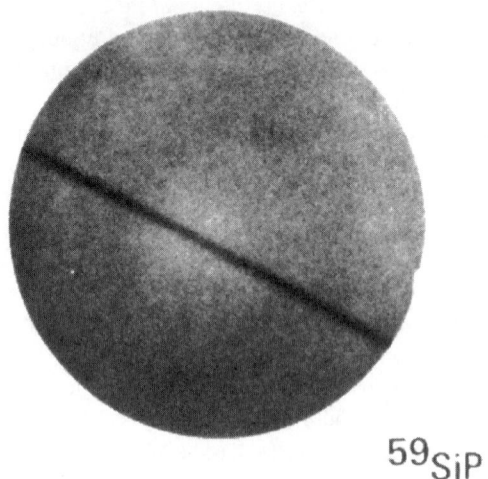

Figure 3.45   Segregation of P at the grain boundaries is similar to that seen in Figure 3.44. This wafer was also diffused in P but at 773 K for 3 h (Pollock *et al.* 1982)

consequence, many impurities and dopants when segregated in this way in silicon are extremely important with respect to the overall electrical properties.

Segregation to the surface of silicon is made difficult by the SiO_2 oxide layer that is usually present unless the surfaces are correctly prepared using ultrahigh vacuum procedures. For a silicon (111) orientated surface obtained by cleavage, the original (2×1) arrangement of atoms transforms to a (7×7) structure when annealed under ultrahigh vacuum by eliminating oxygen contamination from the surface. However, for silicon doped with boron, a $(\sqrt{3} \times \sqrt{3})$ R30 structure is observed (Korobtsov *et al.*, 1988). Scanning tunnelling microscopy (STM) has made a substantial impact on the study of semiconductor surfaces over the past 10–15 years (Chen, 1993). This has been particularly valuable for examining surfaces as they transform under the influence of a driving force and has led to consideration of the response to thermal cycling, electromigration, sputtering and chemical etching (Hoshino *et al.*, 1995; Hibino and Ogino, 1994; Seiple and Pelz, 1994; Feltz *et al.*, 1994). Certainly, homoexpitaxial growth of Si on a (111) (7×7) surface is of considerable interest since this reveals that the initial stages of growth are a result of the formation of three-dimensional islands (Hasegawa *et al.*, 1994; Shigeta *et al.*, 1995). Only after these islands have coalesced does two-dimensional layer-by-layer growth occur. This is because the defects produced assist the rearrangement of surface atoms so that the atomic arrangement of the surrounding matrix and the perfection of the interface are important when either impurity or dopant atoms are introduced (Zheng *et al.*, 1994). STM studies of a boron layer on silicon produced by thermal decomposition of a diborane show three distinct surface reconstructions and pronounced segregation of the boron layer (Wang *et al.*, 1995) (Figure 3.46). (Orr, 1996). In the region of high boron concentration, 50 % of the Si is present in the first layer so that the bulk structure is essentially replaced substitutionally by the boron atoms. This observed heterogeneity at the surface indicates the complexity associated with additions of boron to silicon, particularly at the higher concentrations (Orr, 1996).

In the case of grain boundary segregation in polycrystalline silicon doped with elements such as As, heat treatment at temperatures in the range 973–1273 K (Grovenor *et al.*, 1985), produces enrichment of the grain boundaries in As atoms. These achieve at a temperature of an average saturation concentration of about 12 at. % 973 K. For silicon containing low concentrations of arsenic, this segregation removes dangling bonds since these atoms locate at positions within the boundary to allow pentavalent coordination. However, at high concentrations, the arsenic atoms also segregate to some of the tetrahedral coordination sites in the grain boundaries. These two locations modify the ability of the As to ionise, the former preventing, and the latter retaining this capability. It is the latter that allows an extensive space charge to be created around the grain boundaries, leading to an increase in the bulk resistivity of the silicon. However, this interpretation only takes account of the valence of the segregated atoms, and it has to be recognised that strain relaxation and interfacial tension cannot be neglected (Oppolzer *et al.*, 1985). Arsenic segregation has also been observed, associated with silicon films deposited on {100} orientation silicon wafers, and at various interfaces such as those between the silicon and the

Figure 3.46 Scanning tunnelling microscope image of diborane-exposed surface, showing the formation of reconstructed islands and patchy reconstruction of the substrate terrace for silicon (Orr, 1996)

substrate and between the deposited silicon and surface oxide (Mandurah, 1981). Here, the concentration of the segregated As is of the same order as that measured for the grain boundaries, demonstrating the importance for both types of interface.

Blue emitting light diodes were the first nitride devices to become commercially available (Nakamura and Fasol, 1997; Gaska *et al.*, 1998). The III to V nitrides of main interest are the direct band gap AlN, GaN and InN and their alloys because they span a wide band gap from 6.2 eV for AlN to 1.9 eV for InN. These materials have been reviewed recently by Moneman (1999), and doping data are available mainly for GaN, with little reported for AlN or InN systems. However, AlN shows p-type properties when highly doped carbon layers are present (Li *et al.*, 1997).

3.6.4 Glasses

As indicated in Chapter 2, it is the surface of glass materials that is significant since, it is here that reactions and modifications occur. Such changes in composition and chemistry at the surface are important for characterising the reactivity of glass surfaces. The main constituents of glasses arise from bridging and non-bridging oxygen atoms in the form of Si–O–Si and Si–O, other network forming elements replacing Si such as Al and B, network modifying cations such as Fe and Ca and ion-

Table 3.2 Composition of E-glass as determined from the constituents used in the manufacturing process and by X-ray photoelectron spectroscopy[a]

| | | Composition (at. %) | |
| | | X-ray photoelectron spectroscopy | |
Element	Bulk	Fibre	Plate
O	61.7	66.2	72.7
Si	18.8	21.9	17.9
Ca	7.9	7.1	2.6
Al	5.9	4.3	4.7
Na/K	1.1	0.5[b]	0.4[b]
B	4.4	< 0.05	1.7
Mg	0.3	< 0.05	< 0.05

[a]The XPS percentages were adjusted so that the carbon contamination was taken as zero.
[b]Atomic percentages of sodium only; no potassium was found on the glass fibre or plate surface.

exchangeable cations of Na and K. The structure of glasses centres on the location of alkali and alkaline earth ions in the Si–O structure to give an overall pattern of repetition and hence no characteristic chemical composition or short range order (Smart, 1998; Paul, 1977, and Kruger, 1988). Certainly, it has been demonstrated that compositional analysis of surface layers for both plates and fibres of glass that the surface compositions can be significantly different from the nominal composition of the bulk. Table 3.2 (Fagerholm *et al.*, 1996) shows a depletion of B, Mg, Al and Na at the surface of glass fibres and Ca for glass plate. By contrast, fibres show the highest concentrations of Si and O while plate is enriched in O as the hydroxyl group. These depletions or enrichments arise as a result of the manufacturing history.

3.6.5 Metal–Ceramic Interfaces

The properties of metal–ceramic composite materials are a function of the interface between these two materials. This special interface occurs widely in structural, electronic and functional materials and indeed in catalyst systems (Rühle and Evans, 1989; Evans *et al.*, 1990; Bruley *et al.*, 1994). There are two ways to describe the electronic structure of a solid by assuming

(a) the electrons are primarily localised to particular ions in the solid and
(b) the outer electrons are loosely bound and move freely between atoms.

Unfortunately, metal oxides do not readily fit either approach. As a consequence, oxides are often described according to which metal atoms are in the transition range of the Periodic Table. Non-transition metal oxides such as MgO and Al_2O_3 have cation orbitals that are involved with s and p bonding symmetry. Transition metal oxides, by contrast, have cation bonding orbitals with d symmetry and the cations have two or more valence states of similar energy, so that it is relatively easy to change the valence of the cation. As a consequence, it is important if the properties of metal–ceramic materials are to be obtained to understand the boundary and electronic structure of this heterogeneous interface at the atomic level.

In this respect, the developments of electron energy loss spectroscopy have provided powerful tools to provide information on the chemical composition from the intensity of the ionisation edges. In addition, the near-edge structure is used to obtain information about bonding and electronic configuration. To date, much of the work has centred on the metal–alumina interfaces and, in general, the spatial difference technique has been invoked to enable interpretation of the electron energy spectra (see Chapter 4). Two frequently considered metal–ceramic systems are M–TiO$_2$ and M-Al$_2$O$_3$. In the case of Al$_2$O$_3$, a range of metals on this substrate have been investigated, including Al, Cu, Nb and Ti (Scheu *et al.*, 1995, 1998; Bruley *et al.*, 1994; Dehm *et al.*, 1998). For the Al/Al$_2$O$_3$ interface the Al–L$_{2,3}$ edge spectra recorded for the aluminium, the interface and the alumina together with the difference spectra obtained by subtracting the Al and Al$_2$O$_3$ reference spectra are shown in Figure 3.47. The remaining intensity represents the spectrum obtained from the interfacial atoms having a coordination different from the bulk materials. As a comparison, Figure 3.48 shows the spectrum calculated for differently tetrahedrally coordinated clusters of oxygen atoms with a central Al atom (Figure 3.49). The influence of the type of nearest neighbour atoms is shown in Figure 3.48 for an AlO$_4$ cluster and an Al(O$_3$Al) cluster with four equal bond lengths of 0.17 nm

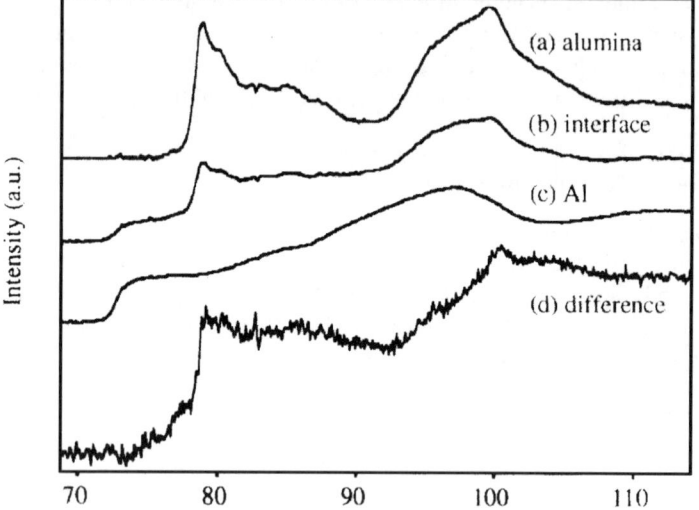

Figure 3.47 Al − L$_{2,3}$ edge energy loss near edge structure (ELNES) from alumina infiltrated with aluminium. The spectra were recorded (a) in Al$_2$O$_3$ near the interface, (b) at the interface and (c) in Al metal near the interface; (d) shows the calculated difference spectrum. Reproduced by permission of Société Française de Microscope Électronique from Scheu *et al.* (1995) *Microsc. Microanalysis Microstruct.*, **6**, 19

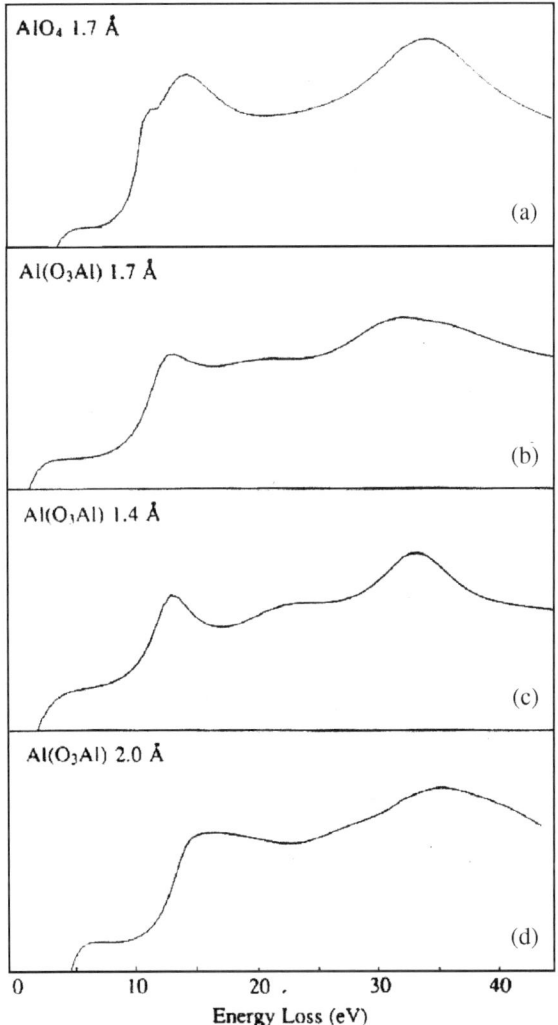

Figure 3.48 Multiple scattering calculations for the Al − L$_{2,3}$ Energy loss near edge spectra of a tetrahedral Al(O$_3$Al) cluster with all three Al–O bond lengths equal to 0.17 nm and the Al–Al bond lengths equal to (b) 0.17 nm, (c) 0.142 nm and (d) 0.2 nm. For comparison, (a) shows the calculations for a tetrahedral AlO$_4$ cluster with all Al–O distances equal to a 0.17 nm spectrum. These are to be compared with measured values in Figure 3.47. Reproduced by permission of Société Française de Microscope Électronique from Scheu *et al.* (1995) *Microsc. Microanalysis Microstruct.*, **6**, 19

which approximates to the amorphous Al$_2$O$_3$ condition. Best agreement with the experimental data is obtained for clusters where the Al–Al bond has a length greater than that for the Al–O bond. Hence, in this case the Al environment at the metal–ceramic interface approximates to a tetrahedral arrangement with a central Al atom

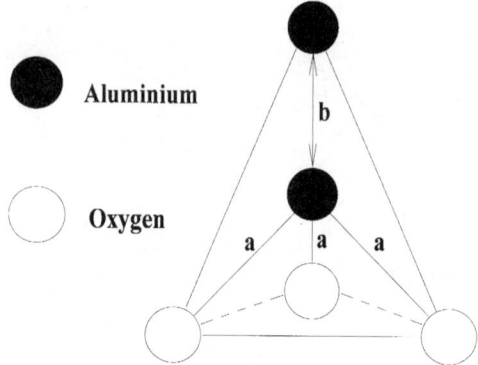

Figure 3.49 Model of the Al(O$_3$Al) cluster used for the multiple scattering calculations. The fixed bond lengths of 0.17 nm are labelled as a, whereas b is the distance varied during the calculations (Scheu *et al.*, 1995)

coordinated to three O atoms and one Al atom with a greater bond length. By comparison, a similar fcc metal, copper, was found to have copper–oxygen bonding across the interfaces and a Cu$^+$ oxidation state at the copper–Al$_2$O$_3$ interface (Scheu *et al.*, 1995). The interfacial width over which bonding occurs is estimated to be $\sim 0.34 \pm 0.06$ nm, whereas for the case of titanium, which has a well established affinity for oxygen, this width is greater, 0.5 ± 1 nm (Dehm *et al.*, 1998).

The relative oxygen affinity is important since it governs the competition of two atoms or cations for any available oxygen at the interface. At one extreme are the metals such as Ca, Y, Mg, Be, Hf, Sr, Ba, Zr, Al and Ti which have a strong affinity for oxygen, the least reactive are Cu, Rh and the noble metals and intermediate between them are V, Ta, Mn, Cr, Zn, Fe, Mo, Sn, Ni, Ru, Pb and Bi. As a first approximation, a metal with a low oxygen affinity produces little interaction; conversely, a reactive metal deposited on to a less reactive metal oxide will generally reduce the substrate (Henrich, 1998). When the affinities of the metal and oxide are similar, e.g. Mg 52 and Al 47.2 (relative units), then Mg would be expected to reduce Al$_2$O$_3$ at an Mg–Al$_2$O$_3$ interface, but, in practice, an atomically sharp interface is formed and Mg does not reduce Al$_2$O$_3$ (Yu and Lad, 1994). For Al and Ti, the oxygen affinities are 47.2 and 44.2 respectively and, although closer than Mg and Al, there is a strong interaction when Al is deposited on to a TiO$_2$ substrate, leading to oxidation of the Al (Dake and Lad, 1993). By contrast, scanning tunnelling microscopy shows that the Al grows as islands on a TiO$_2$ substrate and does not interact (Carrol *et al.*, 1994). Hence, oxygen affinity should be considered as only a starting point for predicting metal–oxide interfacial interactions.

3.7 REFERENCES

Aaron, H. B. and Aaronson, H. J. (1968) *Acta Metall.*, **16**, 789.
Ademek, J., Bartuska, P., Lejcek, P. and Baretsky, B. (1999) *Scr. Metall.*, **40**, 485.

Anthony, T. (1971) *Proc. Symp. on Radiation Induced Voids in Metals and Alloys*, eds J. W. Corbett and L. Cianniello, USAEC Symp. Series 26, Eng. 710601, p. 630.

Aucoututier, M., Rallon, O., Mantref, M. and Belouet, C. (1982) *J. Phys.*, **43**(C1), 17.

Aufray, B., Rolland, A. and Gas, P. (1986) *Surf. Sci.*, **178**, 872.

Baik, S. Fowler, D. E., Blakely, J. M. and Raj, R.(1985) *J. Am. Ceram. Soc.*, **68**, 281.

Bain, E. C. Abor, R. H. and Rutherford, J. J. B. (1933) *Trans. Am. Soc. Steel Treat.*, **21**, 481.

Balasubramaniam, P. V. and Stein, D. F. (1973) *Metall. Trans.*, **4**, 1735.

Bark, J. and White, C. L. (1987) *J. Am. Ceram. Soc.*, **70**, 682.

Bernardini, J., Cebané, R. and Cebané, J. (1985) *Surf. Sci.*, **162**, 519.

Bieber, C. G. and Dekker, R. F. (1961) *Trans. AIME*, **221**, 62.

Bougers, P. F. and Franken, P. E. C. (1981) *Adv. Ceram.*, **I**, 38.

Bourret, A. and Colliex, C. (1983) *Ultramicroscopy*, **9**, 982.

Bowen, P. Hippsley, C. A. and Knott, J. F. (1984) *Acta Metall.*, **32**, 637.

Bremer, S. J. and Ming-Jian, M. (1991) *Scr. Metall.*, **25**, 1243.

Briant, C. L. (1985) *Acta Metall.*, **33**, 124.

Briant, C. L. (1987) *Acta Metall.*, **35**, 149.

Brimhall, J. L., Baer, D. R. and Jones, R. H. (1983) *J. Nucl. Mater.*, **117**, 218.

Brimhall, J. L., Baer, D. R. and Jones, R. H. (1984) *J. Nucl. Mater.*, **122**, 196.

Bruley, J., Brydson, R., Müllejans, H., Mayer, J., Gutekunst, G., Mader, W., Knauss, D. and Rühle, M. J. (1994) *Mater. Res.*, **9**, 2574.

Burton, J. J, and Machlin, E, (1976) *Phys. Rev. Lett.*, **37**, 1433.

Cabané, F. and Cabané, J. (1991) *Interface Segregation and Related Processes in Materials*, ed. J. Nowotry, Trans Tech Publ. (Switzerland).

Cameron, W. E. (1977) *Am. Miner.*, **67**, 747.

Carrol, D. L., Liang, Y. and Bonnel, D. A. (1994) *J. Vac. Sci. Technol.*, **A123**, 2298.

Carter, R. D., Damcott, D. L., Atzmon, M., Was, G. S., Bruemmer, S. and Kenik, E. A. (1994) *J. Nucl. Mater.*, **211**, 70.

Chang, L.-S., Rabkin, E., Hofmann, S. and Gust, W. (1999) *Acta Mater.*, **47**, 2951.

Chastell, D. J. and Flewitt, P. E. J. (1979) *Mater. Sci. Eng.*, **138**, 153.

Chen, C. (1993) *J. Introduction to Scanning Tunnelling Microscopy*, New York: Oxford University Press.

Chiang, Y. M. and Takagi, T. (1990) *J. Am. Ceram. Soc.*, **73**, 3286.

Chiang, Y. M., Henriksen, A. F., Kingery, W. D. and Finello, D. (1981) *J. Aus. Ceram. Soc.*, **64**, 385.

Ciancelli, A. K., Feng, H. C. and Vitek, V. (1978) *Acta Metall.*, **26**, 1317.

Clark, J. B. (1964) *Acta Metall.*, **12**, 1197.

Clarke, D. R. (1978) *J. Appl. Phys.*, **49**, 2407.

Clarke, D. R. (1988) *Boundaries in Semiconductors*, eds J. H. Leamy, G. E. Pike and H. Seager, New York: North-Holland.

Clausing, R. E., Heatherley, L. Faulkner, R. G., Rowcliffe, A. F. and Farrel, K. (1986) *J. Nucl. Mater.*, **141**, 978.

Clayton, J. Q. and Knott, J. F. (1982) *Metal Sci.*, **16**, 145.

Coble, R. L. (1962) *J. Am. Ceram. Soc.*, **48**, 123.

Cornish, A. J. and Day, M. K. B. (1971) *J. Inst. Metals*, **99**, 377.

Cowan, R. L. and Tedmon, C. S. (1973) *Advances in Corrosion Science and Technology*, eds M. G. Fontana and R. W. Stoehler, New York: Plenum Press, Vol. 3, p. 293.

Cundy, S. L., Metherell, A. J. F., Whelan, M. J., Unwin, P. N. T. and Nicholson, R. B. (1968) *Proc. R. Soc. (Lond. A)*, **307**, 267.

Cutler, I. B. (1970) *High Temperature Oxides Part III*, ed. A. M. Alpan, New York: Academic Press, p. 129.

Dake, L. S. and Lad, R. J. (1993) *Surf. Sci.*, **289**, 297.

Dehm, G., Schere, C., Rühle, M. and Raj, J. (1998) *Acta Metall.*, **46**, 759.

Denisov, Y. P., Drachinskiy, A. S., Ivashchenko, Y. N., Kraynikov, A. V., Pronin, A. V., Timofeyev, N. I. and Firstov, S. A. (1987) *Fiz. Metallov Metalloved.*, **63**, 604.

Desu, S. B. and Payne, D. A. (1990) *J. Am. Ceram. Soc.*, **73**, 3398.

Doig, P. and Edington, J. W. (1975a) *Metall. Trans. A*, **6A**, 943.

Doig, P. and Edington, J. W. (1975b) *Corrosion*, **31**, 347.

Doig, D. and Flewitt, P. E. J. (1983) *J. Micros.*, **130**. 377.

Doig, P., Flewitt, P. E. J., Edington, J. W. (1977) *Corrosion*, **63**, 217.

Doig, P. Lonsdale, D. and Flewitt, P. E. J. (1981) *Metall. Trans. A*, **12A**, 1277.

Donald, A. (1976) *Phil. Mag.*, **34**, 1189.

Duffy, D. M. and Tasker, P. W. (1984) *Phil. Mag.*, **50A**, 143.

Dumoulin, P. and Guttman, M. (1980) *Mater. Sci. Eng.*, **42**, 249.

Edwards, B. C., Bishop, H. E., Rivière, J. C. and Eyre, B. L. (1976) *Acta Metall.*, **24**, 957.

El' Rabat, B. and Priestner, L. (1988) *Mater. Sci. Eng.*, **A101**, 117.

Erhart, H. and Grabke, H. J. (1981) *Metal Sci.*, **15**, 401.

Ernst, F., Kienzle, O. and Rühle, M. (1999) *J. Eur. Ceram. Soc.*, **19**, 665.

Evans, A. G., Ashby, M. R. and Hirth, J. P. (1990) *Metal Ceramic Interfaces*, New York: Pergamon Press.

Fagerholm, H. M., Rosenholm, J. B., Horr, T. J. and Smart, R. St C. (1996) *Colloid Surf, A*, **110**, 11.

Faulkner, R. G. and Jiang, H. (1993) *Mater. Sci. Technol.*, **9**, 665.

Feltz, A., Memmert, U. and Behm, R. J. (1994) *Surf. Sci.*, **309**, 216.

Finlan, C. T., Lin, Y. P. and Steeds, J. W. (1986) *Proc. EMAG 85*, Bristol: Institute of Physics Publishing, No. 78. p. 207.

Foiles, S. M. (1987) *MRS Proc.*, **81**, 51.

Fraczkiewicz, A. and Biscondi, M. (1985) *J. Phys. (France)*, **46**(C4), 497.

Fukushima, H. and Birnbaum, H. K. (1984) *Acta Metall.*, **32**, 851.

Gaska, R., Yang, J. W. and Bykhovski, A. D., Shar, M. S., Kaminski, V. V. and Soloviov, S. M. (1998) *Appl. Phys. Lett.*, **72**, 64.

Garjanolli, F., Alter, D., Dewes, P. and Nelson, J. L. (1988) *Proc. 3rd Inst. Symp. on Environmental Degradation of Materials*, eds E. S. Theus and J. R. Weeks, Warrendale: Mat. Sci. p. 657.

Gas, P., Guttman, M. and Barnardini, J. (1981) *Acta Metall.*, **30**, 15.

George, E. P., Lui, C. T. and Padgett, R. A. (1989) *Scr. Metall.*, **23**, 979.

Goldschmidt, H. J. (1971) *J. Iron Steel Inst.*, **209**, 910.

Grovenor, C. R. M. (1989) *Microelectronic Materials*, Bristol: Institute of Physics Publishing.

Grovenor, C. R. M., Smith, D. A. and Wang, C. Y. (1985) *J. Phys.*, **46–64**, 411.

Guttman, M. (1975) *Surf. Sci.*, **53**, 213.

Guttman, M. (1976) *Metals Sci.*, **10**, 337.

Guttman, M. (1980) *Mater. Sci. Eng.*, **42**, 227.

Guttman, M., Krahe, P. R., Abil, F., Ansil, G., Bruineaux, M. and Cohen, C. (1974) *Metall. Trans. A*, **5A**, 167.

Hall, E. L., Imeson, D. and Van der Sande, J. F. (1981) *Phil. Mag.*, **A43**, 1569.

Hampe, W. (1874) *Z. Berg-, Hütten- Salinenwesen*, **22**, 93.

Hasegawa, T., Kohno, M., Hosaka, S. and Hosoki, S. (1994) *J. Vac Sci. Technol.*, **12B**, 2078.

Hashimoto, M., Ishida, Y., Yamamoto, R., Doyama, M. and Fujiwara, T. (1984) *J. Surf. Sci.*, **144**, 182.

He, X. L., Chu, Y. Y. and Ko, T. (1982) *Acta Metall.*, **30**, 11.

Henrich, V. E. (1998) *Ceramic Interfaces. Properties and Applications*, eds R. St C. Smart and J. Nowotny, London: Institute of Materials.

Hibino, H. and Ogino, T. (1994) *Phys. Rev. Lett.*, **72**, 657.

Hippsley, C. A. (1987) *Acta Metall.*, **35**, 2399.

Hippsley, C. A., Knott, J. F. and Edwards, B. C. (1980) *Acta Metall.*, **28**, 869.

Hippsley, C. A., Knott, J. F. and Edwards, B. C. (1982) *Acta Metall.*, **30**, 64.

Hippsley, C. A., Knott, J. F. and Edwards, B. C. (1984) *Acta Metall.*, **32**,637.

Hirschwald, W., Sikora, J. and Stolze, F. (1985) *Surf. Interface Analysis*, **7**, 155.

Hirshwald, W., Loechel, B., Nowotny, J., Oblakowski, J., Sikola, T. and Stolze, F. (1981) *Bull. Acad. Pol. Sci. Sec. Chem.*, **15**, 169.

Holt, R. T. and Wallace, W. (1976) *Int. Metals Rev.*, **21**, 1.

Hondros, E. D. (1967) *Metal Sci.*, **1**, 36.

Hondros, E. D. (1985) in *Reactive Solids*, eds P. Bennett and L. C. Dunfour, Elsevier.

Hondros, E. D. and Seah, M. P. (1977) *Int. Metals Rev.*, **22**, 262.

Hondros, E. D. and Seah, M. P. (1978) *Int. Metals. Rev.*, December.

Hondros, E. D., Seah, M. P. and Lea, C. (1976) *Metals Mater.*, January, 26.

Hoshino, T., Kokubun, K., Fujiwara, H., Kumamoto, K., Ishimaru, T. and Ohdomari, I. (1995) *Phys. Rev. Lett.*, **75**, 2372.

Hway, S. K. and Morris, J. W. (1980) *Metall. Trans. A*, **11A**, 1197.

Jenko, M. and Lucas, M. (1996). *Mater. Sci. Forum*, **207**, 401.

Jiang, H. and Faulkner, R. (1996a) *Acta Metall.*, **44**, 1865.

Jiang, H. and Faulkner, R. (1996b) *Acta Metall.*, **44**, 1857.

Johnson, W. C. (1977) *Metall. Trans.*, **8A**, 1413.

Johnson, W. C. and Stein, D. F. (1975) *J. Am. Ceram. Soc.*, **58**, 485.

Johnson, W. C., Chaukia, N. Z., Ku, R., Bomback, J. L. and Wynblatt, P. P. (1978) *J. Vac. Sci.*, **15**. 147.

Johnson, W. C., Joshi, A. and Stein, D. F. (1976) *Metall. Trans. A*, **7A**, 949.

Johnson, W. C., Stein, D. F. and Rice, R. W. (1974) *J. Am. Ceram. Soc.*, **57**, 342.

Johnson, W. C., Stein, D. F. and Rice, R. W. (1975) *Grain Boundaries in Engineering Materials*, eds J. L. Walter, J. H. Westbrook and D. A. Woodford, Baton Rouge: Claitors Publ. p. 223.

Jorgensen, P. J. (1957) *J. Am. Ceram. Soc.*, **40**, 80.

Joshi, A. and Stein, D. F. (1991) *J. Inst. Metals.*, **99**, 178.

Joshi, A., Shastry, C. R. and Levy, M. (1981) *Metall. Trans. A*, **12A**, 1081.

Jupp, R. S., Stein, D. F. and Smith, D. W. (1980) *J. Mater. Sci.*, **96**, 15.

Kaneko, H., Nislizawa, T., Tamaki, K. and Anifuji, T. (1965) *J. Jap. Inst. Metals.*, **29**, 166.

Kanlea, B. and Schneider, H. (1993) Proc. Hot Isostatic Pressing, Antwerp.

Karlsson, L. and Norden, H. (1984) *Proc. Stainless Steel*, Gotenborg, 1984, London: Institute of Metals, No. B20, p. 85.

Karlsson, L., Norden, H. and Odelius, H. (1988) *Acta Metall.*, **36**, 1.

Kazmerski, L. L. (1982) *J. Vac. Sci. Technol.*, **20**, 423.

Kearns, M. A. and Burstein, G. T. (1985) *Acta Metall.*, **33**, 1143.

Kellar, H., Sukagarmi, R. and Kook, A. (1971) *Arch. Eisenhüttenwes.*, **42**, 293.

Kent, W. B. (1994) *J. Vac. Sci. Technol.*, **11**, 1038.

Kingery, W. D. (1984) *Pure Appl. Chem.*, **56**, 1703.

Kingery, W. D., Mitamura, T., Vander Sande, J. B. and Hall, E. L. (1979a) *J. Mater. Sci.*, **14**, 1766.

Kingery, W. K., Vander Sande, J. B. and Mitamura, T. (1979b) *J. Am. Ceram. Soc.*, **62**, 221.

Korobtsov, V., Lifshits, V. G. and Zotov, A. V. (1988) *Surf. Sci.*, **195**, 466.

Krahe, P. R. and Guttmann, M. (1973) *Scr. Metall.*, **7**, 387.

Kruger, A. A. (1988) in *Surface and Near Surface Chemistry of Oxide Materials*, eds J. Nowotry and L. C. Dufour, Amsterdam: Elsevier, Mat. Sci. Monograph 47, p. 413.

Kizanowski, J. E. (1989) *Scr. Metall.*, **23**, 1271.

Laws, M. S. and Goodhew, P. J. (1991) *Acta Metall. Mater.*, **39**, 1525.

Lea, C. (1980) *Metal Sci.*, **14**, 107.

Lea, C. and Seah, M. P. (1975a) *Surf. Sc.*, **53**, 272.

Lea, C. and Seah, M. P. (1975b) *Scr. Metall.*, **9**, 583.

Lewandowski, J. J., Hippsley, C. A. and Knott, J. F. (1987) *Acta Metall.*, **35**, 2081.

Li, Y. X., Salamanca-Riba, L., Spencer, M. G., Wongchotigal, K., Zhon, P., Tang, X., Talyansky, V. and Venkatesan, T. (1997) *Mater. Res. Soc. Symp. Proc.*, **449**, 555.

Liu, C. T., White, C. L. and Horton, J. A. (1985) *Acta Metall.*, **33**, 1585.

Lejček P. and Hofmann, S. (1995) *Crit. Rev. Solid State Mater. Sci.*, **20**, 1.

McMahon, C. J. (1976) *Mater. Sci. Eng.*, **25**, 233.

McMahon, C. J., Cianelli, A. K. and Feng, H. C. (1977) *Metall. Trans. A*, **8A**, 1055.

McMahon, C. J., Nichols, A. W., James, I. P., English, C. A. and Williams, T. M. (1986) *Radiation Induced Sensitisation of Stainless Steels*, ed. D. I. R. Norris, London CEGB.

Maier, J. (1995) *Solid State Ionics*, **75**, 139.

Mandurah, M. M., Soraswat, K. C., Helms, C. R. and Kamins, T. I. (1981) *J. Appl. Phys.*, **51**, 5755.

Marcus, H. L., Harris, J. M. and Szalkowski, F. J. (1971) *Fracture Mechanics of Ceramics, Vol. 1, Character of Ceramics*, eds L. I. Hinch and R. W. Gould, New York: Marcel Dekker, p. 398.

Martrepierre, P., Rofes-Vernis, J. and Thivellien, D. (1980) *Proc. Int. Symp. Boron in Steels*, eds S. K. Banerji and J. E. Morvial, New York: The Metallurgical Society of AIME.

Marwick, A. D. (1978) *J. Phys.*, **8**, 1849.

Masri, P. and Tasker, P. W. (1985) *Surf. Sci.*, **149**, 209.

Matsuoka, M. (1971) *Jap. Appl. Phys.*, **10**, 736.

Mills, M. J. (1989) *Scr. Metall.*, **23**, 2061.

Misra, R. D. K. (1996) *Acta Metall. Mater.*, **44**, 885.

Misra, R. D. K. and Balasubramanian, T. V. (1989) *Acta Metall.*, **37**, 1475.

Misra, R. D. K. and Rao, P. R. (1997) *Mater. Sci. Technol.*, **13**, 277.

Misra, R. D. K. and Rao, P. R. (1993) *Mater. Sci. Technol.*, **9**, 497.

Misra, R. D. K., Balasubramanian, T. V. and Rao. P. R. (1987) *Acta Metall. Mater.*, **35**, 2995.

Mitamura, T., Hall, E. L., Kingery, W. D. and Van der Sande, J. B. (1979) *Ceram. Int.*, **5**, 1311.

Möller, R., Brenner, S. S. and Grabke, H. J. (1986) *Scr. Metall.*, **20**, 587.

Moneman, B. (1999) *J. Mater. Sci., Mater. Electronics*, **10**, 227.

Morgan, T. S., Little, E. A., Faulkner, R. E. and Titchmarsh, J. M. (1992) *Effect of Radiation on Materials, 15th Int. Symp.*, eds R. Stoller, A. S. Kinso and D. S. Gills, Philadelphia: ASTM STP 1125.

Morrissey, F. H. J. (1997) PhD thesis, University of Bristol.

Morral, J. E. and Cameron, T. B. (1980) *Proc. Int. Symp. Boron in Steels*, eds S. K. Banerji and J. E. Morral, New York: The Metallurgical Society of AIME.

Mukhopadhyay, S. M., Jardine, A. P., Blakely, J. M. and Baik, S. (1988) *J. Am. Ceram. Soc.*, **71**, 358.

Mulford, R. A. (1983) *Metall. Trans. A*, **14A**, 865.

Mulford, R. A., McMahon, C. J., Pope, D. P. and Feng, H. C. (1976) *Metall. Trans. A*, **7A**, 1269.

Muller, D. A. and Mills, M. J. (1999) *Mater. Sci. Eng.*, **A260**, 12.

Muller, D. A., Subramanian, S., Batson, P. E., Silcox, J. and Sass, S. L. (1996) *Acta Metall.*, **44**, 1637.

Muroga, T., Yamaguchi, A. and Yoshuda, N. (1989) *Effect of Radiation on Materials, 14th Symp.*, ASTM, STP1046.

Nakamura, S. and Fasol, G. (1997) *The Blue Laser Diode*, Berlin: Springer-Verlag.

Nettleship, D. and Wild, R. K. (1990) *Surf. Interface Analysis*, **16**, 552.

Norris, D. I. R., Baker, C. and Titchmarsh, J. M. (1986) *Radiation Induced Sensitisation of Steels*, ed. D. I. R. Norris, London: CEGB, p. 86.

Norris, D. I. R., Baker, C., Taylor, C. and Titchmarsh, J. M. (1992) *Effects of Radiation on Material, 15th Int. Symp.*, eds R. Stoller, A. S. Kuman and D. S. Gilles, ASTM, STP1125 p. 603.

Nowotny, J. (1988) *J. Solid State Ionics*, **28**, 1235.

Nowotny, J. and Sloma, M. (1988) *Surface and Near Surface Chemistry of Oxide Materials*, eds J. Nowotny and L. C. Dufour, New York: Elsevier, p. 281.

Nowotny, J. and Wagner, J. B. (1981) *Oxid. Metals*, **15**, 169.

Ogura, T., Watanabe, T., Karashira, S. and Masumoto, T. (1987) *Acta Metall.*, **35**, 1807.

Ohnuki, S., Takahashi, H. and Takeyama, T. (1982) *J. Nucl. Mater.*, **104** 1121.

Ohtami, H. and McMahon, C. J. (1994) *Acta Metall.*, **23**, 379.

Ohtami, H., Ferry, H. C., McMahon, C. J. and Mulford, R. A. (1979) *Metall. Trans.* A, **6A**, 87.

Oppolzer, H., Eckers, W. and Schaber, H. (1985) *J. Phys.*, **46**(C4), 523.

Orr, B. G. (1996) *Solid State Mater. Sci.*, **1**, 11.

Padilha, A. F., Schanz, G. and Anderko, K. (1982) *J. Nucl. Mater.*, **105**, 77.

Paine, D. C., Weatherley, G. C. and Aust, K. T. (1986) *J. Mater. Sci.*, **21**, 4257.

Papazin, J. M. and Besher, D. N. (1971) *Metall. Trans.*, **2**, 497.

Paul, A. (1977) *J. Mater. Sci.*, **12**, 2246.

Pierantoni, M., Aufray, B. and Cabané, F. (1985) *J. Phys.*, **46**, 517.

Pineau, A., Aufray, B., Cabané-Brouty, F. and Cabané, J. (1983) *Acta Metall.*, **31**, 1047.

Polak, M. (1990) *Surface Segregation Phenomena*, eds P. Dowber and A. Miller, Boca Raton, FL: CRC Press, p. 291.

Pollock, G. A., Deline, V. A. and Furman, B. K. (1982) *Grain Boundaries in Semiconductors*, eds H. J. Leamy, G. E. Pike and C. H. Seager, New York: North-Holland.

Powell, B. D. and Mydura, H. (1973) *Acta Metall.*, **21**, 1151.

Powell, B. D. and Woodruff, D. P. (1976) *Phil. Mag.*, **34**, 169.

Ramano, L. T., Wilshaw, P. R., Long, N. J. and Grovenor, C. R. M. (1989) *Supercond. Sci. Technol.*, **1**, 285.

Ranasubramaman, P. V. and Stein, D. F. (1975) *Metall. Trans.*, **4**, 1735.

Rehn, L. R., Okamoto, P. R. and Wiesbech, S. H. (1979) *J. Nucl. Mater.*, **80**, 172.

Rice, R. W. (1966) *Materials Science Research*, eds W. K. Kriegel and H. Palmour, New York: Plenum Press, Vol. 3, p. 387.

Rugg, H. de and Vielfhaus, H. (1986) *Surf. Sci.*, **173**, 418.

Rühle, M. and Evans, A. G. (1989) *Mater. Sci. Eng.*, **A107**, 187.

Rühle, M., Evans, A. G., Ashby, M. F. and Hirth, J. P. (eds) (1990) *Metal Ceramic Interfaces*, New York: Pergamon Press.

Russell, P. E., Herrington, C. R., Burke, D. E. and Holloway, P. (1982) *Grain Boundaries in SemiConductors*, eds M. H. Leamy, G. E. Pike and G. H. Seager, New York: North-Holland, p. 185.

Sakurai, T., Kuk, Y., Birchenall, A. K., Pickering, H. W. and Grabke, H. J. (1980) *Proc. 27th Int. Field Ion Symp.*, eds Y. Yashiro and N. Igata, Tokyo, p. 334.

Scheu, C., Dehm, G., Müllejans, H., Brydson, R. and RÜhle, M. (1995) *Microsc. Micro-analysis Microstruct.*, **6**, 19.

Scheu, C., Dehm, G., Rühle, M. and Brydson, R. (1998) *Phil. Mag. A*, **78**, 439.

Schneider, H., Okada, K. and Pask, J. A. (1994) *Mullite and Mullite Ceramics*, John Wiley & Sons, p. 164.

Schultz, B. J. and McMahon, C. J. (1972) *Temper Embrittlement in Steels*, Philadelphia: ASTM, STP449, p. 104.

Seah, M. P. and Lea, C. (1975) *Phil. Mag.*, **31**, 627.

Seah, M. P., Spencer, P. J. and Hondros, F. D. (1979) *Metals. Sci.*, **13**, 307.

Seiple, J. B. and Pelz, J. P. (1994) *Phys. Rev. Lett.*, **73**, 999.

Shastry, C. R. and Judd, G. (1972) *Metall. Trans.*, **33**, 779.

Shigeta, Y., Endo, J. and Maki, K. (1995) *Phys. Rev. B.*, **51**, 2021.

Smart, R. S. (1998) in *Handbook of Surface and Interface Analysis*, eds J. C. Riviere and S. Myhra, New York: Marcel Dekker.

Stawstrom, M. and Hillert, M. (1969) *J. Iron Steel Inst.*, **207**, 77.

Stickler, R. (1974) in *High Temperature Materials in Gas Turbines*, eds R. R. Salmund and M. O. Spiedel, New York: Elsevier.

Stillen, K., Nilison, J. and Noring, K. (1996) *Metall. Trans. A*, **27A**, 327.

Suzuki, S., Abiko, K. and Kiruma, H. (1980) *Scr. Metall.*, **2**, 455.

Takahashi, H., Ohnuki, S. and Takayama, T. (1981) *J. Nucl. Mater.*, **103**, 1415.

Tanaka, I., Kleebe, H. J. and Cinibulk, M. K. (1994) *J. Am. Ceram. Soc.*, **77**, 911.

Tatsumi, K., Okumura, N. and Funaki, S. (1986) *Proc. 4th Int. Conf. Trans. Jap. Inst. Metals.*, ed. Y. Ishida, vol. 427, p. 27.

Tauber, C., Grabke, H. J. and Bunsenge, B. (1978) *Phys. Chem.*, **82**, 1298.

Taylor, I. T. and Edgar, R. L. (1971) *Metall. Trans.*, **2**, 833.

Taylor, K. A. (1993) *Metall. Trans. A*, **24A**, 1017.

Tedmon, C. S., Vermilyea, D. A. and Rosocauski, J. H. (1971) *J. Elect. Chem. Soc.*, **118**.

Thomas, B. J. and Henry, G. (1980) *Proc. Int. Symp. Boron in Steels*, eds S. K. Banerji and J. E. Morvial, New York: The Metallurgical Society of AIME.

Thorvaldsson, T. and Dunlop, G. L. (1983) *J. Mater. Sci.*, **18**, 793.

Titchmarsh, J. M. and Dumbill, S. (1996) *J. Nucl. Mater.*, **227**, 203.

Toshi, A. (1975) *Scr. Metall.*, **9**, 251.

Tvrdý, M., Seidl, R., Hyspecka, L. and Mazanec, K. (1985a) *Scr. Metall.*, **19**, 51.

Tvrdý, M., Seidl, R., Hyspecka, L. and Mazanec, K. (1985b) *Scr. Metall.*, **19**, 1199.

Turan, S. and Knowles, K. M. (1995) *J. Microsc.*, **177**, 287.

Shigeta, Y., Endo, J. and Maki, K. (1995) *Phys. Rev.*, **51B**, 2021.

Ustino, J. I. and Shchikov, V. (1983) *Acta Metall.*, **31**, 335.

Viswanathan, R. and Toshi, A. (1975) *Metall. Trans. A*, **6A**, 2289.

Walsh, J. M. and Anderson, N. P. (1976) *Superalloys, Metallurgy and Manufacture*, ed. B. H. Kear, Baton Rouge, LA: Claitors Publ.

Wang, C. M., Pan, W. Q., Hoffmann, M. J., Cannon, R. M. and Rühle, M. ((1996) *J. Am. Ceram. Soc.*, **79**, 788.

Wang, Y., Hamers, R. J. and Kaxiras, E. (1995) *Phys. Rev. Lett.*, **74**, 403.

Watanabe, T., Kitamira, S. and Karachine, S. (1980) *Acta Metall.*, **28**, 455.

Way, H. Y., Nafafubadi, R., Srolebitz, D. J. and Lesan, R. (1993) *Acta Metall.*, **41**, 2533.

Westbrook, J. H. (1980) *Residuals and Additives and Materials Properties*, London: Royal Society, p. 25.

Wild, R. K. and Hickey, J. (1997) Proc. ECASIA 97, Eds. I. Olefjord, L. Nyborg and D. Briggs p. 643.

Williams, T. M., Stoneham, A. M. and Harries, D. R. (1976) *Metal Sci.*, **10**, 14.

Wirth, A., Andreoni, I. and Gregory, G. (1986) *Surf. Interface Analysis*, **9**, 157.

Wynblatt, P. and McClure, R. C. (1981) *Mater. Sci. Res.*, eds J. Pask and A. Evans, Vol. 14, p. 83.

Wybblatt, P. and Ku, R. C. (1977) *Surface Sci.* **65**, 511.

Yishi, Y. and Huiliang, Z. (1983) *Acta Metall.*, **19**, 4118.

Yu, J., McMahon, C. J. and Wada, T. (1974) *Metall. Trans.*, **5**, 2235.

Yu, Y. and Lad, R. J. (1994) *Mater. Res. Soc. Symp. Proc.*, **4**, 583.

Yuan, Z. X., Song, S. H., Faulkner, R. G. and Xu, T. D. (1994) *Acta Metall.*, **42**, 127.

Zheng, J. F., Walker, J. D., Salmeron, M. B. and Weber, E. R. (1994) *Phys. Rev. Lett.*, **72**, 2414.

Zhou, Y. X., Fu, S. C. and McMahon, C. J. (1981) *Metall. Trans. A*, **12A**, 959.

Chapter 4

Measurement of Composition

4.1 INTRODUCTION

It is generally those elements of a lower concentration within the overall material system, for example a metal alloy, that are of particular importance with respect to segregation processes. As described previously, elements within Groups IV to VI of the Periodic Table are important impurities in steels (Figure 3.6). As a consequence, it is not sufficient to have a knowledge of just the bulk chemical composition, but rather the distribution is required even to the level of the location of the specific atoms within the overall microstructure. However, it has to be recognised that measurement of the bulk composition can be difficult when it is required at the parts per million or even parts per billion level. This aspect is not considered in this book, and the reader is referred to a series of references that deal specifically with bulk chemical analysis of materials (Ortner, 1983; Adams, 1983; Jenkins, 1977).

As considered in the preceding chapters, segregation of solute and impurity atoms can occur to surfaces, grain boundaries, interfaces and interphase boundaries in materials for a range of equilibrium and non-equilibrium processes. However, the segregated atoms are usually confined in spatial extent to a few atom distances about these regions, even in the case of the more extended composition profiles associated with the non-equilibrium processes. Figure 4.1 provides a simple summary of the small spatial range of such distributions that have to be measured and, typically, the full width at half maximum (fwhm) extends from about 1 to 10 atom distances. It is this aspect that has presented a challenge to quantifying the amount and the spatial extent of such segregation. Moreover, for interpretation of the contribution of the segregated atoms to physical, mechanical and chemical processes, it is necessary to measure the boundary specific and mean concentrations together with, the spatial extent and the distribution of the solute atoms, and establish their chemical state to determine how these atoms are bonded within boundaries. The techniques that are available for making these measurements may be divided broadly into two

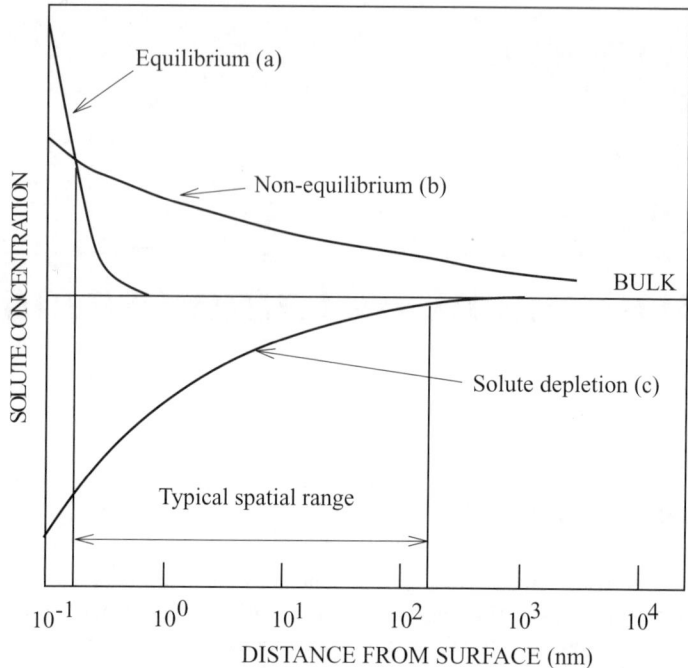

Figure 4.1   Schematic diagram showing spatial extent of common solute distributions near a free surface or an interface including given boundaries for (a) equilibrium segregation, (b) non-equilibrium segregation and (c) solute depletion

categories: indirect and direct. Obviously, it is the direct methods that should offer the ability to provide the information required. However, the indirect methods, which are usually less sophisticated and demanding, should not be neglected.

4.2   INDIRECT METHODS

These methods involve the measurement of the change in a property of the material arising from the change in local composition. They were generally developed prior to the advent of the high-resolution direct methods of measurement, and it is certainly the case that, more recently, less attention has been given to these methods for establishing a measure of segregation. However, although dated, they remain helpful, and the reader is commended the reviews by Westbrook (1964), Inman and Tipler (1963) and Joshi (1978).

4.2.1   Microhardness

In the microhardness technique, an indenter of typically 1–10 μm cross-section, either Vickers or Knoop geometry, is used, and small loads of between 1 and 50 g are applied (Gleiter and Chalmers, 1972). Since the spatial resolution of the technique is large relative to the extent of the segregation profiles that develop about a grain boundary or interface, microhardness measurements would be anticipated, at best, to provide semi-quantitative information about any change in the local composition. It has been used to examine segregation profiles produced under non-equilibrium conditions. For example, hardness changes are found to extend to distances of up to several micrometres from the observed position of an etched grain boundary for non-equilibrium segregations in zinc containing either 100 ppm Al, where there is a hardness increase, or 100 ppm Au, where there is a decrease (Aust *et al.*, 1968) (Figure 4.2). Similarly, microhardness measurements undertaken on a range of intermetallic compounds (Aust, 1968), dilute binary alloys (Jorgenson and Anderson, 1967; Westbrook and Aust, 1963) and commercial alloys such as Type 304 austenitic stainless steel (Westbrook, 1970) show changes in the hardness local to the position of grain boundaries. It is essential with this technique that the specimens are prepared correctly to ensure that any surface flowed layers are removed. Since it is not possible readily to establish the grain boundary orientation in relation to the surface on which the measurement is to be undertaken (Braunovic, 1974), misinterpretation of the origins of the measured change in hardness using this technique can occur.

4.2.2 Electrical Resistivity

In polycrystalline materials the electrical resistivity is the sum of electron scattering in the bulk crystal lattice and at internal interfaces. As a consequence, even simple solute enrichment or depletion at grain boundaries will change the resistivity of both

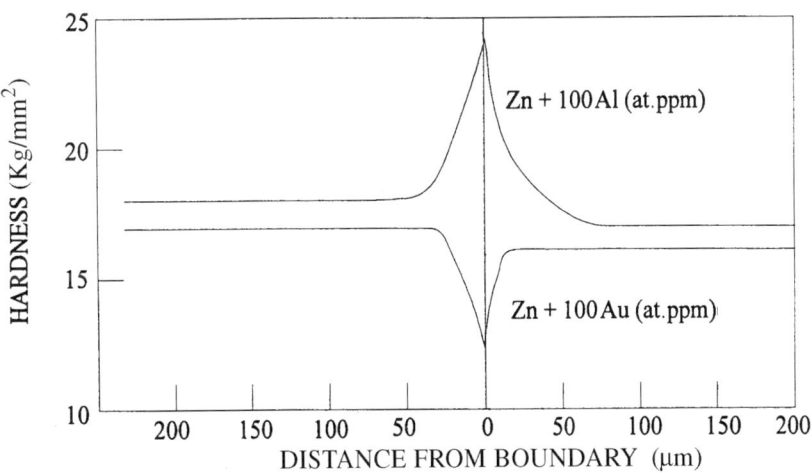

Figure 4.2 Hardness distance profiles near a grain boundary in zinc with (i) 100 at. ppm aluminium and (ii) 100 at. ppm gold (after Aust *et al.*, 1968): the former shows a hardness increase and the latter a decrease

the grains and the boundaries. Unfortunately, it is difficult to use formal theory of electrical resistivity to interpret these changes quantitatively (Seager and Schottky, 1959).

For conducting polycrystals containing small concentrations of solute atoms, differences in the electrical resistivity can be attributed to the segregation of the solute to the grain boundaries (Gleiter and Chalmers, 1972; Seager and Schottky, 1959; Braunovic and Haworth, 1969). There are examples, however, where specific geometry specimens contain a known type of grain boundary, such as bicrystals, have been used for electrical resistance measurements (Beskrovnyi, 1957). Figure 4.3 shows the change in resistance of high purity aluminium containing impurity iron atoms measured using an eddy current decay method at a temperature of 4 K (Kasen, 1970 and 1972). The change in the resistance is $\Delta\rho = \rho_{\text{measured}} - \rho_0$, where ρ_0 is the reference resistivity for a single crystal of homogeneous material. Therefore, changes are a combination of the grain boundary resistivity, which is a function

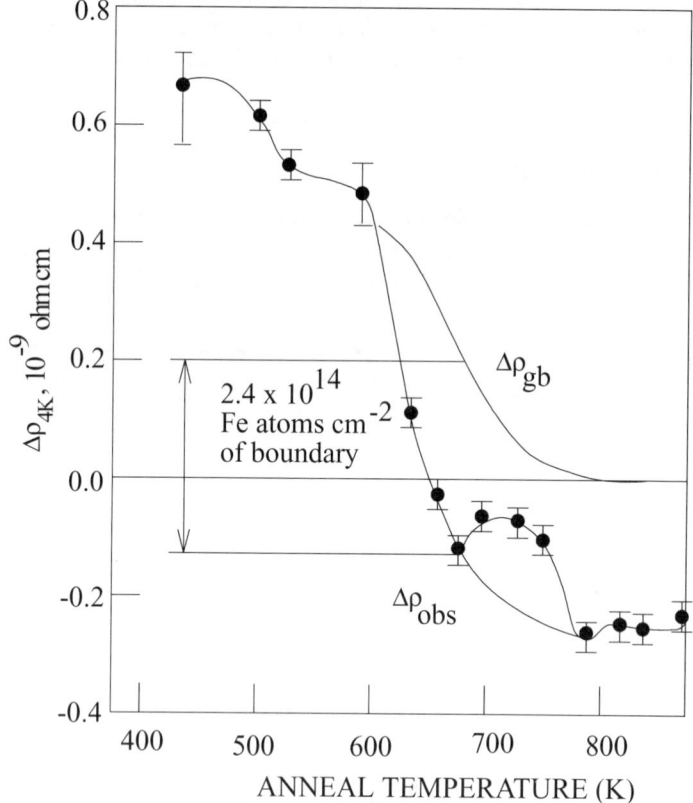

Figure 4.3   Change in residual resistivity ($\Delta\rho$) for a temperature of 4 K as a function of the step-isochronal anneal temperature for a specimen of ultra-pure aluminium doped with iron (courtesy of Kasen, 1972)

of both the number of boundaries present and their composition, and the lattice resistivity, which changes with solute segregation. Clearly, as shown by Kasen (1970), the difference in resistivity between $\Delta\rho_{ob}$ and $\Delta\rho_{gb}$ (Figure 4.3), the latter being for an alloy without segregation to the grain boundaries, can be attributed to segregation of iron atoms to grain boundaries after isochronal heat treatment at different temperatures.

Electrical resistivity theory for semi-conducting materials is well developed and offers the possibility of more detailed analysis (Grovenor, 1989). Models that describe the electrical properties of grain boundaries in semiconductors are based upon distortion of the crystal lattice created by trapping states localised to this region of the microstructure. By analogy with the Schottky barrier, the charge flow from the grains occurs at mid-gap states and will equalise the position of the Fermi levels in grains and grain boundaries to create a symmetrical Schottky barrier about the boundary (Figure 4.4a) (Taylor *et al.*, 1952). Seto (1975) has shown that the height of this potential barrier, ϕ_B, varies with impurity concentration and predicts three regimes of behaviour:

(a) at low concentrations where there are few free carriers trapped and both band bending and the barrier are small,
(b) at intermediate concentration as more carriers are available and the barrier increases and
(c) at higher concentrations there is an excess of carriers needed to saturate all the mid-gap boundary states so that the Fermi level is fixed and approaches the equilibrium level in the grain, thereby reducing the barrier.

The predicted variation in barrier height with grain boundary doping concentration is given schematically in Figure 4.4b. Conduction or current, I, across a grain boundary is modelled by including contributions from thermionic emission over the Schottky barrier at the boundary. The current across the boundary (Rai-Choudhury and Hower, 1973, and Fonash, 1972), which is given by

$$I = A^* T^2 \exp\{(-\phi_B + \xi/kT)[(EV_\alpha/kT) - 1]\} \tag{4.1}$$

where A^* is the effective Richardson coefficient, T is temperature, V_α is the applied potential, ϕ_B is the barrier height which is proportional to composition, E and ξ are defined in Figure 4.4. A typical current, I, versus voltage, V, for a grain boundary is shown in Figure 4.5a where the ohmic and resistive regions are predicted by equation (4.1). Figure 4.5b shows the carrier mobility and the corresponding resistivity for polycrystalline silicon with a concentration of dopant compared with that for a corresponding single crystal. For further details of this extensive and interesting area, the reader is referred to the various models that have been developed to describe these contributions (Gleiter and Chalmers, 1972; Grovenor, 1985, 1989; Matare, 1984).

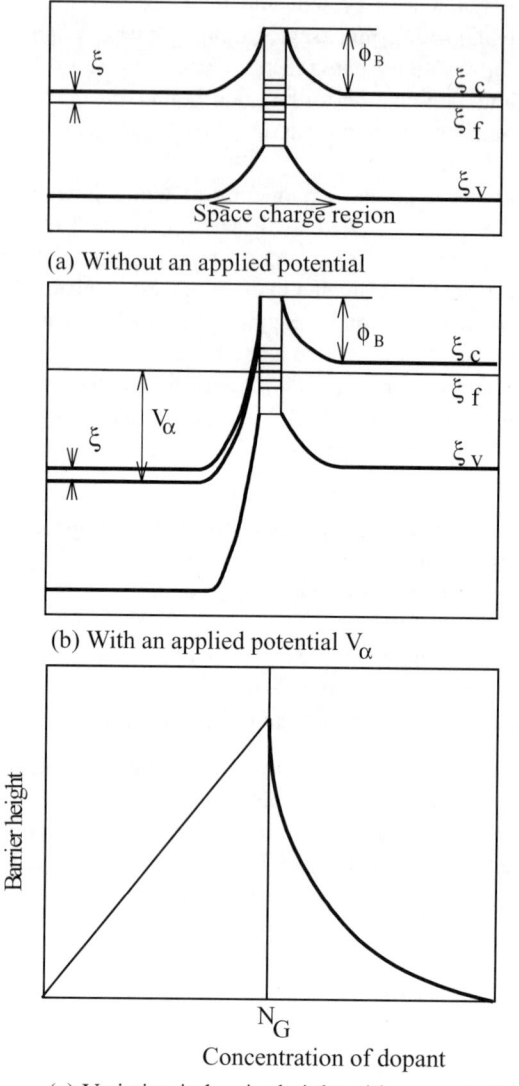

(a) Without an applied potential

(b) With an applied potential V_α

(c) Variation in barrier height with amount of dopant

Figure 4.4   Schematic diagrams showing (a) the proposed form of the band bending around a grain boundary in *n*-type semiconductor material, (b) the same boundary with a potential V_a applied to the material and (c) the variation in grain boundary barrier height with the amount of dopant in the semiconductor (after Seto, 1975)

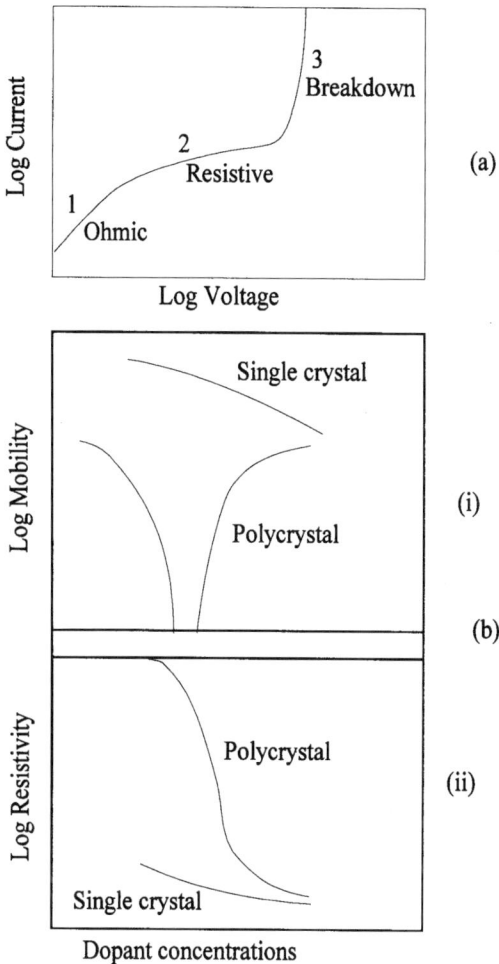

Figure 4.5 (a) Variation in transboundary current with the applied voltage, showing ohmic, resistive and breakdown behaviour, and (b) plots of (i) the variation in carrier mobility and (ii) the resistivity as a function of the dopant concentration in polycrystalline and single crystal silicon (after Seto, 1975)

4.2.3 Electrochemical Methods

Electrochemical methods have been used for some considerable time to reveal compositional changes at grain boundaries. Certainly, as discussed in Section 4.3.2, chemical and electrochemical etching may be used to provide a direct, if qualitative, indication of the presence of segregated species at grain boundaries and interfaces. Doig and Edington (1975) demonstrated that it is possible to undertake measurements of the electrochemical potentials at grain boundaries by using scanning microreference electrodes. These microelectrodes, typically of 25 µm tip diameter, provided a probe that is used to traverse the surface of the specimen at a constant distance ~ 100 µm. By a suitable choice of the electrolyte and the applied potential, the anodic regions within a specimen can be selectively identified. By scanning the microelectrode probe over the surface of the specimen, it is possible to develop a quantitative electrochemical image of the surface and thereby locate and quantify segregation. The technique has been applied to the local chromium depletions at grain boundaries in austenitic stainless steels developed by heat treatment. However, more recently, scanning probe microscopes have been gaining in popularity, and here the scanning electrochemical microscope offers the ability to investigate local electrochemical activity on a surface of a sample in an electrolyte (Band and Farr, 1992; James *et al.*, 1998). This technique scans a microelectrode over the sample to determine local reaction rates. Moreover it provides considerable information particularly if combined with a shear force back feed to maintain the probe at a constant distance from the surface of the material since it then allows larger electrochemical activity to be detected because the probe scans closer to the surface.

4.2.4   X-ray Diffraction

Several studies of grain boundary segregation in materials have been undertaken using X-ray diffraction techniques. The basis of X-ray diffraction is described elsewhere (Cullity, 1963; Flewitt and Wild, 1994), but the following measurements can be made:

(a) lattice parameter (Jorgenson and Westbrook, 1964; Arkharov and Vangengeim, 1957),
(b) scattering intensity (Arkharov *et al.*, 1962) and
(c) absorption (Arkharov *et al.*, 1953).

To obtain helpful data from these techniques, it is necessary to have a sufficient volume of solute segregated to the surfaces, interfaces or grain boundaries, so that there are large differences in the measured parameters. Consequently, X-ray diffraction techniques have been used with limited success and are usually restricted to systems where there is a marked segregation to the grain boundaries. For example, lattice parameter determination depends upon being able to undertake measurements on a system where the solute atom segregation to the grain boundaries may be suppressed and subsequent heat treatment effects segregation. The lattice parameter is measured from the position of an X-ray diffraction peak obtained from the bulk lattice and the change associated with the redistribution of solute or impurity elements to grain boundaries. Thus, the resultant change in the lattice parameters can be extremely small even when appreciable segregation of the solute atoms occurs, so that maximum resolution for the technique is required.

4.2.5 Internal Friction

The dissipation of energy that occurs in a material when stress and strain are out of phase is referred to as the internal friction of the system. It arises because of the presence of time dependent anelastic deformation in addition to purely elastic time independent distortion. Using this technique it is possible to consider the effect of redistribution of certain solute or impurity atoms, particularly interstitial atoms such as carbon, on the change to the overall damping of the system. Essentially, the greatest dissipation of energy is given by the relationship (Gleiter and Chalmers, 1972)

$$\omega t_0 \exp(Q^f/RT_p) = 1 \tag{4.2}$$

where T_p is the temperature for a fixed frequency, ω, where the maximum logarithmic decrement, δ, is observed ($\delta \alpha Q^{-1}$, where Q is the energy dissipation), Q^f is the energy barrier associated with the damping process and t_0 is a function of the relaxation time. Hence, measuring T_p for the material at two fixed frequencies provides a relationship with a gradient of Q^f/R. This technique has limited application, but it has been used to study the distribution of carbon in iron–carbon alloys (Lagerberg and Josefsson, 1955) and the segregation of phosphorus in both steels and α-iron alloyed with either nickel or silicon (Glikman and Cherpatov, 1972).

4.2.6 Summary

As pointed out by Joshi (1978), the choice of indirect methods to evaluate segregation to the interfaces and grain boundaries in materials covers a very broad range of measurable properties that should be sensitive to the changes brought about by the redistribution of the solute or impurity atoms to such preferred locations within the overall microstructure. In addition to those techniques described above, there are surface and grain boundary energy measurements (Odin *et al.*, 1956; Low, 1969) chemical etching (Cohen *et al.*, 1947; Krahe and Guttmann, 1974), including the etching of 'ghost boundaries' arising from pre-existing segregations (Philips, 1963–64), corrosion, liquid metal embrittlement, etc. However, as pointed out earlier in this chapter, caution has to be exercised when interpreting indirect data since they may arise from other intrinsic changes in the material. As a consequence, direct methods are usually to be preferred.

4.3 DIRECT METHODS

By contrast to the indirect methods, the direct methods allow detection, quantification of the amount and spatial extent of the segregated element at a surface, interface, grain boundary or interphase boundary. Moreover, since models have been, and are being, developed to describe the physical and mechanical properties associated with local boundary composition, there is a further need to be able to establish the detailed distribution of the segregated atoms within the boundary region and to have a complete characterisation including the chemical or electronic state. Unfortunately, since the segregation leading to either an enrichment or depletion of atoms at these locations is usually low, even detection can be a significant challenge. We will now consider techniques that cover a range of resolution from optical microscopy with a resolution of ~ 250 nm to high-resolution methods such as the atom probe or scanning tunnelling microscopy which enable individual atoms to be imaged. For a general introduction to many of these techniques, the reader is referred to works by Flewitt and Wild (1985, 1994).

4.3.1 Light Microscopy

The optical, light microscope provides a powerful tool for examining, evaluating and quantifying the microstructure of materials. It has a resolution of about 250 μm, depending on the particular illumination source used, with a similar depth of field. Although originally developed to operate in the transmission or reflected modes, the latter for polished and etched materials, the optical microscope remains the most useful and easily applied technique for establishing the microstructure of a range of materials. However, as with any visual technique, the value of the information that can be derived from a surface, intersecting with grain boundaries and interphase boundaries depends critically upon both the method of preparation of the specimen to be examined. Moreover, the sampling procedure selected is also important since the region viewed represents only a small fraction of the total volume of the material.

Specimens are usually prepared by a mechanical lapping sequence followed, in certain circumstances, by final chemical or electrochemical polishing to remove the 'flowed' surface layer (Samuels, 1968). Although the specimen preparation is an essential part of any optical microscopy undertaken to interrogate boundaries in materials, we do not wish to develop the detail of the various lapping and polishing procedures, but refer the reader to other studies for the underlying theory and for the practical details of mechanical, chemical and electrolytic polishing (Samuels, 1968; Lacombe, 1963; McTegart, 1959; Hoare and Mowat, 1950; Brouillet, 1955; Greaves and Wrighton, 1966). However, subsequent electrochemical or chemical etches are then used to identify any change to the local composition of the grain boundary or interphase boundary. A typical example is selective chemical etching, used to identify 'sensitisation' and associated depletion of the grain boundary in chromium for austenitic stainless steels.

4.3.2 Chemical and Electrolytic Etching

Although chemical and electrolytic etching techniques are used to reveal the microstructure of materials for examination under the light microscope. These techniques cannot be used for unambiguous identification of segregated elements or local concentrations at boundaries. An extension of the technique compared with optical microscopy is chemical or electrolytical etching to effect selective dissolution of the region of interest and then undertake chemical analysis on the resulting solutions. The technique can be applied to dissolution of selected surfaces, interfaces or grain boundaries.

The selection of the chemical reagents for preferential dissolution has to be undertaken with care to ensure that the reagent removes only those elements of interest and then from the particular microstructural feature. Hence, it is necessary to maintain a careful balance between the chemical etching reagent, the etching period and the observed effect on the overall microstructure by undertaking light microscopy comparisons. This applies equally to electrolytic etching, whereby the grain boundary or interface of interest is subject to preferential dissolution, and as such it is essential to establish the potential versus current curve for the 'specific' microstructural feature. In this respect it is preferable to undertake the etching under potentiostatic control to maintain a selected dissolution voltage on the specimen (Lacombe, 1963). The chemical etch technique has been applied to a range of metal alloy systems, including enrichment of tin in Cu–Sn alloys (Gleiter and Chalmers, 1972; Clifton and Smith, 1949), copper in Zn–Cu alloys (Lacombe, 1963) and manganese enrichment in Ni–Cr–Mn steels (Gleiter and Chalmers, 1972). The extension to the dissolution to intergranular fracture surfaces has been applied to a wide range of metal alloy systems including phosphorus, antimony and nickel in temper embrittled Ni–Cr steels (Clifton and Smith, 1949).

Chemical analysis is then undertaken on the resulting solutions, using one of a range of methods including conventional wet techniques, analysis with radiotracers (Dean and Davey, 1938), neutron activation combined with gamma ray spectroscopy and atomic absorption techniques. Neutron activation analysis has a high detection sensitivity for selected elements, where the stable nuclei within the residues obtained from drying extracted solutions are irradiated with a neutron flux to form unstable radionuclides (Schultz, 1959). It is these radionuclides that are quantified by gamma ray spectroscopy.

For many materials, apart from metals, it is difficult to evaluate grain boundary composition, and here advances have been made using chemical etching techniques, for example, detection of solute segregation to the grain boundaries in ceramics. Grain boundary enrichment has been measured in manganese–zinc spinel ferrites doped with Ca, Y and Mg. Moreover, the technique has been applied to sodium chloride doped with calcium chloride using acetone as the etchant to show surface enrichment in Ca in specimens quenched from different temperatures in the range 423–773 K (Kummer and Youngs, 1963).

In general, chemical and electrochemical etching techniques provide an indication of a change in chemical composition at an interface, grain boundary or interphase boundary, but the results can be open to misinterpretation. They are normally used only if the results are considered in conjunction with data from other supportive techniques.

4.3.3 Autoradiography

The term autoradiography is applied to any technique that demonstrates the macroscopic or microscopic location of a radioactive source in a sample (Priestley, 1992). The macroscopic location is usually achieved by applying a film, such as X-ray film, to the section of the specimen that contains the radioactive source and, after a defined period, the film is developed to reveal the distribution of radioactivity. Detection of microscopic locations requires a procedure with a higher spatial resolution, and this can be achieved by using a nuclear emission detector rather than a film, although, depending upon the size and distribution of the feature, in many cases the latter is still used (Rogers, 1979).

The autoradiographic technique is direct and has been applied with success to a range of metallic and ceramic systems. For example, α or β emissions from segregated species have been used to identify segregation to grain boundaries in a range of systems including polonium in a Pb–Bi alloy, tin in α-iron, silver in Cu and Sn, nickel in Al_2O_3, titanium in Sn, phosphorus in α-iron and boron in type 316 stainless steel (Thomas and Chalmers, 1955; Thomas and Winegard, 1952; Tiller and Winegard, 1955; Ainsle *et al.*, 1960; Coulomb *et al.*, 1959; Miller *et al.*, 1960; Jorgenson and Westbrook, 1964; Weinberg, 1963; Yukawa and Sinnot, 1955; Harris and Marwick, 1980). The optical autoradiographs in Figure 4.6 show the distribution of boron in Type 316 austenitic stainless steel specimens containing 3 and 90 ppm boron in the bulk respectively and following heating to a temperature of 1323 K and air cooling (Wild, 1980); they reveal non-equilibrium segregation of boron to the grain boundaries. To detect these segregations by this technique, theoretical considerations show that the concentration of the radioactive tracer element at the grain boundary must be between 10^2 and 10^3 greater than the concentration in the parent grain (Joshi, 1978; Weinberg, 1963) and the wavelength emitted should be long. The visibility of segregation is also a function of the penetration depth of the emitted radiation. Hence, the sensitivity for detection of a particular atom species at a grain boundary is a function of both the concentration and the range of the emitted radiation. Since the radioactive emissions are a function of the concentration, it is possible to quantify the amount of segregation using either microdensitometer measurements from the film or digitised images with quantitative image analysis procedures (Seger *et al.*, 1992; Flewitt and Wild, 1994). As a consequence, this technique has the advantage of quantification of composition but, more importantly, provides the ability to examine the distribution of a particular atom species within the overall microstructure.

4.3.4 Electron Optical Techniques

As we have described previously, the chemical composition of grain boundaries, interfaces and interphase boundaries is related to the bulk chemical composition, the thermal, mechanical and physical history and the structure of the individual boundaries and interfaces. Hence, the wide range of electron optical techniques now available are well suited to assist with determining details of segregation

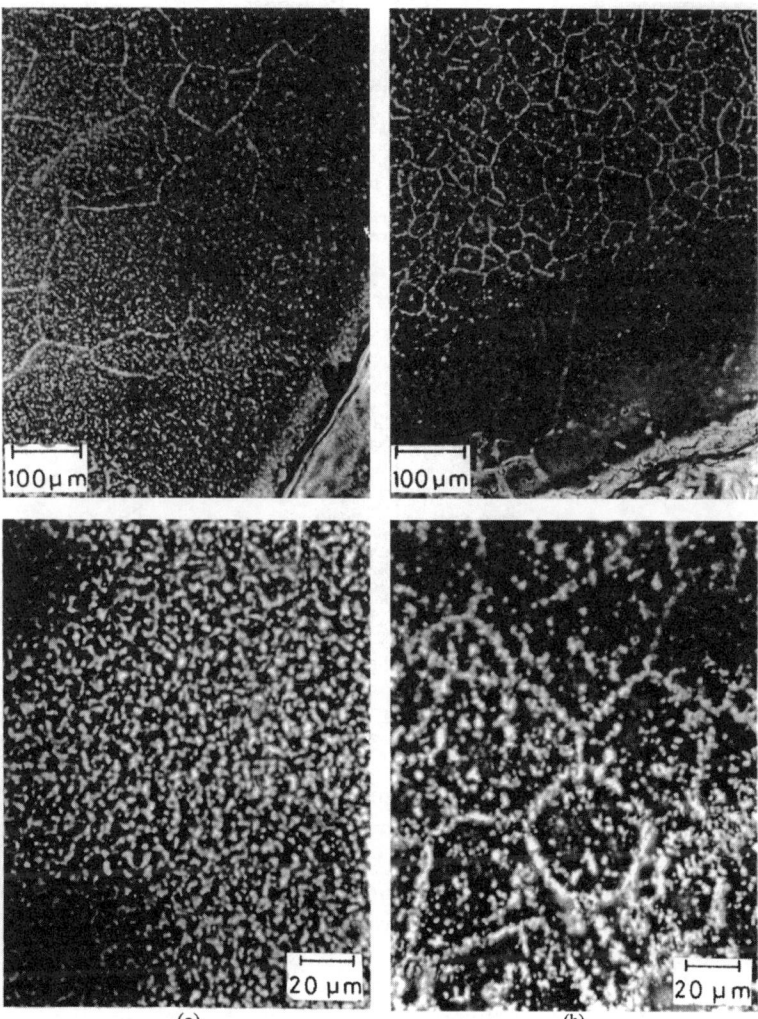

Figure 4.6 Autoradiographs of boron in Type 316 steel specimens with (a) 3 ppm and (b) 90 ppm bulk boron and following heating to a temperature of 1323 K and air cooling (after Wild, 1980)

combined with the crystallographic form of these features. We will now consider the range of electron optical techniques available for this application, with particular attention focused on analytical electron microscopy.

INTERACTION OF ELECTRONS WITH MATERIALS

The penetration depth of electrons varies with the energy of the incident electrons and the atomic number of the material examined. Figure 4.7a shows the variation in

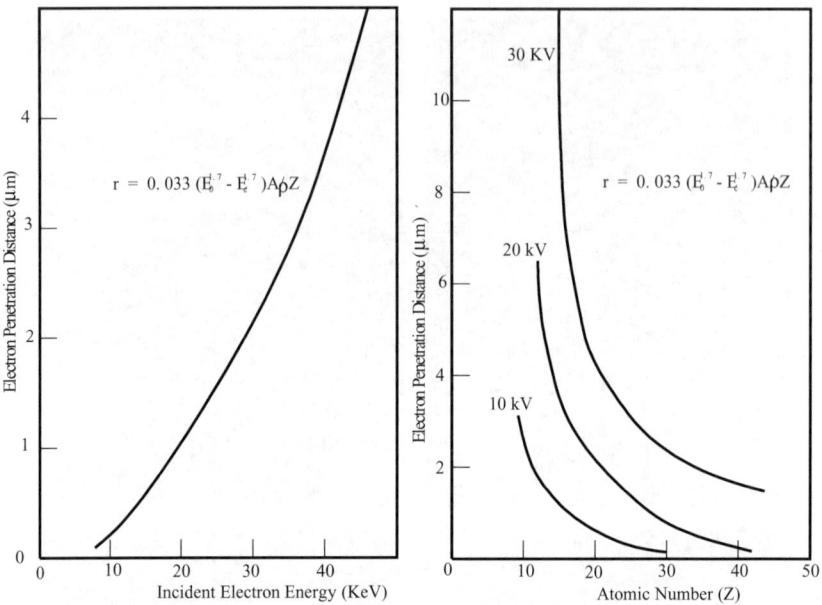

Figure 4.7 (a) Mean free path length of electrons in stainless steel as a function of electron energy and (b) of 10, 20 and 30 keV as a function of the atomic number of the material being probed

the mean free path of electrons in a stainless steel as a function of incident beam energy (Castaing, 1960) which increases from a fraction of a micrometre, at energies in the region of 10 keV, up to 2 µm at 30 keV. In Figure 4.7b the mean free path of electrons is plotted as a function of atomic number for the three incident electron energies 10, 20 and 30 keV. Thus, we can conclude that the mean free path of electrons in elements of low atomic number is very large and can be as great as 10 µm for elements with an atomic number below 20, while elements with high atomic numbers (greater than 40) have short electron mean free paths generally less than 2 µm. This atomic number dependence is important when characterising the microstructure of materials. However, at low electron kinetic energies, in the range 10–2000 eV, the electron mean free path is independent of atomic number and increases in all solids from approximately 1 nm at 10 eV to 10 nm at 2000 eV.

When an incident high energy electron beam penetrates a material, it is scattered by the atoms to give a distribution within the body (Figure 4.8) so that the resolution is limited by the spread of these electrons around the incident beam. Figure 4.9 is a plot of the intensity of the secondary electrons as a function of the distance from the centre of a 5 nm and 50 nm diameter incident electron beam on aluminium and gold. The majority of the electrons are associated with the incident beam, but a fraction emanate from the adjacent volume of material (Seah, 1980) as a result of electron

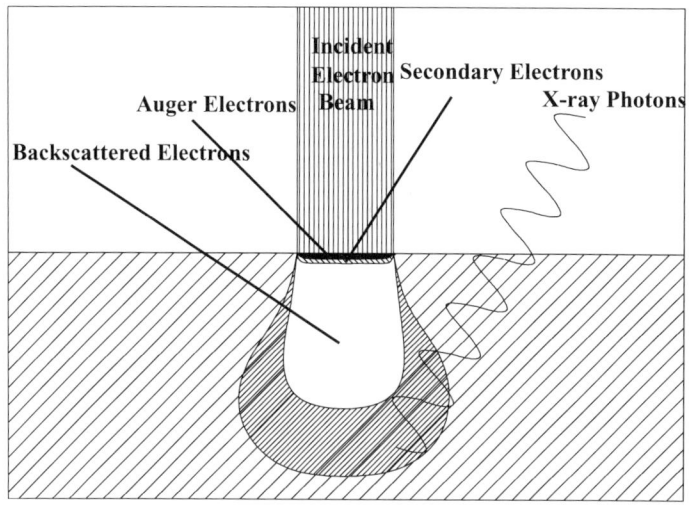

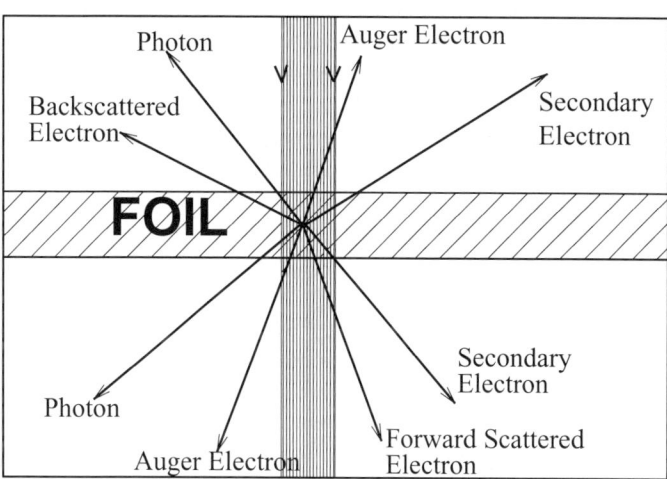

Figure 4.8 Schematic diagram illustrating (a) the volume of material that is probed by an incident electron beam together with the volumes from which X-ray and backscattered, Auger and secondary electrons emanate and (b) the corresponding thin foil diagram where there are now signals in the forward direction.

processes taking place within the material; scattered secondary electrons are detected at a distance of up to 2 μm from the centre for the gold. The intensity of these scattered electrons is lower in aluminium than in gold, but generally they do not significantly degrade the image except where their intensity makes a significant contribution to the total. It is the backscattered and secondary electrons (Figure 4.8a) that are used to form images in the scanning electron microscope, and the associated

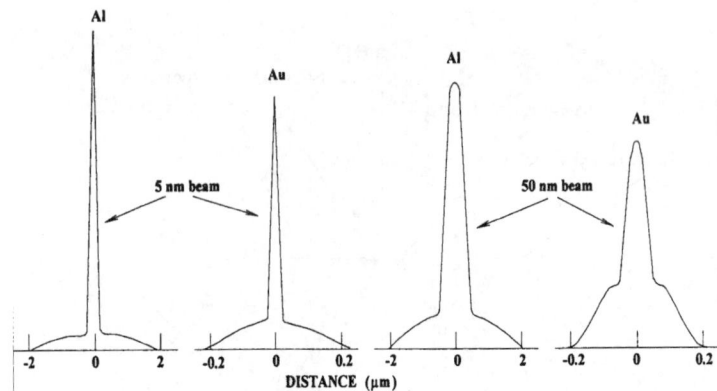

Figure 4.9   Scattered secondary electron distribution as a function of distance following bombardment of gold and aluminium by electron beams of 5 and 50 nm diameter (after Seah, 1986)

characteristic X-rays are interrogated in the electron microprobe analyser. Problems associated with the electron scattering within a material, which places an ultimate spatial resolution limit on both the image and the analytical microscopy, are reduced by the use of thin foil specimens, typically ~ 100 nm thick, which may be examined in the transmission electron microscope (Figure 4.8b). The electron signals transmitted and diffracted through these specimens can be used to produce transmission electron images and provide crystallographic characterisation by electron diffraction. Since, within this thin cross-section of material, there is little spread of the incident electron beam, microanalyses can be undertaken to a high spatial resolution so that this approach is well suited to investigating the segregation of elements to grain boundaries, interfaces and interphase boundaries. In addition to the emission of characteristic X-rays as shown in Figure 4.8b, there are a range of other signals emitted, for example Auger electrons, which provide near surface composition information, and inelastically scattered electrons (electron energy loss spectroscopy) can be used to provide local composition information.

The X-ray microanalytical techniques require efficient detection and discrimination of the characteristic radiation emitted from a specimen bombarded with high-energy electrons. The X-radiation may be divided into (a) characteristic X-rays related to the constituent elements and (b) continuous background X-radiation produced by electrons decelerated within the specimen. The characteristic X-rays emitted by a specific element may be identified from either wavelength, λ, or characteristic energy, E, since:

$$E = hc/\lambda \qquad \qquad (4.3)$$

where h is the Planck constant and c is the velocity of light. This forms the basis of techniques that use characteristic X-rays for microanalysis. Other methods require either a measure of the energy distribution of electrons that have interacted with the specimen or a quantitative evaluation of the lattice spacing using high resolution imaging.

ELECTRON PROBE MICROANALYSIS

The electron probe microanalyser (Cosslett and Duncumb, 1956) detects and quantifies characteristic X-rays emitted when electrons interact with a bulk specimen (Figure 4.8a). Indeed, current generation instruments essentially combine a medium-performance scanning electron microscope with spectrometers to detect and discriminate the emitted X-rays. Thus, electron probe microanalysis is a diagnostic technique that uses bulk specimens to determine chemical composition to a spatial resolution of between 0.1 and 1 µm corresponding to an analysed volume of 10^{-21}– $10^{-18}\,\mathrm{m}^3$ (Cosslett and Duncumb, 1956; Goldstein, 1961; Beaman and Isasi, 1972; Poole and Martin, 1961, 1969; Duncumb and Sheilds, 1963; Reed, 1975; Duncumb, 1979; Aaronson and Domain, 1966). The principal features of these instruments are:

(a) an electron optical system forming an incident electron beam of between 0.1 and 1 µm diameter,
(b) a specimen translation stage,
(c) optical and electron optical imaging,
(d) X-ray detectors and
(e) computers to control and process data.

The design of the electron probe microanalyser accommodates the low efficiency of X-ray production and, therefore, collects a large fraction of the emitted X-rays (Figure 4.10). Various designs of commercial instruments have been produced to

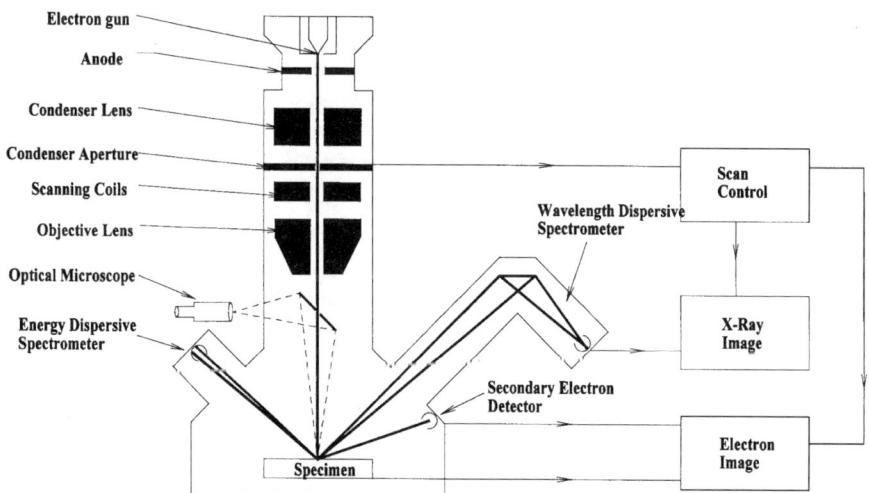

Figure 4.10 Schematic diagram showing the main features of an electron probe microanalyser

optimise X-ray yield which is further improved by inclusion of a high intensity electron source such as an LaB_6 or cold or hot (Schottky) field emission gun in place of a tungsten thermal filament. The emergent X-rays are detected using various combinations of wavelength crystal and energy dispersive spectrometers to allow a range of elements to be analysed simultaneously to reduce errors in the finally evaluated chemical composition of a particular feature.

The interaction of the incident electron beam with the bulk specimen (Figure 4.8a) can be described experimentally, analytically and by Monte Carlo calculations (Figure 4.11). The magnitude of the scattering, which defines the resolution, depends upon the mean atomic number of the phase analysed and the incident electron beam conditions. There is a minimum energy or excitation potential to effect K, L and M X-ray spectra, which increases with atomic number for a given element on passing from M to L to K lines. The intensity, I, of a characteristic X-ray emitted by an electron beam increases with both beam current, i, and the overvoltage: the amount by which the voltage, V_0, of the incident electrons exceeds the critical value for excitation, V_E. This intensity, I, is given by

$$I = iK*(V_0 - V_E)^n \tag{4.4}$$

where $K*$ is a constant dependent upon atomic number and X-ray spectra series and n is approximately 1.67 for $V/V_E \geqslant 3$; as voltage increases, n tends to unity. The intensity of the continuous background emission at wavelengths of 0.1–$0.128\,n$ is approximately

$$I\lambda = (\bar{Z}/\lambda^2)(aV - b/\lambda) + (N_x\bar{Z}^{-2}/\lambda^2) \tag{4.5}$$

where a, b and N_x are constants, V is the applied voltage and $\bar{Z}$ is the mean atomic number for the specimen. Thus, the background intensity at a fixed value of λ and $\bar{Z}$ is directly proportional to the electron beam voltage, and it is desirable to operate the instrument with a large overvoltage to increase the peak-to-background ratio. However, since overvoltage increases electron penetration into the specimen and correspondingly reduces spatial resolution, operating conditions are selected for $2 < (V/V_E) < 5$. The theoretical maximum electron probe current, I_m, is given by

$$I_m = 3\pi^2/16B[eV/kT(d^{8/3}/C_S^{2/3})] \tag{4.6}$$

where B is the electron source brightness, V is the applied voltage, kT is the thermal energy of the electron, d is the diameter of the incident electron beam and C_S is the spherical aberration coefficient.

The maximum useful count rate for energy dispersive spectrometer systems operating at about optimum resolution is approximately 2–$3 \times 10^3\,\mathrm{cs}^{-1}$ over the entire energy range, whereas, for wavelength dispersive spectrometers sct to a specific element, count rates in excess of $5 \times 10^3\,\mathrm{cs}^{-1}$ can be accepted without a loss of resolution. Thus, it is important to select an electron probe of a size to optimise the overall analysis. Using a heated tungsten filament, the electron beam current varies as (beam diameter)$^{8/3}$, equation (4.6), so that, at 20 keV, incident

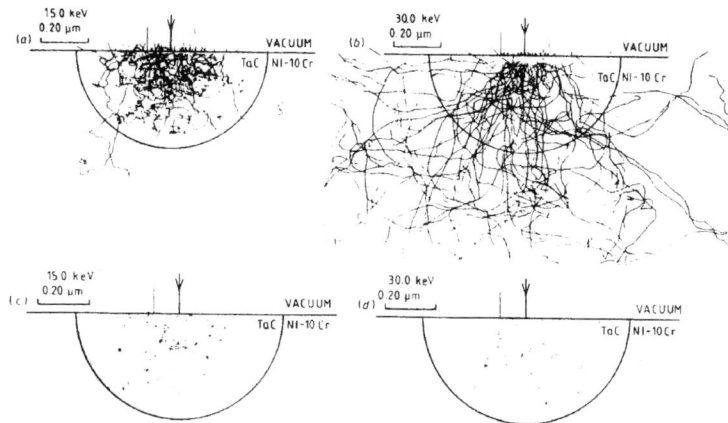

Figure 4.11 Monte Carlo simulation interaction of a 0.2 µm diameter beam with a hypothetical 1 µm diameter hemispherical TaC inclusion in a NiCr matrix: (a) electron trajectories, 15 keV; (b) electron trajectories, 30 keV; Ta Mα X-rays at (c) 15 keV and (d) 30 keV

electron beams of 0.2–2 µm diameter produce a current that is typically in the range 10^{-10}–10^{-6} A. For the bulk specimens used in the electron probe microanalyser, spectral resolution of the chemical analysis does not improve for an electron beam with a diameter much less than 1 µm; the volume of X-ray production is controlled by electron scattering and penetration rather than probe size (Figure 4.11). Here, the electron trajectories and the region of X-ray production at the higher voltage exceeds 1 µm or four times the incident probe diameter. In such cases, the analysis can be optimised by increasing the electron beam current and using wavelength dispersive spectrometers, where the benefit of the higher count rate and energy resolution is invoked.

As a consequence of the above, electron probe microanalysis using bulk specimens has restricted application to those specific areas of interface segregation where it is long range and/or large, and in recent times it has been largely superseded by techniques of more appropriate spatial resolution that make use of thin foil specimens. However, Aaronson and Domain (1966) reported enrichment of nickel at the boundaries between ferrite and austenite in a low-carbon nickel steel, and extended profiles have been observed in aluminium base alloys (Slastry and Judd, 1972; Clark, 1964).

HIGH-RESOLUTION TRANSMISSION ELECTRON MICROSCOPY

An extension to conventional transmission electron microscopy has been to increase the resolution of the images and achieve this at higher accelerating voltages. The first high-resolution images achieved by Allpress *et al.* (1969) gave intensity peaks that correspond to the atom positions in two dimensional lattices. High-resolution

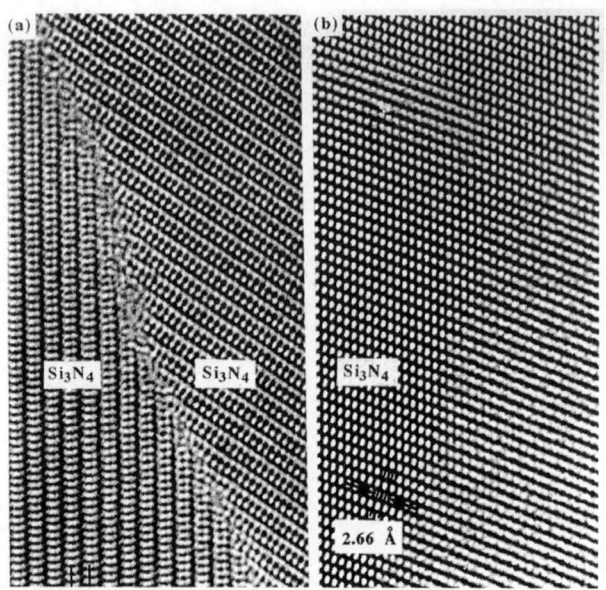

Figure 4.12 Two grain boundaries between two β-Si$_3$N$_4$ grains imaged under high resolution (Turan and Knowles, 1995): (a) shows an asymmetrical high-angle tilt grain boundary with the electron beam direction close to [011] while (b) shows a low-angle grain boundary with the electron beam direction close to [211]. Reproduced by permission of Blackwell Science Publishers from Turan, S. and Knowles, K. M. (1995) *J. Microsc.*, **177**, 287

imaging of crystalline materials requires the highest possible structural resolution defined by the first zero of the plane contrast transfer function at optimum focus (Erickson and Klug, 1971; Hofmann, 1978). To achieve this, it is necessary to have an instrument with the smallest possible values for the objective lens focal length and aberration coefficients combined with maximum mechanical stability. In practice, this is achieved using top entry stages and miniaturised objective lens pole pieces. If the transmitted beam and one diffracted beam from a very thin area of a foil are selected using the objective aperture of a microscope fitted with a high resolving power objective lens, a periodic fringe pattern is formed by phase interference between the two beams. Under carefully controlled conditions, the periodicity of the fringes in the image corresponds to the spacing of the original lattice planes in the specimen.

Figure 4.12 shows two grain boundaries imaged under high resolution (Turan and Knowles, 1995). Figure 4.12a shows an asymmetrical high-angle tilt grain boundary, while Figure 4.12b shows a low-angle grain boundary between two β-Si$_3$N$_4$ grains. Although atomic resolution images have been possible for a number of years for some complex oxides with a large unit cell, it is only with higher voltage (up to 400 keV), high-resolution instruments that the stringent conditions required to produce such images in a range of materials, including semiconductors, has

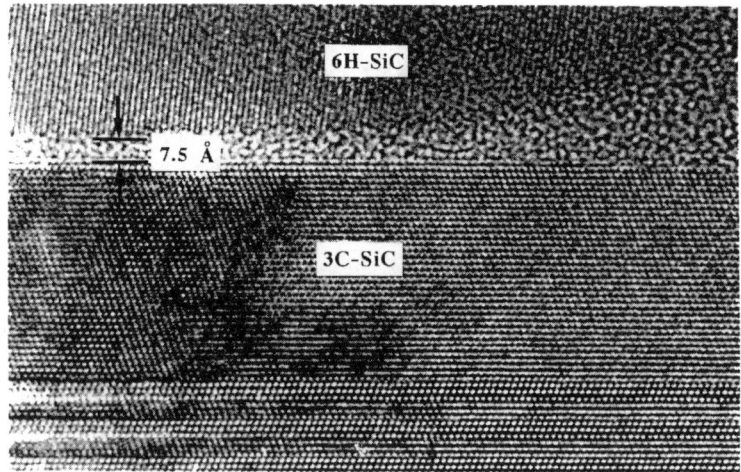

Figure 4.13 High-resolution image of the interface between a 6H–SiC and a 3C–SiC grain, showing the presence of an amorphous intergranular film approximately 0.75 nm thick between the 6H–SiC and 3C–SiC grains. A triple junction is located to the right of the micrograph and the film thickness increases from left to right as a result. Reproduced by permission of Blackwell Science Publishers from Turan, S. and Knowles, K. M. (1995) *J. Microsc.*, **177**, 287.

become possible. Figure 4.13 provides an example of a high-resolution image, in this case the interface between a 6H–SiC and a 3C–SiC grain. This image reveals the presence of an amorphous intergranular film approximately 0.75 nm thick between the 6H–SiC and 3C–SiC grains. A triple junction is located to the right of the micrograph, and the film thickness increases from left to right.

In weak beam microscopy, a high-resolution, dark-field image is formed with a reflection that is weakly excited: the deviation parameter, $\underline{s}$, is large (Cockayne, 1973; Stobbs, 1975). This specialist technique is an extension of the conventional matrix strain field conditions, $\underline{s} = 0$, the image width formed by strain field contrast is large and can be used since, as the value of $\underline{s}$ is increased, the image width decreases and, at large values of $\underline{s}$, approaches the true width of the feature. Although weak beam microscopy is used to decrease the image width of precipitates, it is of particular value when observing dislocations, grain boundaries and interfaces, since under these conditions the image width corresponds to that of the dislocation core. Obviously, higher-resolution transmission techniques are an essential component for correlating the crystallographic and atomic detail with the chemical composition of grain boundaries, interfaces and interphase boundaries.

ANALYTICAL ELECTRON MICROSCOPY

An important development for electron optical systems within the past 30 years has been in scanning transmission electron microscopy (STEM). Two separate but unrelated lines have been followed: the first produced a dedicated instrument of the type devised by Crewe (1968), and the second extended existing conventional transmission electron microscopes (Cowley, 1969). By strict definition, the former is an STEM instrument and is dedicated to providing high spatial resolution scanning transmission images. The latter is a conventional transmission electron microscope fitted with scan coils in the illumination system, with the specimen located at the centre of the objective lens, which is then used as a third condenser to form a small-diameter electron probe (~ 2 nm), and an electron detector is placed below the specimen to record the transmitted image. Both instruments have significantly influenced the understanding of microstructures by providing the opportunity to interface various electron and X-ray detectors so that high-resolution chemical information may be obtained to a similar spatial resolution (Carpenter, 1982). This is to be compared with the poor resolution achieved in the electron probe microanalyser (see the section on electron probe microanalysis). The reciprocity theorem may be applied to compare images formed in the conventional transmission and STEM instruments (Figures 4.14a and b). This theorem states: if a signal is detected at point A when a source is placed at point B, then the same signal in amplitude and phase will be detected at B if a source is placed at A. Replacing the electron source and condenser of the conventional microscope in Figure 4.14a with a

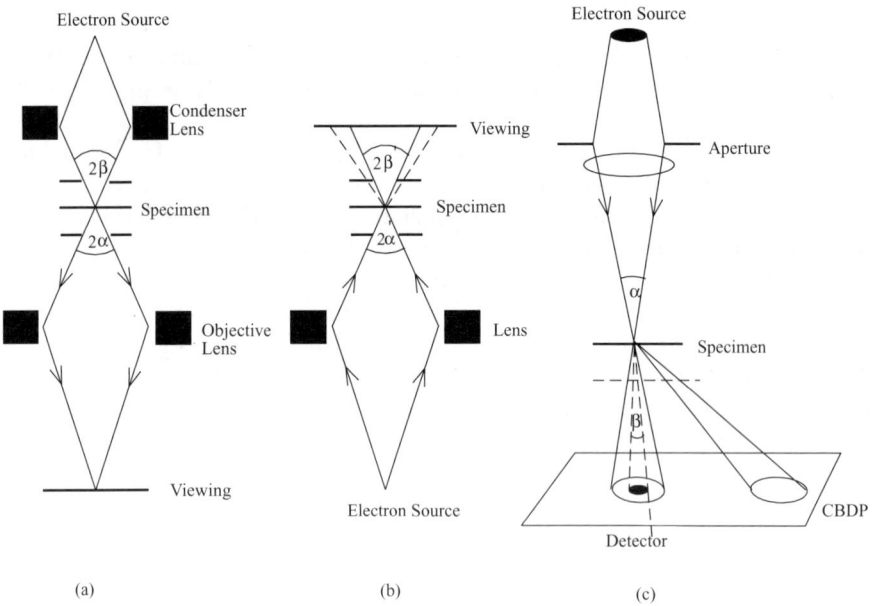

(a)   (b)   (c)

Figure 4.14 Ray diagrams for (a) conventional and (b) scanning transmission electron microscopy, illustrating the reciprocity principal; (c) formation of a convergent beam diffraction pattern in the detector plane of an STEM instrument, showing the position of the bright field detector aperture

detector, and the photographic plate with an electron source, gives a STEM optical system (Figure 4.14b). Since the electron paths are identical for both the instruments then, from the reciprocity theorem, the images should be identical, with a resolution determined approximately by the direction of the incident electron beam. Therefore, in the STEM system shown in Figure 4.14c, the incident beam in the specimen has a convergence defined by the objective aperture.

The high-resolution microanalytical scanning transmission electron microscopes currently available operate at accelerating voltages between 100 and 400 keV and the electron beam can be focused down to ~ 1 nm diameter on the specimen surface. Moreover, when an energy dispersive X-ray spectrometer (Figure 4.15) is interfaced to the microscope local composition within the foil specimens can be determined to a resolution approaching that of the transmission image (Carpenter, 1982; Doig and Flewitt, 1984). Since the dedicated STEM is fitted with a field emission source to form small-diameter (<1 nm), high-intensity electron beams, it is appropriate for undertaking microanalyses at grain boundaries, interfaces and interphase boundaries. Generally, these instruments combine higher brightness field emission sources,

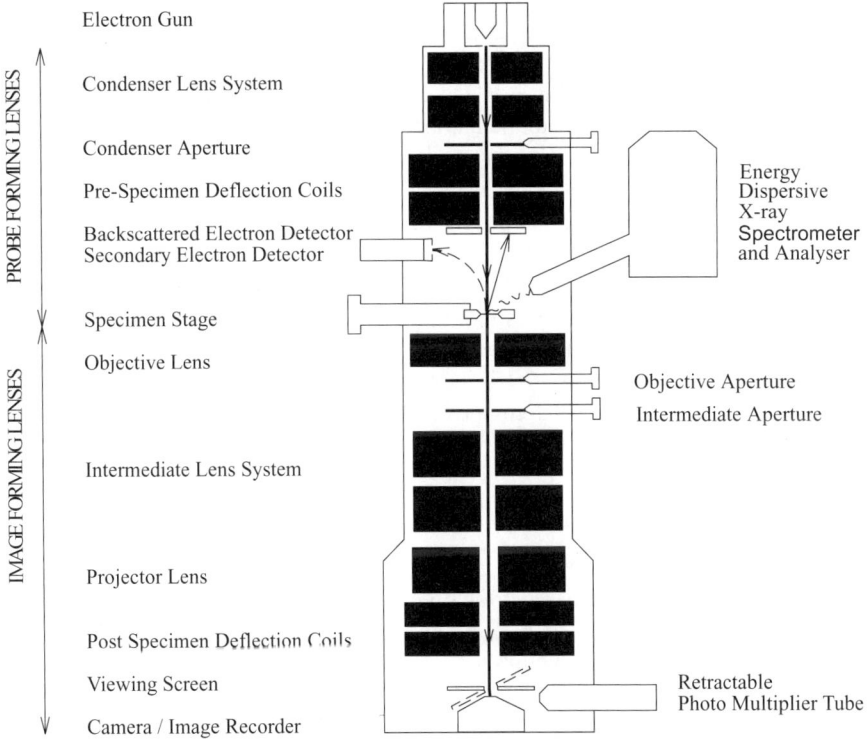

Figure 4.15 Schematic diagram of a scanning transmission electron microscope interface with an energy dispersive X-ray spectrometer for undertaking high spatial resolution chemical analysis

high vacuum systems to reduce contamination, higher accelerating voltages for better specimen penetration, smaller electron beam diameters and improved X-ray detection systems.

Spatial resolution and the minimum quantity of an element detectable are interrelated in microanalysis since any improvement in spatial resolution is balanced by a corresponding decrease in detection limit. The higher spatial resolution gives a smaller analysed volume and reduces the signal intensity so that the acquired energy dispersive X-ray spectrum will be more noisy and the small peaks arising from low concentrations of elements are less readily detected. Spatial resolution is a combination of the electron beam diameter at the surface of the specimen (Williams, 1988) and spreading of this beam as it passes through the thickness of the foil specimen (Figure 4.16a). A quantitative description of spatial resolution by Goldstein *et al.* (1977) considers the spatial extent of an elastically scattered electron beam emerging from the bottom surface of a foil specimen. The resolution is related to the diameter of a cylindrical volume, parallel to the axis of the incident electron probe, within which an arbitrarily selected, fixed proportion (usually 90%) of characteristic X-ray events occur (Figures 4.16a and b) (Doig *et al.*, 1982a). Descriptions of electron scattering and, therefore, spatial resolution of microanalysis in thin foils have been achieved using analytical approximations and Monte Carlo calculations (Doig *et al.*, 1982b; Hall *et al.*, 1981; Stephenson *et al.*, 1981). Regardless of the particular description of spatial resolution, it is generally agreed that scattering of the electron within the foil increases with atomic number, material density and foil thickness and decreases with increasing electron accelerating voltage. Moreover, the spatial resolution degrades as the incident electron beam diameter is increased.

To undertake microanalyses on planar interfaces and grain boundaries that extend throughout the foil specimen volume, it is important to derive a true composition and the spatial distribution of the elements. Therefore, the volume distribution of incident electron beam intensity is required to establish the emitted X-ray intensity (Figures 4.16a and b). The measured characteristic X-ray intensity derived from a given microstructural feature, I', is a convolution of the volume electron flux intensity, $I(v)$, and the solute composition, $X(v)$, distributions within the total sampled region of the foil:

$$I' = K \int_v I(v)X(v) \, dv \tag{4.7}$$

where K is a constant that describes the efficiency of X-ray generation, emission and detection for the particular element of interest. The analytical expression for the electron intensity distribution within the foil, for a total incident electron flux of I_e, is

$$I(r, t) = I_e \left[\pi(2d^2 + \beta_0 d_t^3)\right]^{-1} \left[-r^2/(2d^2 + \beta_0 d_t^3)\right] \tag{4.8}$$

where $I(r, t)$ is the electron flux at a distance r from the centre of the electron beam and depth d_t in the foil, d is the diameter of the incident electron beam,

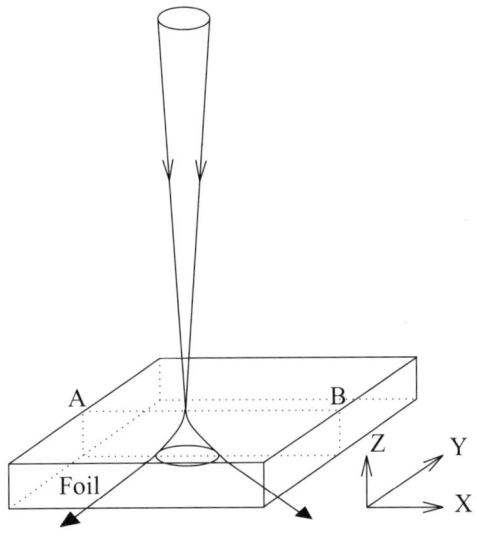

(S) T. E. M. Image

Electron Probe

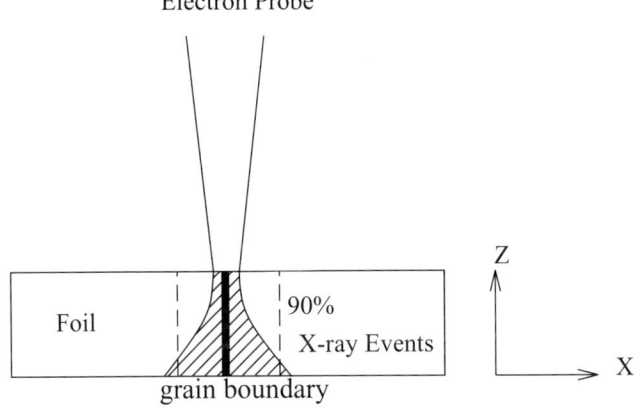

Figure 4.16 STEM-EDS-X-ray microanalysis: (a) interaction of an electron probe with a thin foil; (b) section along A–B in (a), showing the volume of foil from which X-rays are generated together with the cylinder section (dotted) within which 90 % of the X-ray events occur

where $d(\text{fwhm}) = 2.35d$ and β_0 defines the electron scattering characteristic of the foil material:

$$\beta_0 = (4Z/V_0)^2/(\rho/A \times 500) \tag{4.9}$$

where Z, A and ρ are the atomic number, atomic weight and density of the material and V_0 is the electron accelerating voltage in eV; β_0 has units of nm^{-1}. The objective of microanalysis is to optimise the measurement of I' in equation (4.7) for the particular microstructural feature defined by the term $X(v)$. This can be achieved by controlling the term $I(r, t)$ in equation (4.8). Here, the electron flux, I_e, is related to the electron source current, I_0, the incident probe size, d, and the recording time for data acquisition, t, such that (Doig and Flewitt, 1983a)

$$I_e = K'/I_0 d^{8/3} t \qquad (4.10)$$

where K' is a constant dependent on the geometry and design of the individual microscope illumination system. This or similar procedures allow the experimental variables of electron beam diameter, specimen and electron accelerating voltage to be examined, their influence on $I(v)$ and thereby I' to be considered and the microanalysis conditions to be optimised (Doig and Flewitt, 1983a and 1983b).

To determine the composition of a grain boundary or interface, the detectability of the particular element being analysed in the X-ray spectrum is determined by the number of the characteristic X-ray quanta contained in the peak, N_p, compared with those in the background of the spectrum N_b (Doig and Flewitt, 1984; Williams, 1988). For detection, the magnitude of the X-ray peak must exceed the error in its estimation such that

$$N_p > (N_p + N_b)^{1/2} + (N_b)^{1/2} \qquad (4.11)$$

Equation (4.10) relates the electron probe size d to the total incident electron flux I_e for a given probe source. As a consequence, the total intensity of the emitted X-rays increases with electron beam diameter and the background X-ray intensity in the spectrum is proportional to the beam current, I_e, the foil thickness, d^*, and the recording time, t, so that

$$N_b = K'/d^{8/3} d^* t \qquad (4.12)$$

where K' is a constant for a particular X-ray energy and illumination system that characterises the 'effective' electron intensity for X-ray generation and detection. The number of counts in the X-ray peak, N_p, for the element contained in the localised microstructural feature, such as a segregation to a grain boundary, is given by the volume integral:

$$N_p = K'/R d^{8/3} t \int_v I(V) X(V) \, dV \qquad (4.13)$$

where R is the characteristic X-ray peak-to-background ratio obtained from a homogeneous specimen of unit composition and V is the foil volume. The inequality in equation (4.11) defines the condition for detecting a characteristic X-ray energy peak from the segregated species in a recorded X-ray spectrum. Therefore N_b and N_p are calculated from equations (4.12) and (4.13) for given values of specimen and microscope operating conditions to give the optimum conditions for detecting the concentration of a particular element at a grain boundary or interface. The overall behaviour is shown schematically in Figure 4.17 where the electron beam diameter range for detection is given as a function of the characteristic width of a grain

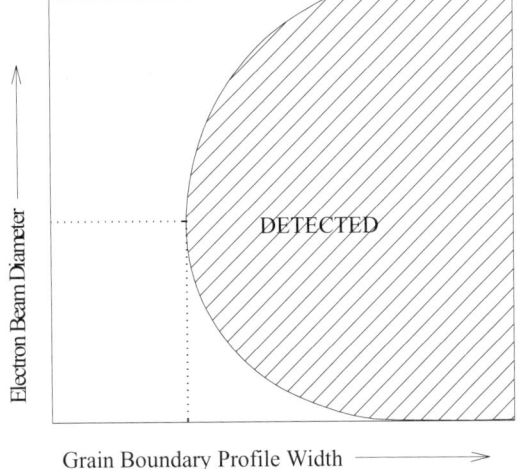

Electron Beam Diameter

DETECTED

Grain Boundary Profile Width ⟶

Figure 4.17 Conditions for detecting a characteristic X-ray energy peak from the segregated species in a recorded X-ray spectrum. The electron beam diameter range for detection is given as a function of the characteristic width of a grain boundary profile (····) indicates maximum sensitivity

boundary segregation profile. For small incident electron beams, the overall X-ray count rate is low such that statistical noise in the recorded spectrum precludes detection, whereas for large beams the volume integral term decreases and the measured N_p/N_b is reduced to a level below that at which increasing statistical confidence in the data compensates. The form of the curve defining the bound for detectibility displays a minimum which provides a measure of the ultimate sensitivity for microanalysis.

Since in the analytical electron microscope the specimens are sufficiently thin to allow electron transmission, few electrons are backscattered and, indeed, they lose only a small proportion of energy within the specimen. Therefore, the characteristic X-ray intensity, I_A, emitted from the thin specimen is given by (Goldstein, 1979)

$$I_A = \mathrm{const}\, X_A W_A \Phi_A \mu_A d^* / A_A \qquad (4.14)$$

where X_A is the concentration of element A, W_A is the fluorescence yield for element A, μ_A is the absorption of element A, Φ_A is the ionisation cross-section and the probability per unit path length of a given energy electron effecting ionisation in a particular K, L or M shell for the atom A, d^* is the specimen thickness and A_A is the atomic weight of element A. If the specimen is infinitely thin, there will be no effect from absorption and fluorescence.

A number of methods have been developed to quantify the chemical analysis of thin foil specimens (Tixier, 1979). The composition of the analysed volume is derived from equation (4.14) by measuring the emitted intensity, I_A, and calculating the constant and other terms. The specimen thickness can vary from one position to another on the foil specimen, and it is not easy to measure this value continually.

Several investigators (Tixier, 1979; Cliff and Lorimer, 1975) overcame this difficulty by basing the analyses upon the intensity ratio of two elements in the foil measured simultaneously and relating this directly to the mass concentration ratio. This is now referred to as the Cliff–Lorimer method, where the ratio of two characteristic X-ray intensities, I_A/I_B, are related to the corresponding weight fraction ratio, X_A/X_B, by

$$X_A/X_B = K_{AB}I_A/I_B \tag{4.15}$$

where K_{AB} is a constant at a given accelerating voltage that is independent of both specimen thickness and composition. A normalisation procedure, $\sum X_n = 1$, is used to convert the weight fractions into weight ratios. The accuracy of this ratio method depends upon the calibration of the K_{AB} values which can be established either experimentally or by calculation. It is unnecessary to establish K_{AB} values for all elemental pair combinations since it is practice to relate each element to a reference, such as silicon.

The relationship between K factors and K_{AB} factors is given by

$$K_{AB} = K_{ASi}/K_{BSi} = K_A/K_B \tag{4.16}$$

Although silicon provides a basis for ceramics and semiconductors, in the case of metals, K factors are referenced to iron. Figures 4.18a to c show the $K_{A_{Si}}$ and $K_{A_{Fe}}$ obtained by Wood *et al.* (1981) at an accelerating voltage of 120 keV. For the iron bases homogeneous alloys are relatively easy to obtain making direct determination of K_{AFe} possible.

$$K_{AFe} = K_{AB}/K_{FeB} \tag{4.17}$$

There are corresponding values for the L series lines. Although this thin specimen criterion neglects effects of X-ray absorption and fluorescence, unfortunately, this is not always possible, and Goldstein *et al.* (1977) offer a correction of K_{AB} for the preferential absorption of X-rays from elements A and B:

$$K_{AB} = K_{AB/TF}\left[\int_0^1 d_B \exp(-\mu_B/\rho_b)\mathrm{cosec}\,\alpha\,dt\right] \bigg/ \left[\int_0^1 d_A \exp(-\mu_A/\rho_A)\mathrm{cosec}\,\alpha\,dt\right] \tag{4.18}$$

where $K_{AB/TF}$ is the absolute value of K_{AB} when there is no absorption or fluorescence (zero thickness), $d_{A,B}$ is the depth distribution of X-ray production from element A or B as a function of mass thickness (ρt), $\mu_{A,B}/\rho_{A,B}$ is the mass absorption coefficient for X-rays from element A or B in the specimen and α is the X-ray take-off angle.

X-ray absorption is important for thin foils when considering X-ray emission from elements of atomic number up to about 11 and/or thick specimens. To a first approximation, absorption is accommodated by assuming that the average X-ray path is half the foil thickness:

$$I = I_0 \exp[\mu\rho d^*/(2\mathrm{cosec}\,\alpha)] \tag{4.19}$$

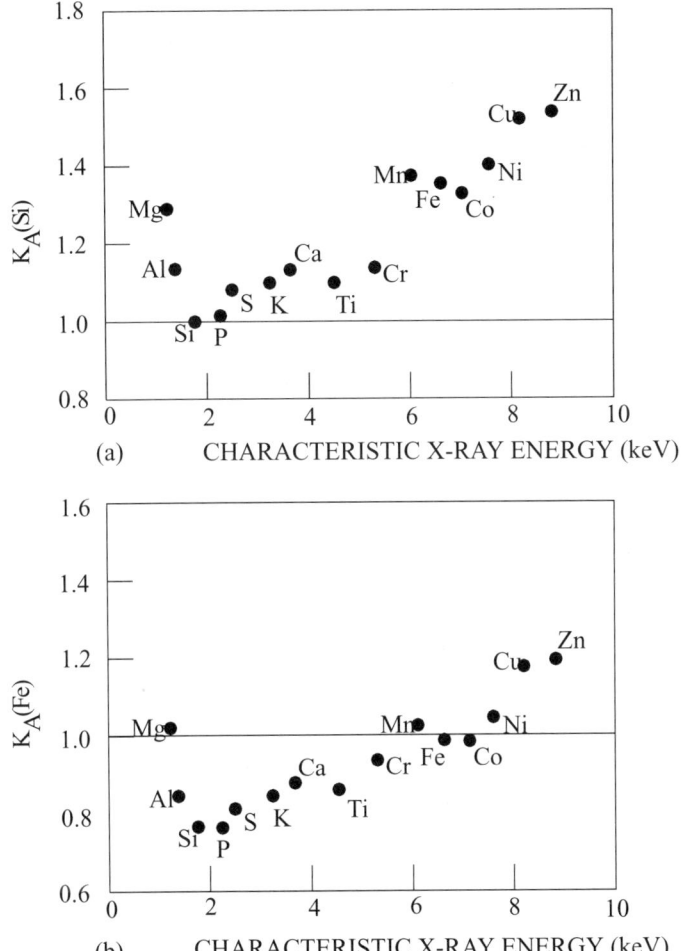

Figure 4.18 Comparison of measured K factors for K_α lines (a) relative to Si (Wood *et al.*, 1981). (b) relative to iron (Wood *et al.*, 1981) for a 120 keV operating potential. Below a characteristic X-ray energy of 3.2 keV, the KA_{Fe} factor is calculated by using the intensity of the L_α and L lines of element A

where an X-ray travels an average distance equal to $d/2$ cosec α, d is the thickness and α is the angle between the spectrometer and the specimen surface. The observed intensity ratio I_A/I_B compared to an infinitely thin specimen becomes

$$I_A/I_B = \left(I_0^A/I_0^B\right) \exp[(\mu_A - \mu_B)\rho d A/(2\text{cosec }\alpha)] \tag{4.20}$$

where μ_A and μ_B are the mass absorption coefficients. An iterative method is necessary to deduce a mean value of the density, ρ, for the specimen. Therefore, it is important to measure the foil thickness at the position where the microanalysis is undertaken. Following an analysis by Tixier (1979) for the specimen fluorescence correction, Nockolds *et al.* (1979) derived an analysis where X-ray generation is assumed to be uniform through the specimen along the line of the incident beam. The ratio of fluorescence intensity I^A to the primary intensity I_A is given by

$$\frac{I^A}{I_A} = X_B W_B[(r_A - 1)/r_A] \frac{A_A}{A_B} \left(\frac{\mu}{\rho}\right)_{AB} \frac{E_{AC}}{E_{BC}} \left(\frac{\ln E_0}{E_{BC}}\right) \Bigg/$$

$$\left(\frac{\ln E_0}{E_{AC}}\right) - \frac{\rho d^*}{2} \left[0.932 - \ln\left(\frac{\mu_B}{\rho_B \rho d}\right)\right] \sec \alpha \qquad (4.21)$$

where W_B is the fluorescence yield of element B, r_A is the absorption edge jump ratio of element A, $(\mu/\rho)_{AB}$ and $(\mu/\rho)_B$ are the mass absorption coefficients of X-rays from element B in element A and the specimen, A_A and A_B are the atomic weights of elements A and B and E_{AC} and E_{BC} are the critical excitation energies for the characteristic radiation of A and B. Twigg *et al.* (1981) showed that equation (4.21) provides the basis for correcting data, for example a foil specimen of Fe–10.5 wt % Cr up to a thickness of $\sim 600\,\text{nm}$.

For a rigorous microanalysis, it is necessary to measure the thickness of the foil specimen to accommodate absorption corrections. A number of techniques have been developed to determine the thickness of thin foil specimens, including the use of thickness fringes (Edington, 1976), trace analysis (Twigg *et al.* 1981) and parallax methods (Chester, 1985) and convergent beam diffraction (Kelly *et al.*, 1975; Allen, 1981; Ecob, 1986). The periodicity of extinction contours is a maximum at the exact Bragg condition $\underline{s} = \phi_B$ and is equal to the extinction distance ξ_g. In a bright-field image, a light and dark sequence of fringes occurs at integral values of t/ξ_g and the extinction distance can be calculated provided the structure factor is known. For some commonly used materials, values can be obtained from tables. Measurements must be made at the exact Bragg condition so that the extinction fringe spacing gives the foil thickness to an accuracy of $\pm 20\,\%$. Alternatively, the projected width of a crystallographically characterised planar feature, X, such as a stacking fault (Figure 4.19a), passing through a foil to intersect the top and bottom surfaces, can be used to determine the thickness. A similar technique involves using parallax separation of objects such as gold particles located on the top and bottom surface of foils. An alternative is to use contamination features (Figure 4.19b) introduced on to the foil surfaces during their examination, where the thickness, d^*, is given by

$$d^* = x/M \tan \alpha \qquad (4.22)$$

where α is the angle between the feature and the foil surface and M is the magnification. The major error with this and all trace techniques is establishing the tilt of the foil, since a departure of $5°$ from the horizontal can lead to an error of between 5 and $10\,\%$ in the calculated thickness. The convergent beam electron

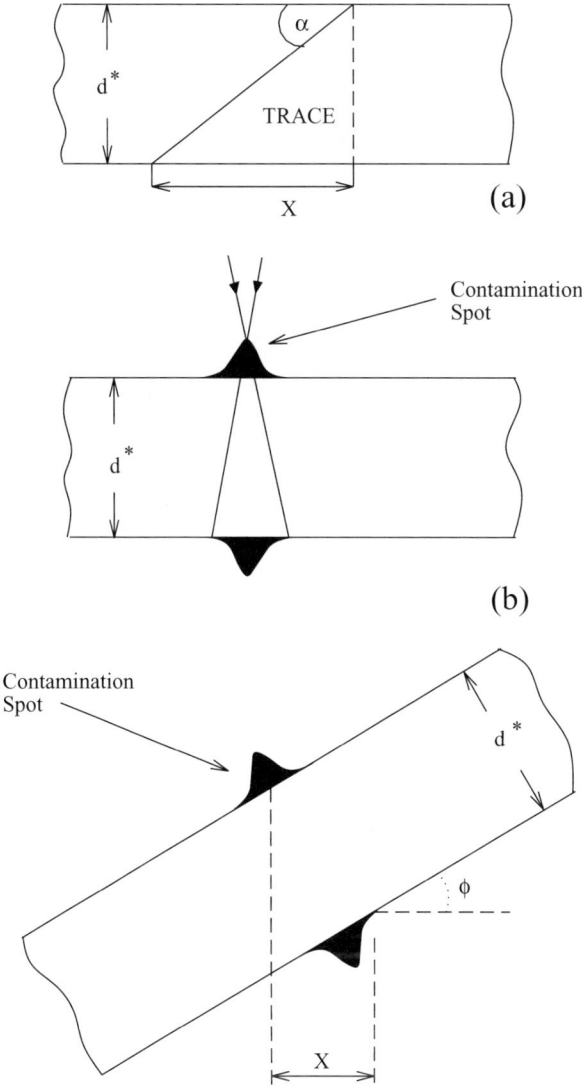

Figure 4.19 Measurement of foil thickness, d^*, from (a) a trace of projected width and (b) surface features such as contamination spots formed on an untilted foil and the same foil tilted through an angle ϕ

diffraction method relies on the two-beam solution of the dynamical equations that describe the diffracted intensity as a function of deviation parameter at given values of d^* and ξ_g. The positions of the minima are described by

$$(\underline{s}/N_i)^2 = -(1/N_i\xi_g)^2 + (1/d^*)^2 \tag{4.23}$$

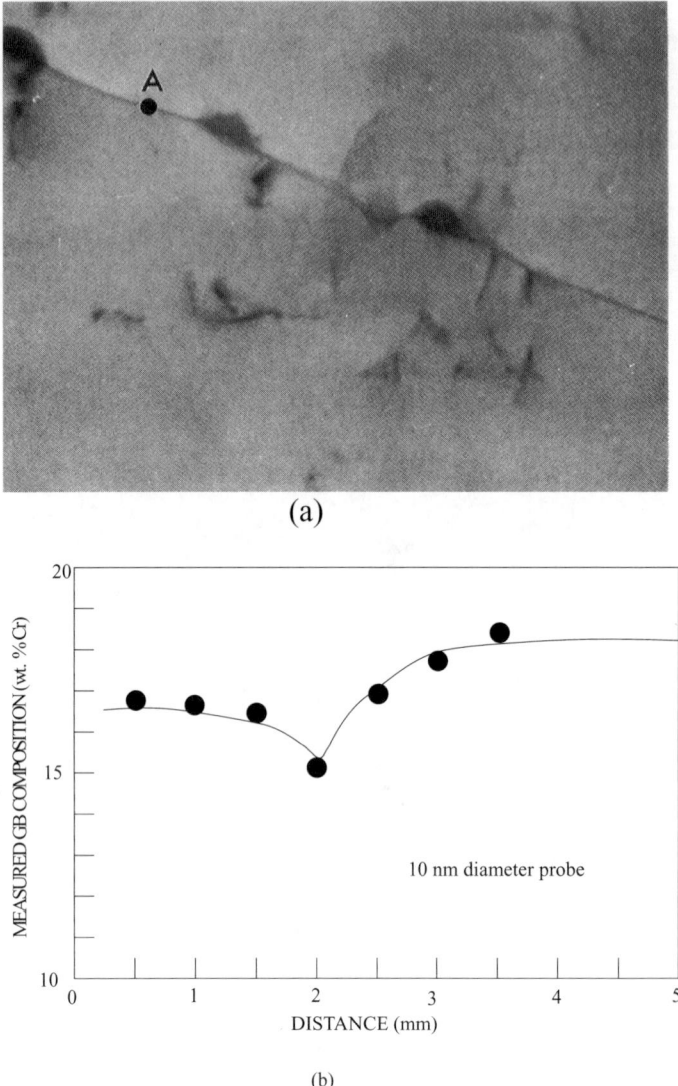

(a)

(b)

Figure 4.20 (a) Transmission electron micrograph from a thin foil taken from the weld fusion boundary of a manual metal arc weld in type 316 austenitic stainless steel. (b) Experimentally measured grain boundary chromium concentration with distance from the weld fusion interface superimposed on that theoretically predicted (solid line) for a 10 mm dia. electron probe. Reproduced by permission of CEGB from Doig and Flewitt (1988) CEGB Report OED/STB/87/0006/R/

where s is the deviation parameter at the ith minimum, and N_i are integers that correspond to the order of the minima. Equation (4.23) is defined for $N_i > t/\xi_g$ and the order of the minima. There is an analogous equation for the maxima in the intensity profile:

$$(s/K_k)^2 = -[(1/\xi_g)(1/K_k)^2] + (1/d^*)^2 1 \tag{4.24}$$

where K_k is a constant. Thus a plot of $(s/N_i, K_k)^2$ against $(1/N_iK_k)^2$ yields values of ξ_g from the gradient and d^* from the intercept. For further details of the analytical procedure, the constants K_k, the nature of the extremum that always occurs at $s = 0$ and the sequence of extreme values of N_i and K_k expected at various thicknesses, the reader is referred to the original paper of Kelly *et al.* (1975) and to the very complete description given by Allen (1981). Ecob (1986) has addressed the errors associated with this method and, apart from measurement errors, these can arise from multi-beam scattering and anomalous absorption, because neither is incorporated into the two-beam, zero absorption equations that form the basis of the method. These difficulties can be overcome for the purpose of quantitative analysis of microstructure by recording all required convergent beam diffraction patterns under identical conditions, and by measuring the effective extinction distance of the given conditions using a preliminary analysis of at least three patterns recorded at widely varying foil thickness.

An example of this application is the development of chromium depletions at the grain boundaries of an austenitic stainless steel weldment as a result of sensitisation introduced by a welding heat treatment cycle (Doig and Flewitt, 1988). Thin foil specimens removed from the heat affected regions within 2.5 mm of the fusion boundary show fine scale grain boundary $M_{23}C_6$ carbide precipitation (Figure 4.20a). X-ray microanalyses of the grain boundaries position A reveal depletion in the element chromium consistent with the preferential precipitation of chromium-rich $M_{23}C_6$-type carbides. Figure 4.20b shows the chromium compositions at grain boundaries located at various distances from the weld fusion boundary. Here, the spatial resolution of analysis is limited by the size of the incident electron probe, the scattering of electrons within the thin foil and the geometrical orientation of the grain boundary with respect to the incident probe. The application of analytical procedures described earlier in this chapter allows estimates of true composition to be established. In general, the measured grain boundary composition is displaced from the true value because the analysing electron beam samples within the volume regions with less depletion; the measured value therefore depends on the spatial extent of the depletion. Using an approximate analytical description of the scattering for an electron probe of 8 nm diameter, a foil thickness of 100 nm and an electron accelerating voltage of 100 keV, corrections can be carried out for the range of composition profiles predicted to result from the welding thermal transient (Doig and Flewitt, 1988), and the values for the grain boundary chromium composition are shown in Figure 4.20b). The predicted values of the grain boundary composition are significantly lower than the bulk value of 18 wt % chromium, showing that detection of the chromium depletion is theoretically possible where, at a distance of about 2 mm from the fusion boundary, the predicted value that would be measured is 15 wt % Cr.

4.4   ELECTRON ENERGY LOSS SPECTROMETRY

Energy losses occur when electrons are reflected or scattered by a solid and this provides the basis for electron energy loss spectrometry (EELS) as an analytical tool (Joy and Maher, 1978). Measurement of these losses allow the composition local to grain boundaries and interfaces to be established to high spatial resolution covering all elements, but in particular those of low atomic number, for example, C, N, O and B which contribute significantly to the properties of steels. An energy loss spectrometer of the type shown schematically in Figure 4.21 can be interfaced to most commercial STEM instruments, thus complementing the chemical analysis systems which measure characteristic X-ray emissions (Loretto *et al.*, 1988). The magnetic prism analyser positioned below the specimen focuses electrons leaving the object point to an image point (Figure 4.21). Electrons with an energy loss, diverging from the object point, are deflected through an angle $\pi/2$ and then focused at the image point, *I*, whereas those with an energy difference $E_0 - \Delta E$ focus at a position displaced by a distance Δx from position *I* to form a line dispersion. Three parameters define the performance of this type of spectrometer:

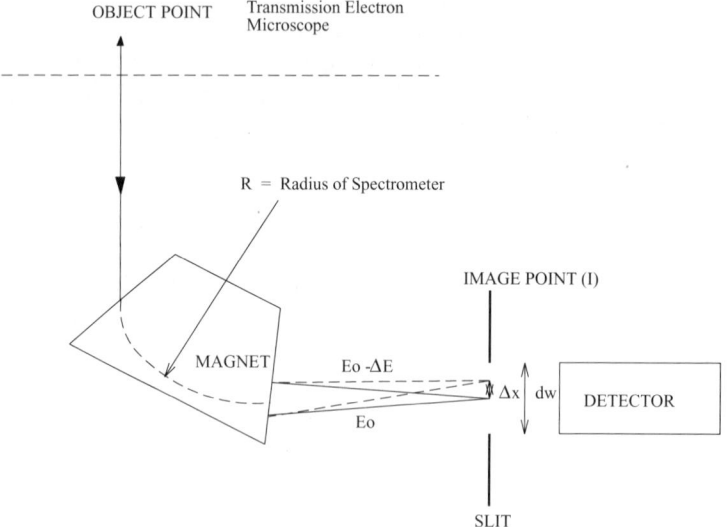

Figure 4.21   Basic analyser layout for the electron energy loss spectrometer. Electrons leaving the object point from the specimen in the transmission electron microscope are deflected through $\pi/2$ by the magnetic field and form a dispersed line image at the position of the detector slit. The dispersion $\Delta E/\Delta x$ is typically a few $\mu m/eV$

(a) the dispersion, $\Delta E/\Delta x$,
(b) the solid angle of acceptance, π/β^2, and
(c) the energy resolution.

For electron beams of energy difference ΔE and separation Δx in the image plane, the dispersion, D^*, for the spectrometer is

$$D^* = 2R/E_0 = \Delta E/\Delta x \tag{4.25}$$

where R is the radius of the electron path through the spectrometer. The minimum energy difference selected by the acceptance slit of width d_w at the imaging position is given by

$$\Delta E = d_w/D^* \tag{4.26}$$

It is this that limits the energy resolution of the system, which for 100 keV incident electrons is typically 1–2 eV for an acceptance angle of ~ 10 mrad. The electrons emerging from the slit are detected by a scintillator-photomultiplier or an array of solid state diodes so that the signals are collected either serially or in parallel. Parallel detection electron energy loss spectrometers based upon solid state detectors have improved detection efficiency by at least two orders of magnitude compared with the serial detection spectrometers. This increased detection both reduces damage to radiation sensitive materials and improves collection of extended energy loss fine structure and increased speed of data acquisition.

A typical electron energy loss spectrum (Figure 4.22) contains three identifiable energy regions for loss of momentum and energy—(a) no loss, (b) low loss and (c) core loss:

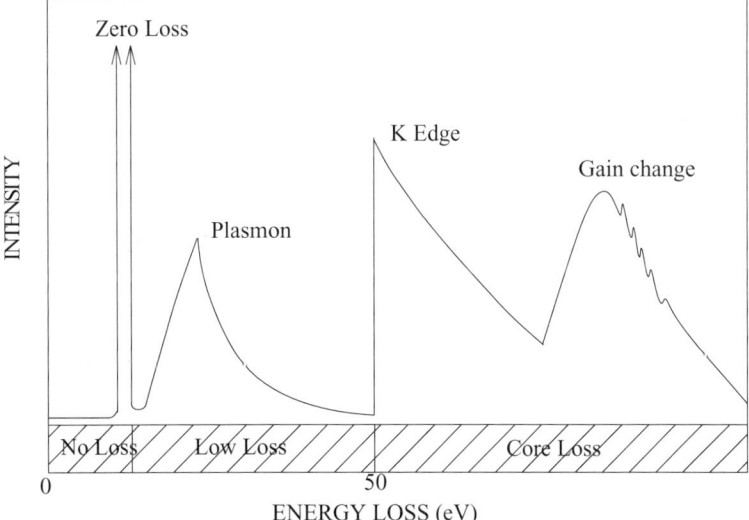

Figure 4.22 Schematic diagram of a typical electron energy loss spectrum showing the main characteristic features: no loss, low loss and core loss

(a) *The no loss region* contains the zero-loss peak. The incident electron beam has a finite width because it has an energy spread, and the electron spectrometer has an energy resolution so that electrons with an energy difference less than this are not separated. The zero-loss peak contains contributions from electrons that are not scattered on passing through the specimen, elastically scattered electrons and those electrons that generate a phonon excitation.

(b) *The low loss region* extends over an energy range from the edge of the zero-loss peak to about 50 eV and contains the plasmon loss peak; the energy losses arise from electrostatic interactions with the electrons. The energy loss structures are caused by either the excitation or ionisation of electrons from various bound states. However, there are the plasmon excitations that occur in metals, because they have 'free' electrons excited by fast incident electrons. The frequency of the plasmon oscillation is proportional to the root of the number of free electrons per unit volume of plasma. The plasmon energy loss has the potential for providing an identification of a material, but, unfortunately, metals and alloys have plasmon peaks of a similar electron energy. With careful calibration, it is possible to measure changes in composition from small shifts in the plasmon peak position.

(c) *Core losses* occur with energies extending from 50 eV upwards: the energy losses arise from inelastic interactions with the inner atomic shells of the specimen atoms, whereas the background intensity results from valence shell excitations. This produces the characteristic edges used for elemental analysis. The energy and momentum distribution of higher energy electrons (30–200 keV) that have interacted with a foil specimen (Figure 4.23) provide information on local chemical composition (Goldstein *et al.*, 1977; Egerton and Cheng, 1987; Egerton, 1989). Since momentum is a vector quantity, a complete description of the interaction requires a knowledge of both the energy change, E, and the angular displacement, θ (Figure 4.23) (Egerton, 1989). Therefore, the signal $I(E)$ detected at some energy loss E by a spectrometer collecting electrons scattered through angles up to β is given by

$$I(E) = IN\eta\sigma_0(\beta, I, E_0) \tag{4.27}$$

where I is the intensity of the incidence electron beam of energy E_0, N is the number of atoms in the irradiated area, η is an efficiency factor and σ_0 is the interaction cross-section. The latter represents the probability that any incident electron suffers an energy loss E as it is scattered into a solid angle less than β (Figure 4.23). A minimum energy E_k is needed to ionise a particular inner shell of an atom such that for $E > E_k$ a finite cross-section of ionisation exists to increase $I(E)$ when this condition is satisfied. This will produce a discontinuity or 'edge' in the spectrum at E_k. Since the energy E_k is approximately the binding energy of the particular atom shell ionised, it characterises the atom, and a single measure of the energy loss identifies the element. Any energy loss by an electron of $E > E_k$ causes ionisation so that the edge extends from E_k to E_0. Since the cross-section falls as E^{-k}, where k is

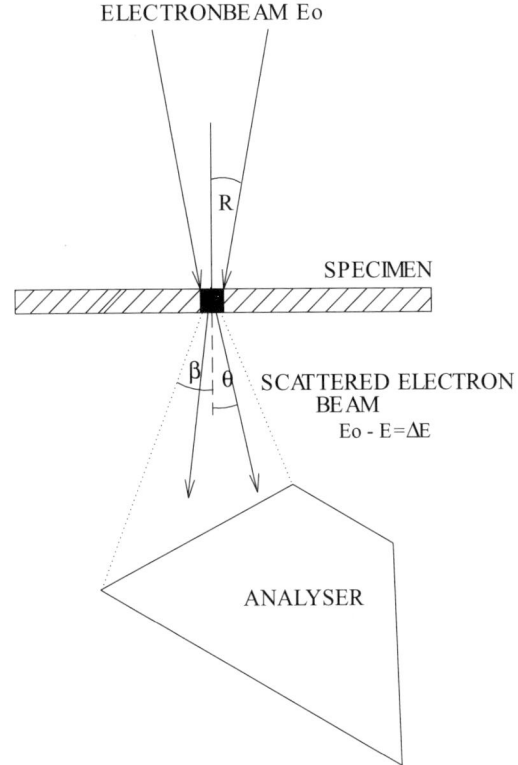

Figure 4.23 Energy loss spectroscopy illustrating the incident electron beam convergence angle α, the scattering angle θ and the spectrometer acceptance angle β

approximately equal to 4, the inner shell edge approximates to a triangle in the spectrum, and for K shell ionisations this is observed experimentally. For L shells the situation is more complex since additional energy terms arise from the angular momentum associated with the 2p orbital electrons. This increases the apparent energy required to ionise the atom, giving a delayed maximum edge of the type observed in the L_{23} shell edge from silicon and other elements in the first row of the periodic table. For elemental analysis, both the energy and the shape of the edge in the spectrum are used. The background is substantial under the edges to be measured in the energy loss spectrum. The cross-section term, σ_0, in equation (4.27) contains contributions from other interactions together with the required edges that combine to produce a signal:

$$I(E) = IKE^{-k} \tag{4.28}$$

where K and k are constants for limited ranges of a spectrum whose values depend upon E, β and the material. Indeed, it is the magnitude of this background relative to that of an identified peak (Figures 4.24a and b) that make peak deconvolution procedures necessary for quantitative evaluation of elemental concentration (Collet *et al.*, 1981). The background under each peak has to be subtracted, using equation (4.28), to fit to the background shape at a position before the edge positions, thus,

$$N = I_k/I\sigma_p\eta \qquad (4.29)$$

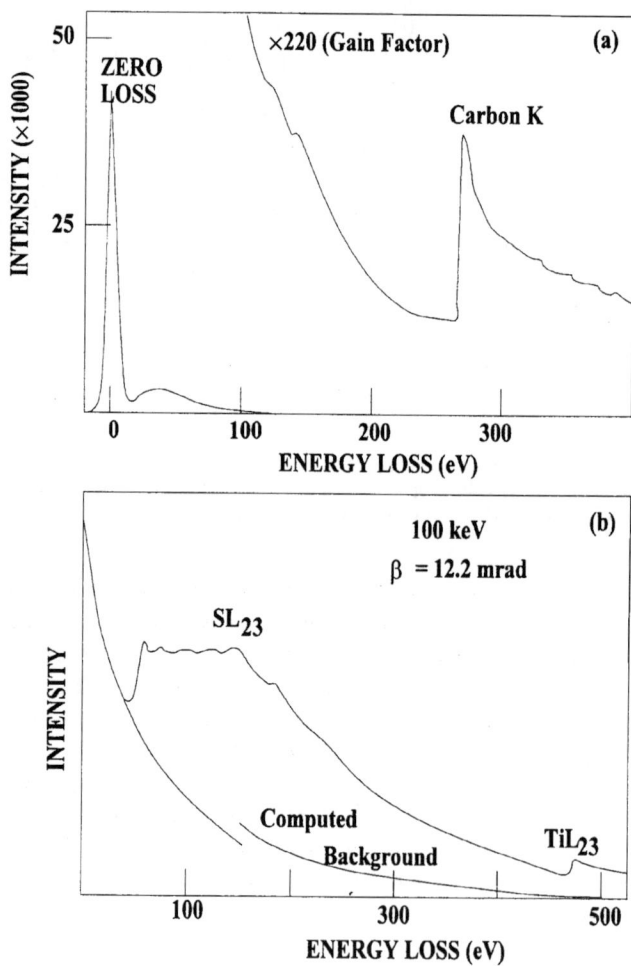

Figure 4.24 Energy loss spectra: (a) spectrum from 20 nm thick carbon, including zero-loss and plasmon peaks (gain change makes edge visible); (b) spectrum from TiS_2 together with computer modelled background fit to AE^{-r}

In practice, the edge integral, the area under the peak, can be obtained for an energy window E so that

$$N = I_K(\beta, \Delta)/I\sigma_p(\beta\Delta)\eta \tag{4.30}$$

where the variables β and σ_p indicate that the integral and the cross-section relate to scattering angles up to β and energy η losses in the window E to $E_k + \Delta$. For quantitative microanalysis, I is conventionally replaced by I_0 $(\beta\Delta)$, the integral in the energy window under the zero-loss peak Δ for the acceptance angle β. This allows backscattering and plasmon scattering in the specimen and integral counts from the element analysed to be evaluated directly from the spectrum. However, when undertaking a microanalysis, elemental ratios are usually required such that

$$N_A/N_B = I_{KA}(\sigma_p\beta, \eta_b)/I_{KB}(\sigma_p A_A \eta_A) \tag{4.31}$$

giving the mass concentration ratios

$$X_A/X_B = (A_A N_A/t_p/(A_B N_B d^*) = (A_A N_A)/(A_B N_B) \tag{4.32}$$

where d^* is the specimen thickness, ρ is the density and A_A is the atomic weight of element A. These relationships are functionally identical to the 'K factor' formulation used for STEM X-ray microanalysis described previously. This technique offers a potentially powerful procedure for undertaking quantitative measurements for low atomic number elements in thin foil specimens, particularly at grain boundary regions.

An important extension to this technique has been the adoption of the subnanometre electron probes to enable high spatial resolution microanalyses to be undertaken where the chemical composition change is at the atomic level. Here, as with the energy dispersive X-ray spectrometer, the process is carried out by sequential point-to-point analysis across the selected grain boundary or interface. For this, the improved performance of the parallel array electron energy loss spectrometers is invoked. Figure 4.25a shows part of an energy loss spectrum from a grain boundary in α phase Zr–2.50 wt % Nb containing 1000 ppm iron that has segregated under the influence of neutron irradiation. The Fe $L_{2,3}$ edge is at a loss energy of $\sim 700\,\text{eV}$ caused by the transitions from the full 2p states to the empty states in the bands at the 4s and 3d atomic levels (Figure 4.25b). Here the transitions are dominated by the 3d states as they are more localised and have larger matrix elements than the 4s states. In addition, the so called white lines are also present and relate to the spin–orbit splitting left in the 2p states: the L3 and L2 lines from the ?p3/2 and 2p1/2 states respectively (Dray *et al.*, 1995). As described by Okamato *et al.* (1992), the white lines vary systematically across the series of transition elements to provide a measure of the number of holes in the d band (Figure 4.26). To quantify these data, two interpretations can be adopted:

(a) the Pearson ratio which is the ratio of the total white line counts to the background and

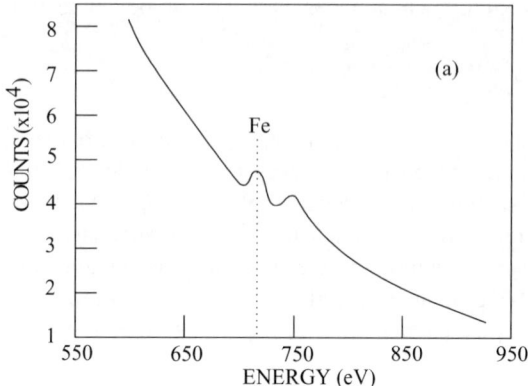

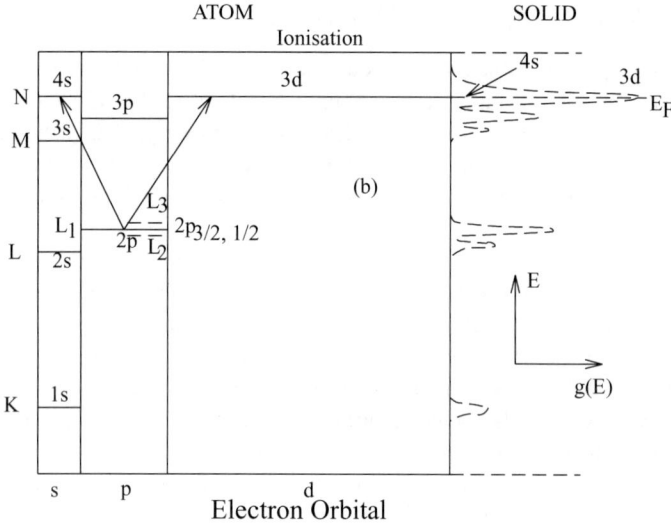

Figure 4.25 (a) Spectrum from an alpha phase grain boundary in Zr −2.5 wt% Nb alloy containing 1000 ppm Fe. The feature rising at 710 eV is due to segregated iron. (STEM in spot mode, channel width 0.83 eV, acquisition time 90 s). (b) The origin of the white lines in transition elements. Electrons from 2p states are promoted into the empty 3d bands. In the solid, the sharp atomic levels are broadened into bands, but the 3d bands are narrow by comparison with the 4s band and give rise to sharp features in the density of states and thus in the energy loss spectrum. The 4s band contributes a relatively featureless continuum, rising under white lines

(b) the ratio of the intensities of the L3/L2 peaks in the two white lines.

Provided these ratios are determined in a prescribed manner (Dray et al., 1995) normalisation against the transition series (Figure 4.27) provides an estimate of the

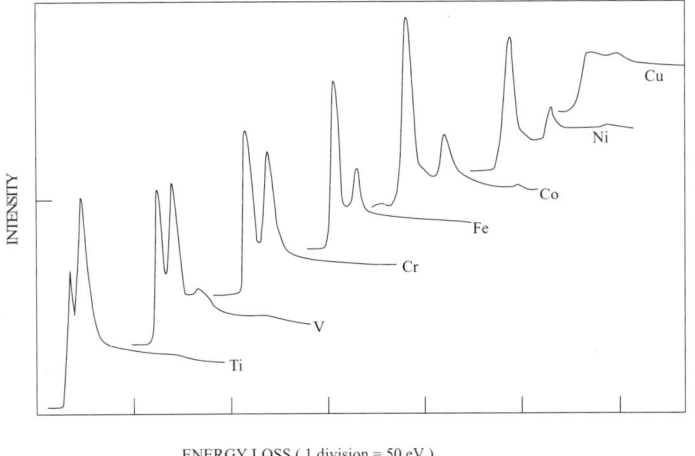

ENERGY LOSS (1 division = 50 eV)

Figure 4.26 Variation in white line intensity across the first series of transition metals (after Okamoto *et al.*, 1992)

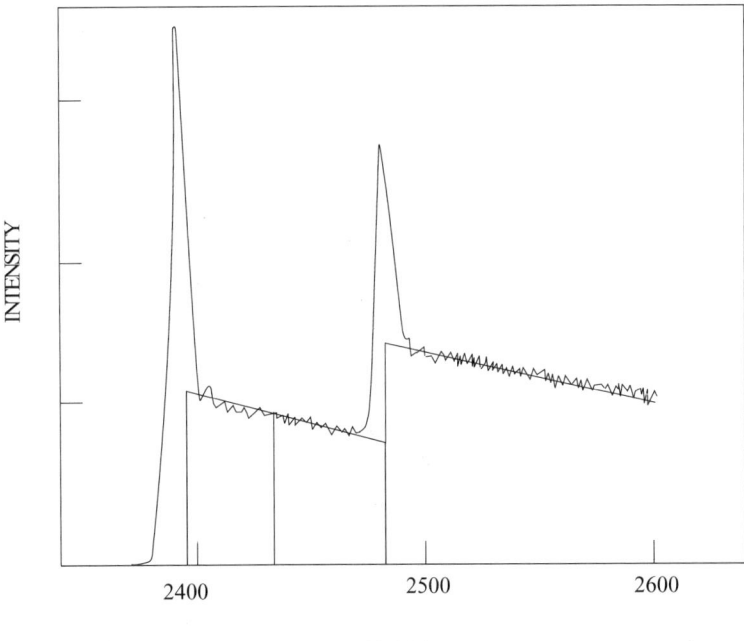

ENERGY LOSS (eV)

Figure 4.27 Schematic diagram of the separation of the white line features from the continuum (after Okamoto *et al.*, 1992)

number of electrons in the d-band of electrons. Oxidation of the iron increases with the Pearson ratio because electrons are transferred from the iron. Figure 4.28 shows that the quantitative use of this technique involves extensive numerical simulation of the background limits in the spectra under the white lines so that a measure of the white line intensity can be obtained. Hence, by use of reference spectra obtained from known intermetallic compounds of Fe, Zr and Nb, the segregated iron has been shown to be in an anionic state as a result of a gain of about two electrons. This state resembles iron in Zr–Nb–Fe alloys. By this means it is possible to consider both the composition and chemical state of species segregated to the grain boundaries (Brown

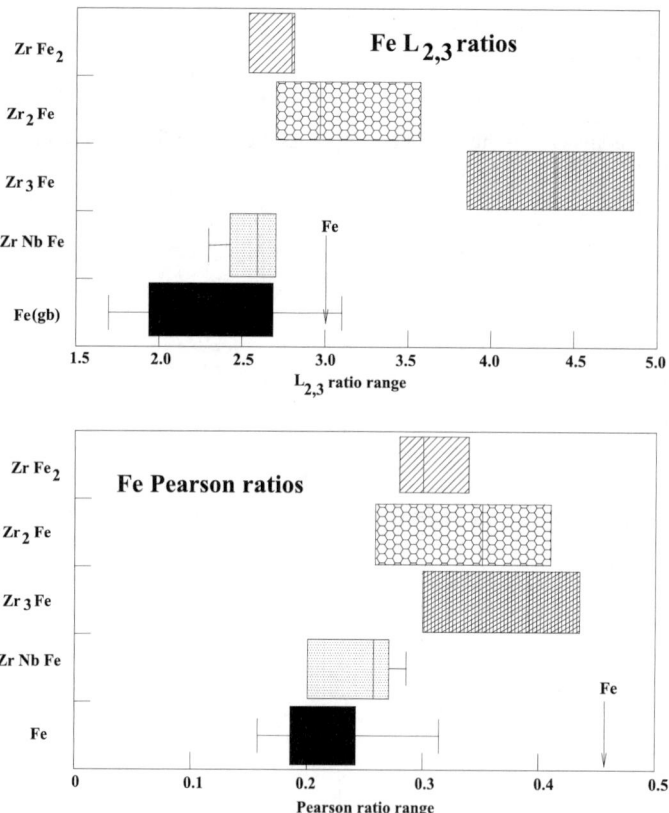

Figure 4.28 (a) The L_3/L_2 intensity ratio and (b) the ratio of intensity in the white lines to that in the continuum, the 'Pearson ratio', for iron in various intermetallic compounds and (bottom row in each diagram) for iron segregated to alpha-phase boundaries. The vertical line in each box is the mean, and separate r.m.s. deviations for readings above and below the mean are shown. Scatter bands are shown where there are significant outliers, particularly for segregated iron. The arrow shows a value for pure metallic iron

et al., 1995). This general approach has been applied to the investigation of segregation of nitrogen to planar defects in diamond (Fallon *et al.*, 1995), radiation-induced segregation in Ni_3Al (Muller and Silcox, 1995) and phosphorus segregation to grain boundaries in a Fe–0.4 wt% P alloy (Ozakaya *et al.*, 1995).

In the latter case, Figure 1.18a shows the $L_{2,3}$ edges obtained from a grain boundary and the matrix of this Fe–0.4 wt% P alloy after a furnace cool from a temperature of 1273 K. In addition, the difference spectra are also shown which are corrected for the thickness changes across the grain boundary, so that the difference between these spectra relates directly to an increase in a shell filling by electrons at the grain boundary. In the spatial difference method, a spectrum from an analysed volume containing a grain boundary is compared with that from the matrix. For this method, two spectra are recorded from an area across the interface and the second displaced from the interface (Figure 4.29) (Bruley *et al.*, 1994). This latter spectrum provides a reference to model the energy dependent background of the first spectrum. After subtracting a smooth power law curve from both these spectra fitted in the region preceding the edge, the numerical difference between the two is obtained by subtracting a scaled version of the latter spectrum from that of the former. The difference spectrum provides the detail of the information associated with the interface. This technique can be applied to a range of metal, ceramic and metal–ceramic systems with the following advantages:

(a) noise produced by detector channel to channel gain variations is removed,
(b) the resultant signal intensities and accuracies are limited only by counting statistics and
(c) any surface artifact differences are removed.

From Figures 1.18a and b it is possible to establish the approximate magnitude of the charge transfer since the ratio of the area under the stripped Fe $L_{2,3}$ edge to a window taken after an edge is a direct function of the d shell occupancy (Okamoto *et al.*, 1992). The iron in the grain boundary has fewer holes in the d shell than the matrix iron, since the d shell relates to the electrons transferred from the phosphorus to the iron. Assuming six electrons in the d shell, this gives four holes which translates to about four-eighths or one-half of an electron transfer. Use of a calibration curve established by Okamoto *et al.* (1992) gives 0.4 of an electron transfer. As a consequence, it is clear that this technique offers a powerful way forward for establishing the chemical state of segregation species to grain boundaries and interfaces.

An extension of the electron energy loss technique is elemental mapping. Here, maps are formed by imaging with electrons that have lost energy corresponding to inner-shell ionisation energies characteristic of the particular elements. This has the advantage that distribution images can be produced to a nanometre-scale resolution with a short acquisition time (Krivanek *et al.*, 1995). More recently, Hofer *et al.* (1999) have extended this to provide quantitative information from the image intensity at any point.

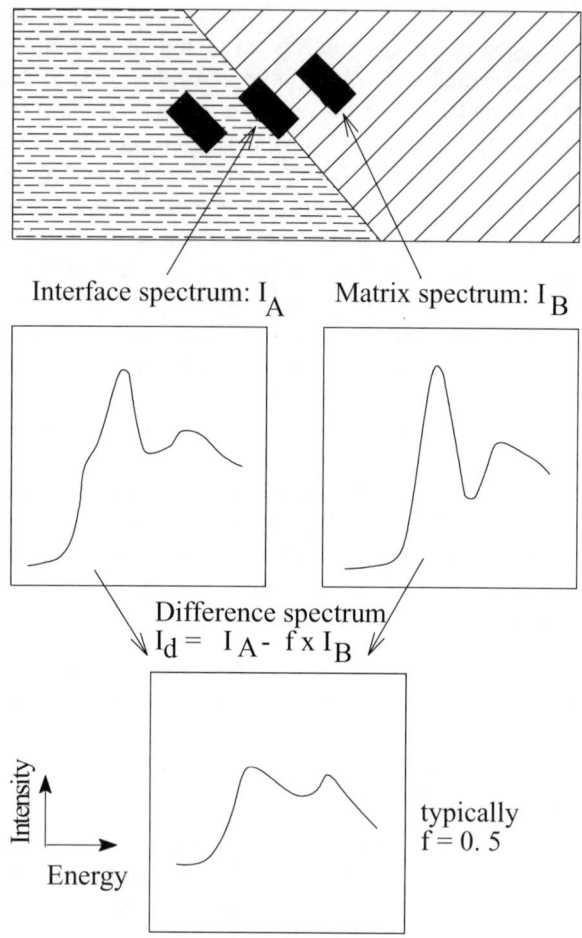

Interface spectrum: I_A Matrix spectrum: I_B

Difference spectrum
$I_d = I_A - f \times I_B$

Intensity

Energy

typically
f = 0. 5

Figure 4.29 Schematic diagram showing the basis of the spatial difference technique. With a beam located at A and B, the measured intensities are I_A and I_B, respectively. When properly scaled by factor f, I_B is used to model the background of I_A. The spatial difference is shown as I_d

4.5   CONTRAST ANALYSIS

4.5.1   Fresnel Contrast

In recent years, both the Fresnel method and high-angle annular dark-field imaging in the STEM instrument have been used to characterise grain boundary segregation in metals, thereby making use of the high spatial resolution these techniques offer. Ozakaya *et al.* (1993) applied both approaches to characterise the distribution of tin at high angle grain boundaries in an Al–0.09 wt % Sn alloy aged at 473 K to obtain mean equilibrium segregation levels of tin to these boundaries. A through focal Fresnel series of data for a high-angle boundary are shown for three thicknesses of foil in Figures 4.30a to c, with the boundary being nearest to the edge in Figure 4.30b. The segregation is localised because this series shows a thin dark absorption line in the mean focus images and absorption-enhanced double dark fringes in the underfocused images. Here, the contrast is affected by the magnitude and form of the potential changes associated with the segregation of the tin and the retained rigid body displacements for the boundary as well as the different absorption behaviour of aluminium and tin. By matching these contrast profiles with those obtained from image simulations, it is possible to obtain a measure of the width (0.3–0.8 nm) and magnitude of the tin segregation (~ 2.0 wt %) at the boundary.

4.5.2 Atomic Number Contrast

Recently, attention has been given to strategies for examining grain boundaries that are appropriate for multicomponent oxides, involving a combination of atomic resolution, atomic number, referred to as Z contrast imaging (Browning *et al.*, 1995; Pennycook and Boatner, 1988) and electron energy loss spectroscopy (Egerton and Cheng, 1987). The Z contrast image, unlike conventional phase contrast images, is incoherent and thereby allows columns of different atomic number atoms to be distinguished and located directly from the image (Figure 4.31). If this is supported by maximum entropy image analysis techniques, it is possible to deduce atom positions accurately (Gull and Daniell, 1978). This provides a valuable tools for examining grain boundaries and interfaces in materials since two columns separated below the resolution limit will image as a single, bright feature that can be deconvolved into the constituents by image analysis.

In practice, Z contrast images are formed by collecting high-angle scattered electrons from the foil specimen on to an annular detector and synchronously displaying the integrated output while the incident electron beam is scanned across the specimen (Figure 4.31) (McGibbon *et al.*, 1996). At high angles in the range 75–150 mrad the detected intensity is mainly the result of thermal diffuse scattering. Lateral coherence between atomic columns in the specimens is averaged by detecting over a large angle range, and coherence between atoms in a given column is reduced by thermal vibrations to residual correlations between near neighbours. Since these latter correlations are a second-order effect, each atom can be considered as scattering independently with a cross-section close to a dependence of Z^2. It is this mass section that forms an object function that is strongly peaked at the position of the atom. Hence, for a specimen where there is little or no dynamic diffraction, the detected density is a convolution of this object function with the electron beam intensity profile. The small width of the object function, which is ~ 0.02 nm, gives a spatial resolution that is limited by the incident electron beam size of about 0.2 nm. Therefore, as the electron beam is scanned over the specimen, an atomic resolution compositional map is generated where the intensity depends upon the atomic number of the atoms from which the columns are compared.

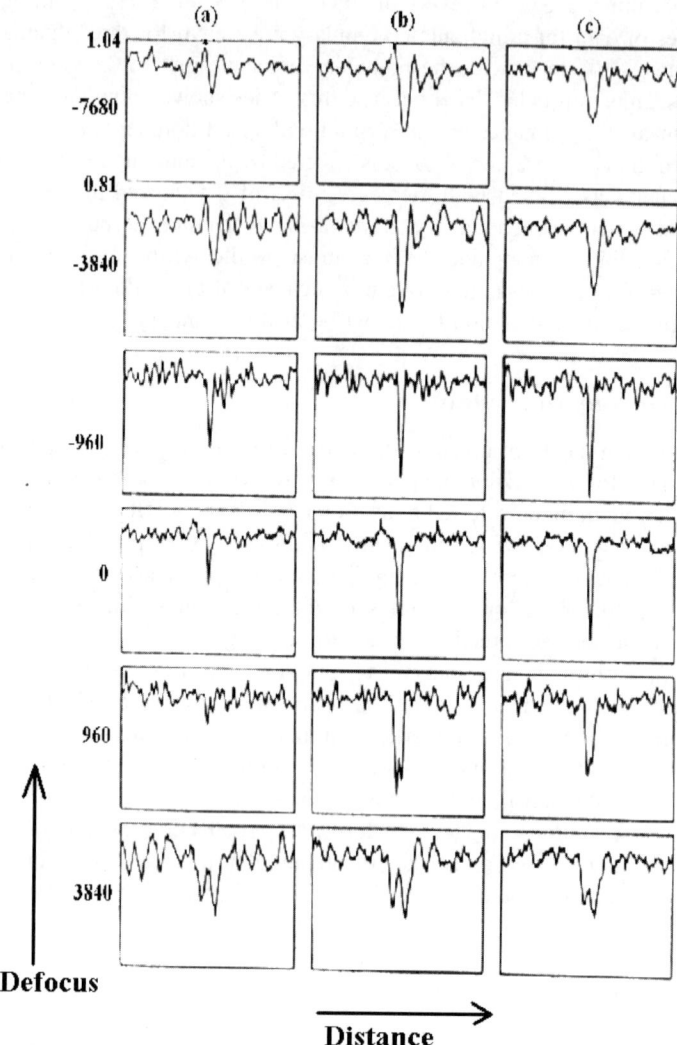

Figure 4.30 Fresnel profiles of a grain boundary in Al–0.09 wt. % Sn for three thicknesses (contrast levels and the defocus in nm are given on the profiles). Reproduced by permission of Blackwell Science Publishers from Ozakaya, D. *et al.* (1996) *J. Microsc.*, **180**, 300

Following this procedure, a *Z* contrast image for a symmetric [001] tilt boundary (36.8°) in SrTiO$_3$ is shown in Figure 4.32a. The brighter spots within the image correspond to the lighter TiO columns ($Z = 22$ for Ti and 8 for O); the pure O columns are not visible. Maximum entropy image processing is used to deconvolute the positions and intensities of the Sr and Ti atoms in each column. Figure 4.32b shows the corresponding grain boundary structural model for this 36.8° symmetric

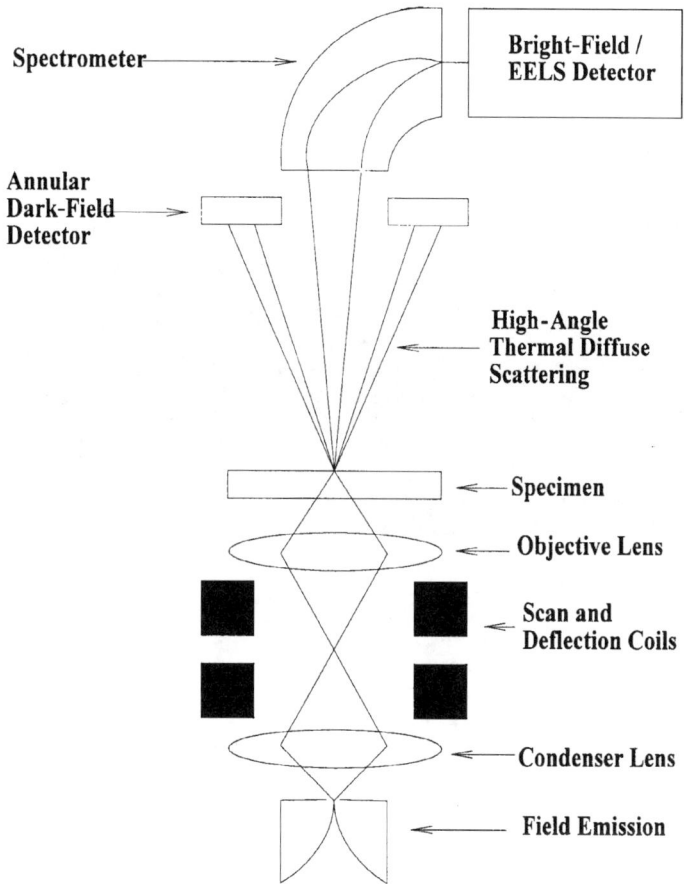

Figure 4.31 Schematic diagram of the detector arrangement in a dedicated STEM, showing that the atomic resolution *Z* contrast image and atomic resolution electron energy loss spectrum can be acquired simultaneously

grain boundary, and Figure 4.32c shows the corresponding energy loss spectra. Therefore, the technique has the potential for establishing the position and type of atoms distributed within the region of a grain boundary or interface.

4.6 AUGER ELECTRON SPECTROSCOPY

4.6.1 Introduction

In 1923, Pierre Auger (1923) discovered that an ionised atom could rearrange with an electron falling from an outer shell into the initial hole, with the energy released being transmitted to an outer shell electron that is ejected with an energy given by

$$E_{\text{Auger}} = E_1 - E_2 - E_3 - \phi \qquad\qquad (4.33)$$

where E_1, E_2 and E_3 are the electron binding energies of the electrons and ϕ is the work function. The energy of the ejected electron is characteristic of the atom and is not dependent on the source of ionisation. In the early 1970s, the technique of Auger electron spectroscopy (AES) was developed in which an incident electron beam of

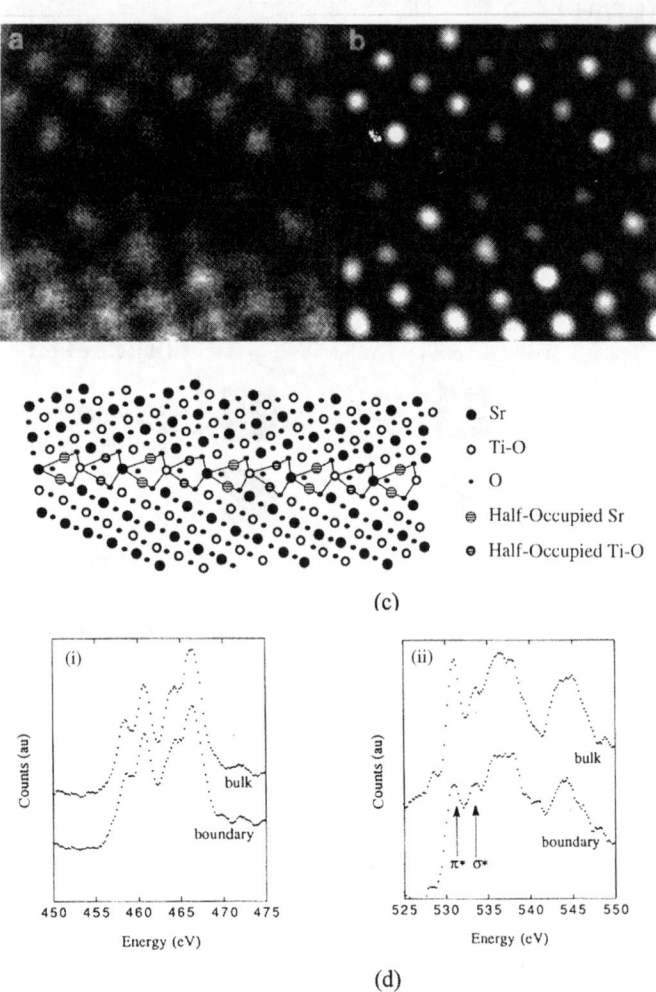

Figure 4.32 (a) Z contrast image of a $36.8°$ symmetric tilt grain boundary in $SrTiO_3$. (b) Maximum entropy image providing scattering intensities and coordinates of the Sr and Ti atomic columns directly from the image. (c) Grain boundary structural model for a $36.8°$ symmetric grain boundary in $SrTiO_3$, where the grain boundary structural units are outlined. (d) Comparison of (i) Ti $L_{2,3}$ and (ii) O K edge spectra acquired from the bulk and boundary of a $SrTiO_3$ bicrystal, showing that the octahedral Ti–O coordination is maintained across the boundary (reproduced with permission from McGibbon *et al.*, 1996)

1–15 keV was used to ionise atoms in a surface and the Auger electron energies were determined using a variety of standard analysers (Riviere, 1990; Briggs and Seah, 1990). The energy of the Auger electrons detected in this technique is normally 0–2000 eV. These electrons have a mean free path in most solids of between 0.5 and 2 nm, and so the detected electrons must originate in the top few atom layers. The incident electron beam can be focused to less than 20 nm diameter by using field emitting electron sources and good-quality electron optics which allow it to be rastered over the sample surface. Thus, images and microanalyses can be obtained from the top few atom layers of a surface with a spatial resolution in the region of 20 nm. Hence, the technique is ideally suited to the study of surfaces in conducting media, but for grain boundary analysis the surface has to be exposed by intergranular fracture, and must be undertaken in an environment that will not contaminate the specimen within the time-scale of the experiment. Grain boundary exposure is usually performed within the ultrahigh vacuum (UHV) of the Auger electron spectrometer and will be in the region of 10^{-8} Pa. If the surface is exposed to air, then much of the information will be lost. To demonstrate this, Seah (1975) coated a clean iron surface with 1 monolayer of tin in an Auger electron spectrometer and produced the spectrum shown in the top part of Figure 4.33a and this was then exposed to air in the preparation chamber before reanalysing. The spectrum recorded after air exposure (the bottom part of Figure 4.33b) shows that the tin peaks

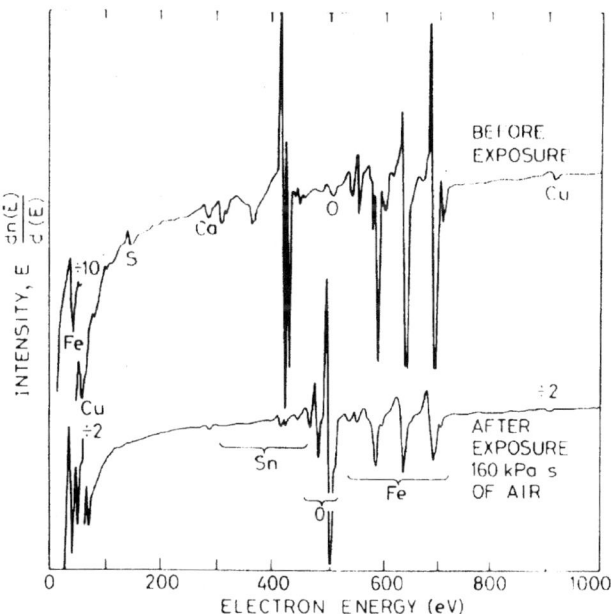

Figure 4.33 Auger electron spectra recorded from the surface of iron with a (a) monolayer of tin deposited on the surface before exposure to air and (b) after exposure to 160 kPa s of air (after Seah, 1975)

at 430 eV have almost disappeared. Tin is an element that is frequently encountered when studying segregation in steels, and it is therefore essential to obtain a grain boundary fracture surface in a proper manner if vital information is not to be lost. There are essentially two approaches to fracture within the spectrometer; in one, the sample is cooled to liquid nitrogen temperature and then fractured by impact, while in the other it is embrittled outside the spectrometer and then fractured in the UHV environment using a slow tensile fracture at room temperature. These two approaches will be described below.

4.6.2 Grain Boundary Analysis

Ferritic steels, molybdenum, tungsten, and many other metals and alloys are suitable for analysis by AES following impact fracture at low temperature. The specimens may either be machined to standard forms, usually rods of 3 to 5 mm diameter and 25 to 30 mm in length with a notch of about 1 mm in depth at the mid-point. Small specimens down to $1 \, \text{mm} \times 1 \, \text{mm} \times 10 \, \text{mm}$ can be fractured by mounting in prepared sleeves. In most systems, fracture is achieved by transmitting a sharp impact to the specimen when it has been cooled to 77 K. A typical fracture surface from an Fe/3 wt% Ni steel, containing 530 ppm of phosphorus and 110 ppm of tin and heat treated at 853 K for 48 h, is shown in Figure 4.34. The fracture path is almost 100% intergranular with very little ductile or cleavage facets. This is an extreme example and most materials fracture with considerably greater proportions of cleavage and ductile facets. Not all metals and alloys can be persuaded to fracture

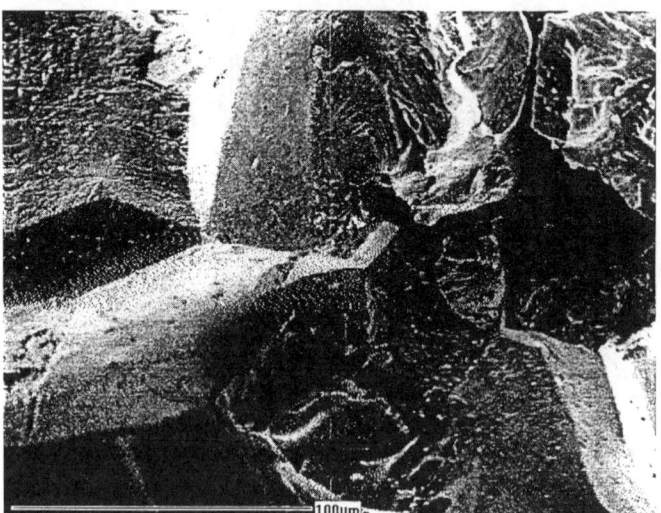

Figure 4.34 Secondary electron image of a fracture surface of a sample of Fe–3 wt% Ni containing trace levels of tin and phosphorus segregated to grain boundaries with the result that fracture is almost 100% intergranular

in an intergranular manner by impact at low temperature. It is possible to extend the group of metals and alloys that will fracture in an intergranular manner if they are first embrittled by charging with hydrogen. Some will then fracture intergranularly by impact but others require to be fractured by a slow tensile pull. This group includes the austenitic stainless steels (Briant, 1985a,b), nickel (Airey, 1985), the nickel based superalloys (Allen and Wild, 1986, 1987; Wild, 1980) and nickel aluminium alloys (Choudhury *et al.*, 1992). There are several methods for charging a material with hydrogen. Probably the most popular and well tested method is cathodic charging (Hall and Briant, 1985). In this method, the specimen is the anode in a solution of sulphuric acid of strength less than 1 N and containing a small quantity (50 mg/l) of sodium arsenite, through which a current is passed. The sodium arsenite is added to increase the hydrogen absorption by acting as a hydrogen recombination poison (Streir, 1970). Figure 4.35 shows an example of Inconel 690 alloy fractured by a slow tensile pull following hydrogen charging.

Auger electron spectroscopy detects all elements from the top few atom layers and with homogeneous samples can give a quantitative analysis. However, care must be exercised when attempting to quantify the levels of a segregant at a grain boundary since it is a single atom layer at this location. A typical Auger spectrum recorded from the grain boundary surface of a chrome nickel steel is shown in Figure 4.36, where peaks from Fe, Cr, Ni and P are visible. The extent of segregation at a grain boundary will be determined by the heat treatment that the material has received and the relative orientations of the adjacent grains. Phosphorus segregation has been found to increase with increasing tilt angle of grain boundary misorientation (Tatsumi *et al.*, 1986), whereas others have found that high index planes have a high level of segregation, and low index planes a low level of segregation (Suzuki *et al.*, 1983). In practice, when a material is fractured intergranularly there will be a distribution of levels of segregation on the fracture facets reflecting the fracture path and grain boundary orientation contributions. It is therefore essential that a number of grains be analysed on each fracture surface in order to obtain meaningful statistics, a recommended number being between 16 and 25. The time required for each analysis in practice determines the upper limit of grains that can be analysed. For a homogeneous material the composition of element i, X_i, is given by

$$X_i = (X_i/S_i) \sum X_j/S_j \tag{4.34}$$

where X_i, X_j are the *i*th and *j*th element peak heights and S_i and S_j are the sensitivity factors for the *i*th and *j*th elements. In most cases of segregation to a grain boundary, the segregating species is present as a single atom layer on the fracture surface and it is assumed that on average half the species remains on each exposed surface following fracture. To calculate the amount of segregant, A, on the surface of matrix, B, it is necessary to know the escape deaths of electrons in the segregating species, λ_A, and the matrix, λ_B, and the energy of the incident electron beam, and to determine the amount of electron backscattering, Γ. This is treated in some detail by

Figure 4.35 Example of Inconel 600 alloy fractured by a slow tensile strain in ultrahigh vacuum following hydrogen charging.

Seah (1977) who derived an equation which allows the amount of segregant to be calculated from:

$$X_i = Q_{AB} \frac{I_A/I_A^{Inf}}{I_B/I_B^{Inf}} \tag{4.35}$$

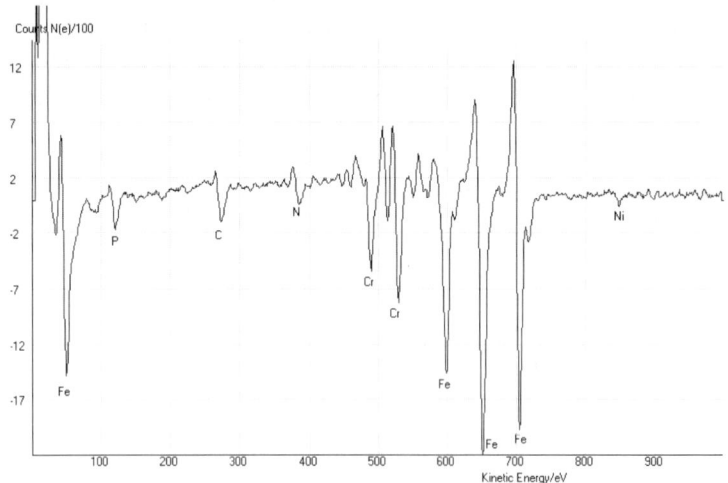

Figure 4.36 Auger electron spectrum recorded from the surface of a grain exposed following impact fracture of a chrome nickel steel in which trace levels of phosphorus have segregated to the grain boundaries.

where

$$Q_{AB} = \left[\frac{\lambda_A(E_A)\cos\Theta}{d_A}\right]\left[\frac{1 + \Gamma_A(E_A)}{1 + \Gamma_B(E_A)}\right] \tag{4.36}$$

where d_A is the size of atom A, and Θ is the angle of the incident electron beam relative to the surface normal. For specific segregants, such as phosphorus, on standard matrices, such as iron, using known incident beam energies it is possible to derive Q_{AB}. This may then be applied to the standard equation to give the segregation in monolayers.

There has been a large number of Auger electron spectroscopy studies of the segregation of the trace elements phosphorus, sulphur, tin and antimony to grain boundaries in iron and ferritic steels at elevated temperatures (Oku *et al.*, 1984; Briant, 1988; Wei and Grabke, 1986; Klug *et al.*, 1985; Choudhury and Padgett, 1990; Jones *et al.*, 1988; Bruemmer *et al.*, 1983; Grabke *et al.*, 1983, 1987; Tatsumi *et al.*, 1988; Viefhaus *et al.*, 1983; Bernardini *et al.*, 1982; Lee and Morris, 1983; Ishida *et al.*, 1985a,b). Similar studies have been reported for stainless steel (Briant, 1985a,b) and for nickel (Mulford, 1983; Bruemmer *et al.*, 1983; Larere *et al.*, 1982; Lassila and Birnbaum, 1987) and nickel based superalloys (Allen and Wild, 1986; Allen *et al.* 1996; Mulford, 1983; Lassila and Birnbaum, 1987; Nettleship and Wild, 1990, 1992; Funkenbusch *et al.*, 1982; Guttmann *et al.*, 1981; Was *et al.*, 1981). There have been many studies concerning the segregation of sulphur to grain boundaries in nickel and its resultant effect on embrittlement. Larere *et al.* (1982) studied the kinetics of sulphur to both grain boundaries and free surfaces in nickel and the effect of other additions such as carbon and calcium on this segregation. Here the segregation free energy of sulphur to grain boundaries (98 kJ mole^{-1}) is almost half that to a free surface (180 kJ mole^{-1}) and C and Ca alter the segregation kinetics with S–C being repulsive and S–Ca being attractive resulting in respectively retarding and accelerating the sulphur build up. Mulford (1987) studied the kinetics of sulphur segregation in nickel and in nickel with Cu, Al, Cr, Mo, W and Hf additions. No evidence for co-segregation was found but these elements influence the precipitation of sulphides which, in turn, effect the grain boundary segregation. The effectiveness of sulphur, phosphorus and antimony in promoting the inter granular embrittlement of nickel was examined by Bruemmer *et al.* (1983). Sulphur was shown to be the most effective due to its large enrichment (10^4–10^5 times the bulk value). Phosphorus was shown actually to reduce embrittlement due to hydrogen by segregating to grain boundaries and reducing the sulphur segregation. Bricknell *et al.* (1982) showed that nickel could be embrittled following high temperature air oxidation. Oxygen was shown to convert internal sulphides to oxides thus releasing sulphur which is able to segregate to grain boundaries causing severe embrittlement. Manganese sulphide particles were shown to be oxidised to a depth of 250 μm in 200 hours at 1273 K air exposure. Lassila and Birnbaum (1987) studied the effect of sulphur on hydrogen embrittlement and found that an increase in sulphur at the grain boundary resulted in lower levels of hydrogen, causing 100% intergranular fracture arising from synergistic effects of the sulphur and hydrogen.

Most ceramics are good insulators and as such present problems when attempts are made to analyse their surfaces by AES due to the build up of charge. This can sometimes be overcome by using low beam currents and rastering over as large an area as possible. Despite these difficulties, many ceramic grain boundaries have been analysed. For example, grain boundaries of sintered SiC have been studied by AES (Hamminger *et al.*, 1983) where they found that SiC doped with boron and carbon showed no grain boundary enrichment but samples doped with aluminium and carbon resulted in a 1 nm layer of aluminium on the grain boundary. Sherman (1984) has studied a calcium partially stabilised zirconia. Here fracture surfaces showed the presence of cavities in which calcium had segregated. Calcium was also found to be present at grain boundaries in concentrations greater than the bulk level (3.7 wt%). Others have used both AES and STEM to study grain boundaries in zirconia containing yttrium and Bi_2O_3 (Winnubst *et al.*, 1983) with enrichment of both yttrium and bismuth at the grain boundaries.

The segregation of boron to the surface and grain boundaries in a range of materials has been reviewed by White and Choudhury (1987); segregation of this small atom has been shown to increase ductility. Liu *et al.* (1985) and Koch *et al.* (1985) studied the effect of boron additions on grain boundary chemistry and tensile properties in Ni_3Al containing 24–26 at. % Al. The ductility was shown to depend critically on the deviation from stoichiometry. As the aluminium content decreased below 25 at. % so the ductility increases and the fracture changes from intergranular to transgranular. This corresponds to an increase in grain boundary boron and a decrease in grain boundary aluminium. In addition boron is shown to segregate more strongly to grain boundaries than to cavity surfaces whereas sulphur, an embrittling agent, tends to segregate more strongly to cavity surfaces than grain boundaries and thus the boron acts to reduce the sulphur at the grain boundary as well as strengthening it. However, unlike embrittling elements, boron does not segregate to ordered free surfaces and it is thought that the presence of Al–Al near neighbours in ordered regions is responsible for inhibiting this segregation. Chaung (1991) has also investigated the effect of small zirconium additions and found that these do not have a similar effect on ductility.

The segregation of oxygen in molybdenum has been studied by AES. In one study of the oxygen segregation to grain boundaries in molybdenum containing 32 wt-ppm of oxygen, it was shown that, by observing the Auger peak positions, the oxygen was present on the grain boundary as pure oxygen and was not combined with molybdenum or any other element (Oku *et al.*, 1986). The addition of aluminium to potassium and silicon doped molybdenum is shown to have a strong gettering effect on the grain boundary oxygen and to also lead to higher K and Si contents (Setti, 1987).

Thorium additions to iridium + 0.3 wt.% tungsten are known to increase ductility during high temperature impact. Alloys containing between 5 and 1000 ppm thorium were studied. Samples containing 5 ppm bulk thorium were shown to have 10 at. % thorium at the grain boundaries after one hour at 1773 K vacuum anneal. However, the grain boundary concentration did not increase when the bulk thorium content was increased further (White *et al.*, 1983).

Tungsten is known to fracture in a brittle mode and that this is caused by the segregation to grain boundaries of nickel (Hoffmann *et al.*, 1974) and phosphorus (Joshi and Stein, 1970). Nieh (1984) demonstrated that nickel would segregate to grain boundaries in tungsten when plated with nickel and that this resulted in severe embrittlement. However, Seah (1981) suggested that carbon should have a beneficial effect in strengthening the grain boundaries. Hofmann and Hofmann (1984) found that the competitive effect of carbon and phosphorus for grain boundary sites was responsible for the strength of Ni/Fe activated sintered tungsten. Studies on the segregation of impurity elements to fracture surfaces of liquid phase sintered W–Ni–Cu and W–Ni–Fe alloys have shown that phosphorus will segregate to between 0.2–0.4 monolayers (Lea *et al.*, 1983). They showed that in furnace cooled specimens the phosphorus adheres to the matrix phase on fracture. Annealing to 1623 K followed by water quench reduces the phosphorus segregation.

An AES study of the segregation to grain boundaries and cavities of antimony in a Cu–0.5% Sb alloy showed that there was an enrichment ratio of 16 with antimony homogeneously distributed on all surfaces (Yu *et al.*, 1983).

4.6.3 Comparison of Grain Boundary Composition by FEGSTEM and AES

The techniques of Auger electron spectroscopy (AES) and field emission gun scanning transmission electron microscopy (FEGSTEM) each have important advantages and drawbacks in the analysis of grain boundaries. Provided a fracture surface can be produced within the ultrahigh vacuum of the scanning Auger microprobe (SAM), is capable of providing a quantitative measure of the composition of the grain boundary surface from many grains in a relatively short period of time. The equation for the surface composition in a homogeneous material (4.34) becomes modified for a single atom layer on a substrate. For a normally incident electron, a phosphorus overlayer on iron can be expressed as

$$\frac{I_p/I_p^\infty}{\sum_i I_i/I_i^\infty} = \frac{\Phi_p \left[\dfrac{1 + r_{Fe}(E_p)}{1 + r_p(E_p)}\right]\{1 - \exp[-d_p/\lambda_p(E_p)]\}}{1 - \Phi_p\{1 - \exp[-d_p/\lambda_p(E_{Fe})]\} + \Phi_p \left[\dfrac{1 + r_{Fe}(E_p)}{1 + r_p(E_p)}\right]\{1 - \exp[-d_p/\lambda_p(E_p)]\}} \tag{4.37}$$

Unfortunately, not all materials can be induced to fracture along their grain boundaries. The FEGSTEM can provide, to a first approximation, a quantitative analysis of a cylinder with a diameter of 2 nm and a length equal to the foil thickness. In practice, the beam spreads on travelling through the foil and the emerging beam has a diameter $d + b$, where d is the incident beam diameter and b is given by

$$b = 625\left(\frac{\rho}{A}\right)^{1/2}\left(\frac{Z}{E_0}\right)d^{*3/2} \tag{4.38}$$

where ρ is the foil density (g cm^{-3}), E_0 is the electron energy (keV), d^* is the foil thickness (cm), A is the atomic weight and Z is the atomic number.

For AES measurements, the above equation reduces to $\Phi_p = 0.04 \times$ (measured at. % phosphorus) while for FEGSTEM measurements we have $\varphi_p = 0.17 \times$ (measured wt % phosphorus).

Combining these gives an empirical relationship between these two techniques

$$\text{wt \% P (FEGSTEM)} = 0.23 \times \text{at. \% P (Auger)} \tag{4.39}$$

If a transmission foil can be produced containing a grain boundary and that boundary can be aligned parallel to the incident electron beam, then the grain boundary composition can be inferred (Doig and Flewitt, 1983a). However, FEGSTEM analysis is time consuming and laborious. Attempts have been made to compare the results from the two techniques (Doig *et al.* 1978; Partridge and Tatlock, 1992; Walmsley *et al.*, 1995). In the most recent study, the segregation of phosphorus to grain boundaries in a range of steels has been measured using a VG FEGSTEM and a VG 310F SAM.

Figure 4.37 shows the results of the experimental comparison in which five alloys have been compared from ferritic steel to stainless steel and nickel-base superalloys. The best fit to the data results in a straight line with a slope of 0.21 and a 95%

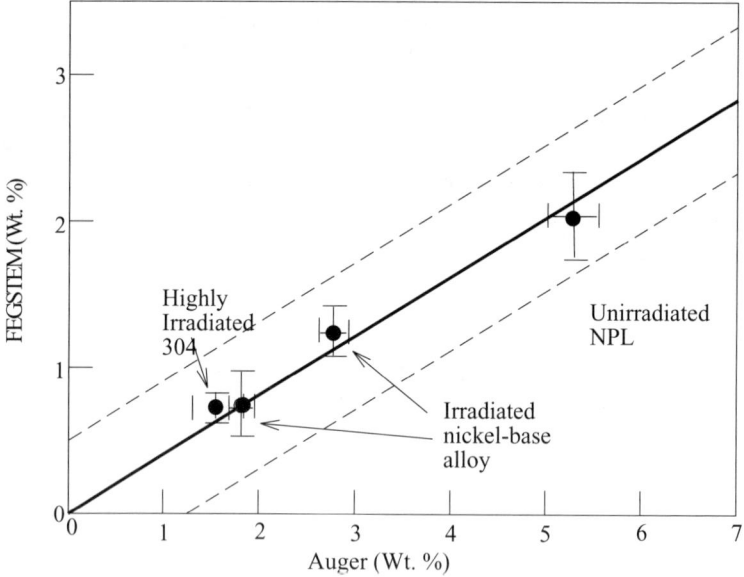

Figure 4.37 Experimental comparison of the quantification of P at grain boundaries using FEGSTEM and Auger spectroscopy in which five alloys have been compared from ferritic steel to stainless steel and nickel-base superalloys. The best fit to the data results in a straight line with a slope of 0.21 and a 95% confidence limit including all the points (Fisher *et al.*, 1996)

confidence limit including all the points. This is remarkably good agreement with the theory and indicates that the two techniques can reliably be used in a complementary manner.

4.7 X-RAY PHOTOELECTRON SPECTROSCOPY

Auger electron spectroscopy has good spatial resolution and can determine quantitatively the presence of most elements, but it is not usually able to give much information on the chemical state of atoms at the surface. X-ray photoelectron spectroscopy (XPS) on the other hand can give chemical state information (Briggs and Seah, 1994; Riviere, 1990). In this technique, photons with energies in the range 1–2 keV are directed on to a surface where atoms are ionised with the ejection of electrons (photoelectrons). The energy of these electrons is measured using a hemispherical electrostatic analyser. The mean free path of electrons with an energy of less than 1 keV is less than 1 nm and, therefore, the analysed volume is less than three atom layers. A major drawback to the technique has been its lack of lateral spatial resolution. The incident X-ray beam illuminates an area several millimetres in size, and the technique could not resolve individual grain boundaries. Recent advances have led to the development of instruments with a lateral resolution of the order of 1 μm and obtain spectra from 900 μm^2. This allows the study of chemical state of the species at a grain boundary, for example an iron–nickel alloy (Hallam and Wild, 1995). An iron–3 % nickel alloy containing phosphorus and tin was heat treated to segregate these elements to the grain boundaries. Figure 4.38 shows the photoelectron distribution map of the tin 3d peak from a fracture surface, together with a secondary electron image of that surface. The individual grains can be identified in the tin element map, and the boundary between the fracture surface and the surface exposed to air can clearly be seen. The narrow scan region of the tin spectrum is reproduced as Figure 4.39. The tin 2p peaks are positioned at 484.9 and 493.3 eV, indicating that the tin is present as elemental tin and not chemically combined with phosphorus.

4.8 SECONDARY ION MASS SPECTROSCOPY

Secondary ion mass spectroscopy (SIMS) is a technique that has been successful in detecting very low levels of elements present on a surface (Benninghoven *et al.*, 1987). The technique bombards the surface with ions and determines the masses of the ejected ions. The incident source is normally a beam of inert gas ions such as argon, although, oxygen ions may be employed. However, gaseous ion sources are difficult to focus and the spatial resolution may be limited to several tens of micrometres if the spatial resolution is determined by rastering the incident beam. In recent years, liquid metal ion sources such as cesium and gallium have been used to provide a fine focused beam of ions. These sources use a metal, with a melting point

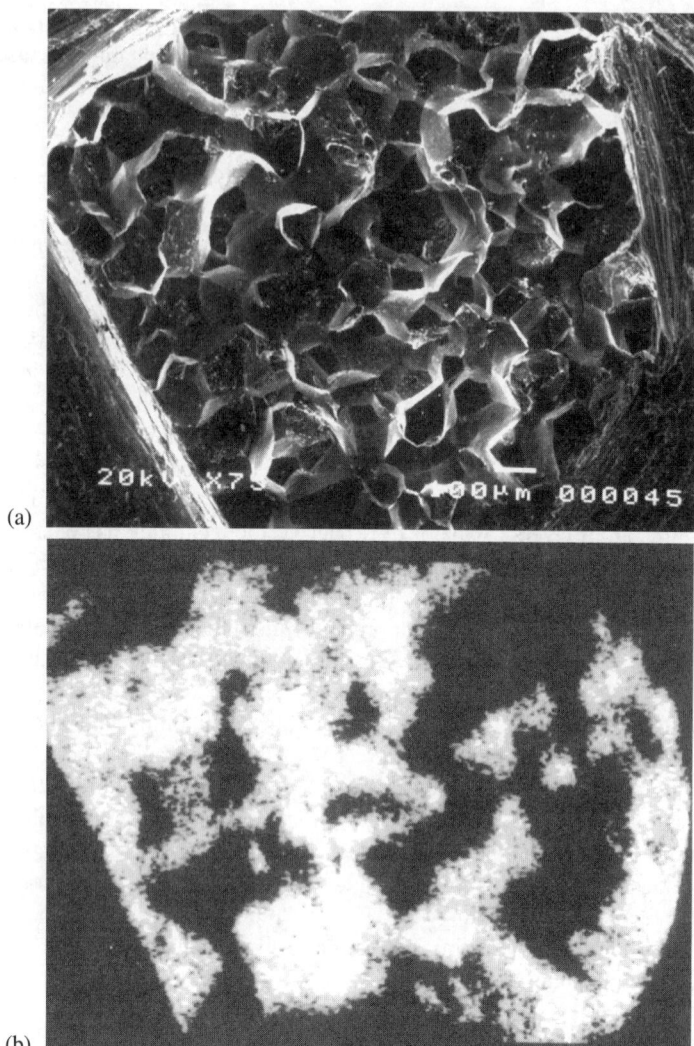

(a)

(b)

Figure 4.38 (a) Secondary electron image from an impact fracture of Fe–3 wt % Ni in which trace levels of tin and phosphorus have segregated to the grain boundaries and (b) the tin 3d map recorded from the surface showing the distribution of tin at the fractured grain boundary surface

close to room temperature, that diffuses to a fine tip where the ions are extracted using a high electrical field. These ion sources can produce an ion beam with a diameter of less than 100 nm and can be used to provide SIMS with high spatial resolution. The ions and ion clusters ejected from the surface are detected and mass analysed using a number of different methods. The simplest uses a quadrupole mass analyser in which the ions travel along a tube while being acted upon by an alternating electrostatic field.

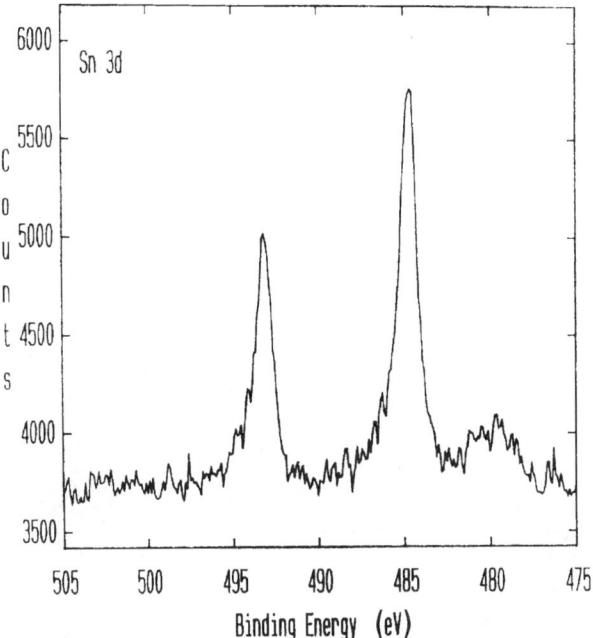

Figure 4.39 X-ray photoelectron spectrum recorded from the grain boundary surface of Fe–3 wt% Ni in which trace levels of tin and phosphorus have segregated to the grain boundaries in the tin 3d region, indicating that the tin peaks are in the position for elemental tin

The ions follow a sinusoidal path with the amplitude determined by the frequency and the ion mass. For a given frequency, only one ion of a given mass will reach the detector. By rastering the frequency of the oscillating field, a mass spectrum can be obtained. An alternative approach is to pass the ejected ions through a magnetic sector or electrostatic sector analyser or a combination of electrostatic and magnetic analysers. In these instruments, the magnetic and/or electrostatic field causes the ions to be deflected around the sector. Only ions with a specific mass will reach the detector for a given field—heavy ions will not be deflected sufficiently while light ions will be deflected too much. A mass spectrum can be obtained by rastering the electrostatic/magnetic fields or by retarding the incident ions. Finally, ions may be mass analysed using a time-of-flight detector. In this arrangement the incident ion source is pulsed. This causes a pulse of ions to be ejected from the surface, and these are accelerated by an electrostatic field into a flight tube. Ions and ion clusters travel down the tube to reach a detector. Light ions are accelerated to the greatest velocity and arrive at the detector first, while heavy ions arrive last, so that by plotting the detected counts as a function of time, a mass spectrum is recorded. The flight tube can be linear but frequently contains a system for reflecting the ions through 180° since

218   *Grain Boundaries*

this has advantages for focusing ions that may not have entered the tube exactly parallel. The time-of-flight system is a parallel detection system, whereas the other systems are serial detectors. It is therefore more sensitive and less damaging to the surface.

The segregation of elements to grain boundaries can be detected using these instruments. In particular, rastering the incident ion beam for a given mass number will produce a spatial map of that mass. Figure 4.40 is an example of the use of high

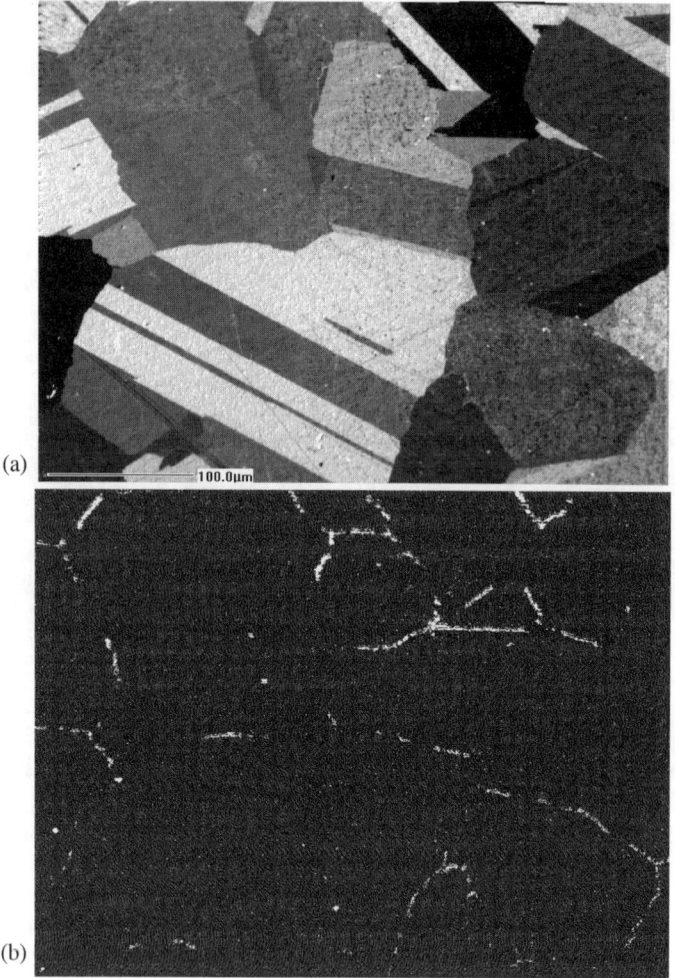

Figure 4.40 Secondary ion map for boron in steel, recorded on a combined electrostatic/ electromagnetic mass spectrometer while bombarding the surface with 25 keV gallium ions from a field emission source

spatial resolution to detect the presence of boron and phosphorus at grain boundaries in Inconel 600 alloy.

4.9 FIELD ION AND ATOM PROBE MICROSCOPY

Muller (1951) developed a technique that was the first to allow atoms to be directly observed. In this technique, a specimen was prepared in the form of a needle with a small radius tip, usually 20–100 nm radius, to which a high potential was applied in the presence of an inert gas. The very high fields in the vicinity of the tip ionised the inert gas atoms which enabled images of the tip to be produced. By further increasing the potential on the tip to between 5 and 20 keV, individual atoms could be induced to evaporate, allowing depth information to be obtained. Grain boundaries present in the tip can be imaged, and the technique has contributed to the understanding of grain boundary structure (Muller, 1963). There were, however, a number of difficulties and limitations with the technique as originally developed. Certainly it was not always possible to distinguish between atoms of different atomic number and, indeed, in certain cases atoms could appear as vacancies (Tsong and Muller, 1966) and no compositional information could be readily obtained.

In 1967, Muller developed the atom probe which surmounted many of the problems associated with the field ion microscope. Ions removed from the tip could be mass analysed using a time of flight mass spectrometer. Essentially, the time taken for the ion to travel from the tip to the detector, t, is determined by the mass-to-charge ratio, M/N, the potential applied to the tip, V, and the distance from the tip to the detector, d_0, by the relationship

$$M/N = \frac{2eVt^2}{d_D^2} \tag{4.42}$$

This technique gives information on the composition within a small area around the tip but does not give spatial resolution. Panitz (1974) made the first attempt to combine the spatial resolution of the field emission microscope with the composition of the atom probe. The advent of position sensitive detectors (Cerezo *et al.*, 1988a, b; Miller, 1992) further improved the capabilities by allowing individual atoms to be identified in space and in mass by the time of flight, and depth information could be obtained as layers were removed. The current state of the art of these instruments is shown schematically in Figure 4.41 (Blavette *et al.* 1993). In this instrument, each pulse removes ions, and all ions are mass analysed and their position is determined from the position of the ion impacts on the detector. As the investigation proceeds, so a three-dimensional picture is built up of the location of all atoms within a cylinder of known dimensions. This method has been used to investigate the grain boundary composition in a nickel-base superalloy (Letellier *et al.*, 1994). In this study, grain boundaries in the nickel-base alloy Astroloy were imaged. The grain

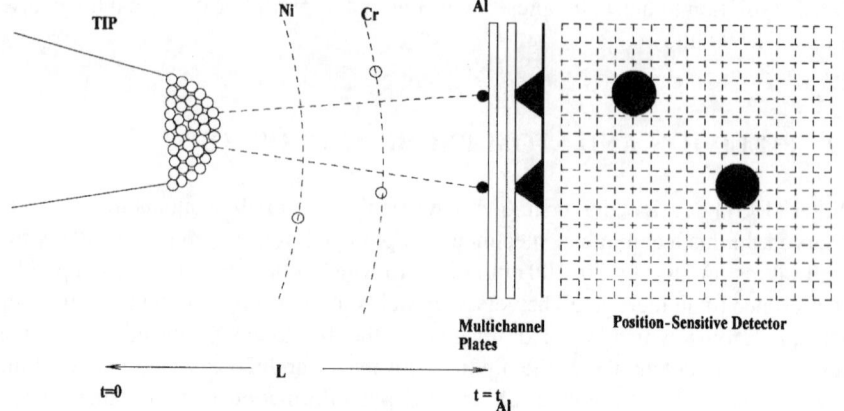

Figure 4.41 Schematic diagram of the tomographic atom probe. The coordinates of atoms on the surface of a sample are deduced from the position of each ion impact on the detector which corresponds to the intensity distribution on the position-sensitive detector. The magnification is $G = L/bR$ (Blavette *et al.* 1993)

size is approximately 40 μm, but the imaging atom probe can only detect a cylinder of 10 nm diameter and it is, therefore, necessary to identify a grain boundary within a prepared tip. This is done by examining the tip in a transmission electron microscope prior to transferring to the imaging atom probe. Figure 4.42 shows the spatial distributions of Al plus Ti, Cr, Mo and B in the vicinity of a serrated grain boundary in this alloy. The total depth that was probed in this study was 120 nm, during which two types of grain boundary were crossed, two γ–γ' boundaries and one γ'–γ' grain boundary.

4.10 SCANNING PROBE MICROSCOPY AND ATOM FORCE MICROSCOPY

A relatively recent development in the study of the morphology of surfaces has been the invention of the scanning tunnelling microscope (STM) and the atomic force microscope (AFM) (Binnig and Rohrer, 1982; Binnig *et al.* 1982; Kuk and Silverman, 1989; Wiesendanger, 1994). The STM is capable of extremely high resolution both in the lateral direction and in the vertical direction, to 0.01 nm resolution. It is, therefore, capable of readily resolving individual atoms and displaying the density of the outer electron shells of the atom and, hence, is ideally suited to the study of grain boundaries, interphase boundaries, interfaces and surfaces.

The basic principles of the technique are illustrated in Figure 4.43. Essentially, a sharp tip, constructed from tungsten or platinum/iridium, is brought to within 0.5–1 nm of the surface to be studied. When the tip is this close the electron wave functions of the surface and the tip overlap and there is a finite probability that an

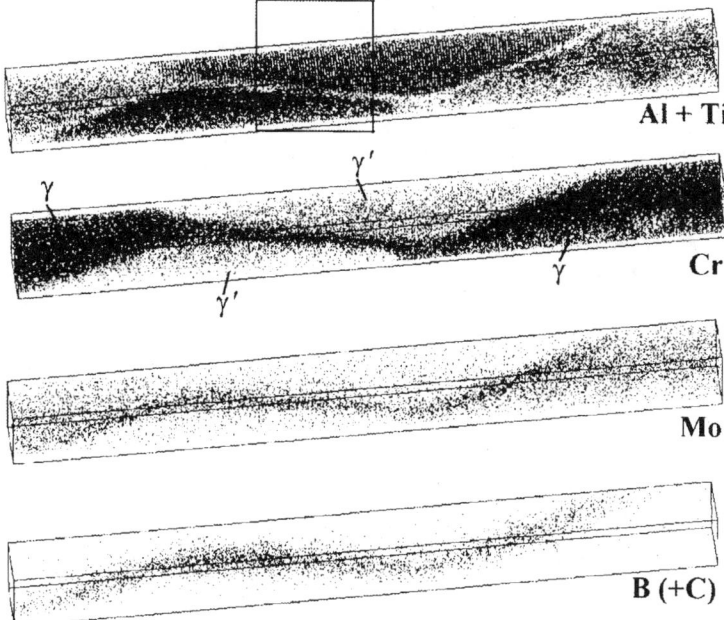

10 nm × 10 nm × 120 nm

Figure 4.42 Grain boundaries in a nickel-base superalloy Astroloy imaged using a tomographic atom probe showing the spatial distribution of Al + Ti, Cr, Mo and B (+ C) in the vicinity of a serrated grain boundary. During analysis, two γ–γ' boundries and one γ'–γ' grain boundary were crossed. Reproduced by permission of Taylor and Francis from Letellier *et al.* (1994) *Phil. Mag. Lett.*, **70**, 189

electron from the surface can tunnel through the potential barrier to the tip and vice versa. The direction of tunnelling will depend on the potential applied between the tip and the surface. Electron tunnelling is a well-established phenomenon and the current that flows depends on the distance between the tip and the surface d_S, the potential between the surface and the tip, V_i, and the tunnel barrier height between the tip and the surface, Φ, and is given by

$$I \propto \frac{V_i}{d_S} \exp(-Kd_S\Phi^{1/2}) \tag{4.43}$$

where $K = 10.25\,(\text{eV})^{-1/2}\,\text{nm}^{-1}$, V_i is usually a few millivolts to a few volts, d_S is 0.5 nm and Φ is a few electron volts which results in a probe current of nanoamps magnitude. The microscope is operated by bringing the tip towards the surface to give a predetermined current I.

Positioning the tip is difficult and usually achieved in two stages. First, a course drive is used to bring the tip within a millimetre of the surface, and then a fine drive takes it to within a nanometre. The tip is then slowly rastered in the x and y directions

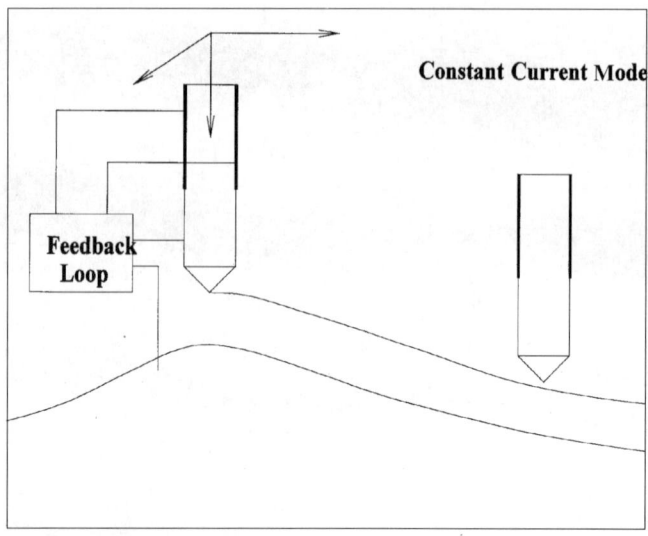

Figure 4.43 Basic principles involved in scanning probe microscopy operating in the constant current mode

while the current I is maintained constant. To do this, the tip must be moved in the z direction as it encounters regions of changing topography. A feedback loop is set to maintain I constant and hence determine changes in z. Changes in the tip position as the result of vibrations and temperature changes have to be accommodated. Early instruments were large constructions, but it was realised that the problems of both temperature and vibration could be minimised by making the units as small as practical. This has the advantage that STM instruments can now be incorporated into scanning electron microscopes and surface analytical instruments such as LEED and scanning Auger microprobes (SAM) (Iwatsuki and Kitamura, 1990). While much of the effort with STM has been to investigate semiconductor device surfaces (Tear, 1990), they can also image metallic surfaces and, therefore, have the potential to study grain boundaries and the segregation of elements to these boundaries. In the region of grain boundaries, abrupt changes are observed in the measured potential distributions. STM studies of the grain boundary region between crystallites in graphite have revealed a disordered region with a width varying between 1 and 10 nm (Albrect *et al.*, 1988), and studies of fullerene films have shown that two-dimensional grain boundaries form and can be imaged (Li *et al.*, 1991a and b). At present there are few reported observations of grain boundaries in metals and alloys using STM. This is not surprising since the raster area of an STM is small in relation to the grain size and studies may need to be carried out in UHV. Figure 4.44 shows an STM image of a C_{60} growth structure of the first three layers containing a grain boundary in the second layer (Li *et al.*, 1991a). It can be seen that the atomic detail of the surface is readily imaged.

(b)

Figure 4.44 STM image of a C_{60} growth structure of the first three layers containing a grain boundary in the second layer (reprinted with permission from Li *et al.*, 1991a. Copyright 1991 American Association for the Advancement of Science. Note the resolution of specific atoms

4.11 REFERENCES

Aaronson, H. I. and Domain, H. A. (1966) TAIMME, **236**, 781.

Adams, F. (1983) in *Analysis of High Temperature Metals*, ed. O. Van Der Biest, London: Applied Science, p. 43.

Ainsle, N., Hoffman, R. F. and Seybolt, A. V. (1960) *Acta Metall.*, **8**, 523.

Airey, G. P. (1985) *Corros.*, **41**, 2.

Albrecht, T. R., Mizes, H. A., Nogami, J., Park, S. I. and Quate, C. F. (1988) *Appl. Phys. Lett.*, **52**, 362.

Allen, G. C. and Wild, R. K. (1986) *Phil. Mag.*, **54A**, L37.

Allen, G. C. and Wild, R. K. (1987) Inst. Phys. Conf. Ser. No. 90; Electron Microscopy and Analysis, p. 13.

Allen, G. C., Flewitt, P. E. J., McIntyre, P., Preece, C., Wild, R. K. and Younes, C. M. (1996) *Mat. Sci. Forum* **207**, 453.

Allen, M. (1981) *Phil. Mag.*, **43A**, 325.

Allpress, J .G., Sanders, J. and Wadsley, D. W. (1969) *Acta Cryst.*, **625**, 1156.

Arkharov, V. I. and Vangengeim, S. D. (1957) *Phys. Metals Metallography*, **4**, 41.
Arkharov, V. I., Bonsor, B. S. and Vangengeim, S. D. (1962) *Fiz. Metallor Metalloved.*, **13**, 86.
Arkharov, V. I., Varskoy, B. N. and Skornyakov, N. N. (1953) *Dokl. Akad. Nauk SSSR*, **89**, 1003.
Auger, P. (1923) *Comptes Rendus Acad. Sci.*, **177**, 169.
Aust, K. T. (1968) *Surface and Interfaces*, Syracuse: Syracuse University Press, Vol. 2, p. 235.
Aust, K. T., Hannenman, R. E., Neissen, P. and Westbrook, J. H. (1968) *Acta Metall.*, **16**, 291.
Band, A. J. and Farr, F. (1992) *Faraday Discussions*, **94**, 1.
Beaman, R. and Isasi, J. A. (1972) Electron Beam Microanalysis, ASTM Spec. Tech. Publ. 506.
Benninghoven, A., Rudenauer, F. G. and Werner, H. W. (1987) *Secondary Ion Mass Spectrometry* New York: John Wiley & Sons.
Bernardini, J., Gas, P., Hondros, E. D. and Seah, M. P. (1982) *Proc. R. Soc. (London) A*, **379**, 159.
Beskrovnyi, A. K. (1957) *Trudy Khar'kov Politekh. Inst.*, **11**, 153.
Binnig, G. and Rohrer, H. (1982) *Helv. Phys. Acta*, **55**, 726.
Binnig, G., Rohrer, H., Gerber, C. and Weibel, E. (1982) *Phys. Rev. Lett.* **49**, 57.
Blavette, D., Deconihout, B., Bostel, A., Sarrau, J. M., Bouet, M. and Menand, A. (1993) *Rev. Sci. Instrum.*, **64**, 2911.
Braunovic, M. (1974) *The Science of Hardness Testing and its Research Applications*, eds J. H. Westbrook and H. Conrad, ASME, p. 329.
Braunovic, M. and Haworth, C. W. (1969) *J. Appl. Phys.*, **40**, 3459.
Bricknell, R. H., Mulford, R. A. and Woodford, D. A. (1982) *Metall. Trans. A*, **13A**, 1223.
Briant, C. L. (1985a) *Acta Metall.*, **33**, 1241.
Briant, C. L. (1985b) *Metall. Trans. A*, **16A**, 2061.
Briant, C. L. (1988) *Acta Metall.*, **36**, 1805.
Briggs, D. and Seah, M. P. (1990) *Practical Surface Analysis by Auger and Photoelectron Spectroscopy*, Chichester: John Wiley & Sons.
Brouillet, P. (1955) *Metaux-Corros.-Ind.*, **30**, 141.
Brown, L. M., Walsh, C. A., Dray, A. and Bleloch, A. L. (1995) *Microsc. Microanalysis Microstruct.*, **6**, 121.
Browning, N. D., Pennycook, S. J., Chisholm, M. F., McGibbon, M. M. and McGibbon, A. J. (1995) *Interface Sci.*, **2**, 397.
Bruemmer, S. M., Jones, R. H., Thomas, M. T. and Baer, D. R. (1983) *Metall. Trans. A*, **14A**, 223.
Bruley, J., Brydson, R., Müllejans, H., Gutekunst, G., Mader, W., Knauss, D. and Rühle, M. (1994) *J. Mater. Res.*, **9**, 2574.
Carpenter, R. W. (1982) *Ultramicroscopy*, **8**, 79.
Castaing, R. (1960) *Adv. Electron Phys.*, **13**, 317.
Cerezo, A., Godfrey, T. J. and Smith, G. D. W. (1988a) *J. Phys. (Paris)*, **C6**(49), 25.
Cerezo, A., Godfrey, T. J. and Smith, G. D. W. (1988b) *Rev. Sci. Instrum.*, **59**, 862.
Chester, R. J. (1985) *Metallography*, **18**, 161.
Choudhury, A., Padgett, R. A. and Brooks, C. R. (1990) *Scr. Metall.*, **24**, 1593.
Choudhury, A., White, C. L. and Brooks, C. R. (1992) *Acta Metall. Mater.*, **40**, 57.
Chuang, T. H. (1991) *Mater. Sci. Eng.*, **A141**, 169.
Clark, J. B. (1964) *Acta Metall.*, **12**, 1197.
Cliff, G. and Lorimer, G. W. (1975) *J. Microsc.*, **103**, 203.
Clifton, D. and Smith, C. S. (1949) *Rev. Sci. Instrum.*, **20**, 583.
Cockayne, D. J. H. (1973) *J. Microsc.*, **98**, 116.
Cohen, J. B., Hurlich, A. and Jacobson, M. (1947) *Trans. ASM*, **39**, 109.
Collet, S. A., Brown, L. M. and Jacobs, M. H. (1981) *Quantitative Microanalysis with High Spatial Resolution*, London: The Metals Society.

Cosslett, V. E. and Duncumb, P. (1956) *Nature*, **117**, 1172.
Coulomb, P., Leymore, C. and Lacombe, P. (1959) *Acta Metall.*, **7**, 691.
Cowley, J. M. (1969) *Appl. Phys. Lett.*, **15**, 58.
Crewe, A. V. (1968) *Science*, **154**, 729.
Cullity, B. D. (1963) *Elements of X-ray Diffraction*, Wiley.
Dean, G. R. and Davey, W. P. (1938) *Trans. Am. Soc. Metals*, **26**, 267.
Doig, P. and Edington, J. W. (1975) *Metall. Trans. A* **6A**, 943.
Doig, P. and Flewitt, P. E. J. (1983a) *Metall. Trans. A*, **124A**, 1943.
Doig, P. and Flewitt, P. E. J. (1983b) *J. Microsc.*, **130**, 377.
Doig, P. and Flewitt, P. E. J. (1984) *J. Phys. (France)*, **54**, 2.
Doig, P. and Flewitt, P. E. J. (1988) *Analytical Electron Microscopy*, London: Institute of Metals.
Doig, P., Flewitt, P. E. J. and Wild, R. K. (1978) *Phil. Mag. A* **37**, 759.
Doig, P., Lonsdale, D. and Flewitt, P. E. J. (1982a) *Metal Sci.*, **16**, 335.
Doig, P., Lonsdale, D. and Flewitt, P. E. J. (1982b) *Scr. Metall.*, **16**, 1201.
Dray, A., Walsh, C. C. and Brown, L. M. (1995) Report for Ontario Hydro.
Duncumb, P. (1979) *J. Microsc.*, **117**, 18.
Duncumb, P. and Shields, P. K. (1963) *Br. J. Appl. Phys.*, **14**, 61.
Ecob, R. (1986) *Scr. Metall.*, **20**, 1001.
Edington, J. W. (1976) *Practical Electron Microscopy in Materials Science, Vol. 3, Philip Tech. Library*, London: Macmillan.
Egerton, R. F. (1989) *Ultramicroscopy*, **28**, 215.
Egerton, R. F. and Cheng, S. C. (1987) *Ultramicroscopy*, **21**, 231.
Erickson, H. P. and Klug, A. (1971) *Phil. Trans., R. Soc. (Lond.) B*, **261**, 1.
Fallon, P. J., Brown, L. M., Barry, J. C. and Bruley, J. (1995) *Phil. Mag.*, **72**, 21.
Fisher, S. B., Scowen, R. S. and Lee, B. J. (1996) Magnox Electric Report TE/GEN/REP/0022/96.
Flewitt, P. E. J. and Wild, R. K. (1985) *Microstructural Characterisation of Metals and Alloys*, London: Institute of Metals.
Flewitt, P. E. J. and Wild, R. K. (1994) *Physical Methods for Materials Characterisation*, Bristol: Institute of Physics Publishing.
Fonash, S. J. (1972) *Electron*, **15**, 783.
Funkenbusch, A. W., Heldt, L. A. and Stein, D. F. (1982) *Metall. Trans. A*, **13A**, 611.
Gleiter, H. and Chalmers, B. (1972) *High Angle Grain Boundaries, Progress in Material Science*, Vol. 16, Oxford: Pergamon.
Glikman, Ye. E. and Cherpatov, Yu. I. (1972) *Fiz. Metallor Metalloved.*, **34**, 90.
Goldstein, J. I. (1961) *Electron Probe Microanalysis, Microanalysis in Interpolation of Extrapolation in Electron Probe Microanalysis*, eds A. J. Tousims and L. Martin, Adv. Elect. Phys., New York: Academic Press.
Goldstein, J. I. (1979) *Introduction to Analytical Electron Microscopy*, eds J. J. Hren, J. I. Goldstein and D. C. Joy, New York: Plenum Press.
Goldstein, J. I., Costly, J. L., Lorimer, G. W. and Read, S. J. B. (1977) *Proc. Workshop in Analytical Electron Microscopy, SEM1*, Chicago: IITI.
Grabke, H. J. R., Erhart, H. and Moller, R. (1983) *Mikrochim. Acta, Supp.*, **10**, 119.
Grabke, H. J. R., Moller, R. Erhart, H. and Brenner, S. S. (1987) *SIA*, **10**, 202.
Greaves, R. H. and Wrighton, H. (1966) *Practical Microscopical Metallography*, London: Chapman and Hall.
Grovenor, C. R. M. (1985) *J. Phys. C, Solid State Phys.*, **18**, 4079.
Grovenor, C. R. M. (1989) *Microelectronic Materials*, Bristol: Institute of Physics Publishing.
Gull, S. T. and Daniell, G. J. (1978) *Nature*, **272**, 616.
Guttmann, M., Dumoulin, P., Tan-Tai, N. and Fontaine, P. (1981) *Corrosion*, **37**, 416.
Guttmann, M., Quantin, B. and Damoulin, P. (1983) *Metal Sci.* **17**, 123.

Hall, E. L. and Briant, C. L. (1985) *Metall. Trans.*, **16A**, 1225.

Hall, E. L., Imeson, D. and Vander Sande, J. B. (1981) *Phil. Mag.*, **A43**, 1569.

Hallam, K. and Wild, R. K. (1995) *Surf. Interface Analysis*, **23**, 133.

Hamminger, R., Grathwohl, G. and Thummler, F. (1983) *J. Mater. Sci.*, **18**, 3154.

Harris, D. R. and Marwick, A. D. (1980) *Phil. Trans. R. Soc. (Lond.)*, **295**, 197.

Hoar, T. P. and Mowat, J. H. S. (1950) *J. Electrodep. Tech. Soc.*, **26**, 7.

Hofer, F., Grogger, W., Kothleitner, G. and Warbichler, P. (1997) *Ultramicroscopy*, **67**, 83.

Hoffmann, J., Hofmann, S. and Tillmann, L. (1984) *Z. Metallkunde*, **65**, 721.

Hofmann, H. and Hofmann, S. (1984) *Scr. Metall.*, **18**, 77.

Hofmann, K. (1978) *J. Phys.*, **19**, 1075.

Inman, M. C. and Tipler, H. R. (1963) *Metall. Rev.*, **8**, 105.

Ishida, K., Yokoyama, S. and Nishizawa, T. (1985) *Acta Metall.*, **33**, 255.

Iwatsuki, M. and Kitamura, S. (1990) *JEOL News*, **28E**, 25.

James, P. I., Garfias-Mesias, F., Moyer, P. J. and Smyrl, W. H., (1998) *J. Electrochem. Soc.*, **145**, L64.

Jenkins, R. (1977) *An Introduction in X-ray Spectrometry*, London: Heyson.

Jones, R. H., Baer, D. R., Charlot, L. A. and Thomas, M. T. (1988) *Metall. Trans. A*, **19A**, 2005.

Jorgenson, P. J. and Anderson, R. C. (1967) *J. Am. Ceram. Soc.*, **50**, 553.

Jorgensen, P. J. and Westbrook, J. H. (1964) *J. Am. Ceram. Soc.*, **47**, 332.

Joshi, A. (1978) *Interfacial Segregation*, eds W. C. Johnson and J. M. Blakely, Metals Park, OH: ASME, p. 39.

Joshi, A. and Stein, D. F. (1970) *Metall. Trans.*, **1**, 2543.

Joy, D. C. and Maher, D. M. (1978) *Ultramicroscopy*, **3**, 39.

Kasen, M. B. (1970) *Phil. Mag.*, **21**, 599.

Kasen, M. B. (1972) *Acta Metall.*, **20**, 105.

Kelly, P. M., Jostons, J., Blake, R. G. (1975) *Phys. Status Solidi*, **A31**, 771.

Klug, R. C., Hintz, M. B. and Rundman, K. B. (1985) *Metall. Trans. A*, **16A**, 797.

Koch, C. C., White, C. L., Padgett, R. A. and Liu, C. T. (1985) *Scr. Metall.*, **19**, 963.

Krahe, P. R. and Guttmann, M. (1974) *Metallography*, **7**, 5.

Krivanek, O. L., Kundmann, M. K. and Kunoto, K. (1995) *J. Microscopy*, **180**, 277.

Kuk,Y. and Silverman, P. J. (1989) *Rev. Sci. Instrum.*, **60**, 165.

Kummer, J. T. and Youngs, J. D. (1963) *J. Phys. Chem.*, **47**, 107.

Lacombe, P. (1963) Metallography 1963, Proc. Sorby Centenary Meeting, Sheffield, ISI Special Report 80, London, p. 50.

Lagerberg, G. and Josefsson, R. (1955) *Acta. Metall.*, **3**, 236.

Larere, A., Guttmann, M., Dumoulin, P. and Roques-Carmes, C. (1982) *Acta Metall.*, **30**, 685.

Lassila, D. H. and Birnbaum, H. K. (1987) *Acta Metall.*, **35**, 1815.

Lea, C., Muddle, B. C. and Edmonds, D. V. (1983) *Metall. Trans. A*, **14A**, 667.

Lee, H. J. and Morris, J. W. (1983) *Metall. Trans. A*, **14A**, 913.

Letellier, L., Guttmann, M. and Blavette, D. (1994) *Phil. Mag. Lett.*, **70**, 189.

Li, Y. Z., Chander, M., Patrin, J. C., Weaver, J. H., Chibante, L. P. F. and Smalley, R. E. (1991a) *Science*, **253**, 429.

Li, Y. Z., Patrin, J. C., Chander, M., Weaver, J. H., Chibante, L. P. F. and Smalley, R. E. (1991b) *Science*, **252**, 547.

Liu, C. T., White, C. L., and Horton, J. A. (1985) *Acta Metall.*, **33**, 213.

Loretto, M. H., Chen, Z. and Burbery, A. J. (1988) *EMAG87, Analytical Electron Microscopy*, London: Institute of Metals.

Low, J. R. (1969) *Trans. AIME*, **245**, 2481.

McGibbon, M. M., Browning, N. D., McGibbon, A. J. and Pennycook, S. J. (1996) *Phil. Mag. A* **73**, 625.

McTegart, W. J. (1959) *Electrolytic and Chemical Polishing of Metals*, London: Pergamon.

Matare, H. F. (1984) *J. Appl. Phys.*, **56**, 2606.

Miller, M. K. (1992) *Surf. Sci.*, **266**, 454.

Miller, T., Bartha, L. and Prohaszka, J. (1960) *Z. Metallkunde*, **51**, 639.

Mulford, R. A. (1983) *Metall. Trans. A*, **14A**, 865.

Muller, D. A. and Silcox, J. (1995) *Phil. Mag.* **71**, 1375.

Muller, E. W. (1951) *Physik*, **131**, 136.

Muller, E. W. (1963) *J. Phys. Soc. Jap.*, **18**, 1.

Nettleship, D. J. and Wild, R. K. (1990) *SIA*, **16**, 552.

Nettleship, D. J. and Wild, R. K. (1992) *Effects of Radiation on Materials: 15th Int. Symp.*, eds R. E. Stoller, A. S. Kumar and D. S. Gelles, Philadelphia: ASTM STP 1125, p. 645.

Nieh, T. G. (1984) *Scr. Metall.*, **18**, 1279.

Nockolds, C., Nasir, M. J., Cliff, F. and Lorimer, G. W. (1979) *Electron Microscopy and Analysis, Institute of Physics Conf. Series No. 52*, Bristol,: Institute of Physics Publishing.

Odin, H., Shaler, A. J. and Wulff, J. (1956) *Trans. AIME*, **185**, 186.

Okamoto, J. K., Pearson, D. H., Ahn, C. C. and Fultz, B. (1992) in *Transmission Electron Energy Loss Spectroscopy in Materials Science*, eds M. M. Disko, C. C. Ahn and B. Fultz, Warrendale: PA: TMS.

Oku, M., Suzuki, S., Abiko, K., Kimura, H. and Hirokawa, K. (1984) *J. Electron Spectrosc. Related Phenom.*, **34**, 55.

Oku, M., Suzuki, S., Kurishita, H. and Yoshinaga, H. (1986) *Appl. Surf. Sci.*, **26**, 42.

Ortner, H. M. 1983) in *Analysis of High Temperature Metals*, ed. O. Van Der Biest, London: Applied Science, p. 1.

Ozakaya, D., Stobbs, W. M. and Brown, L. M. (1993) *Electron Microscopy and Analysis Series*, Bristol: Institute of Physics Publishing, Vol. 138, p. 413.

Ozakaya, D., Yuan, J., Brown, L. M. and Flewitt, P. E. J. (1995) *J. Microscopy*, **180**, 300.

Panitz, J. A. (1974) *J. Vac. Sci. Technol.* **11**, 206.

Partridge, A. and Tatlock, G. J. (1992) *Surface and Interface Analysis* **18**, 713.

Pennycook, S. J. and Boatner, L. A. (1988) *Nature*, **336**, 565.

Philips, W. L. (1963–64) *J. Inst. Metals*, **92**, 94.

Poole, D. M. and Martin, P. M. (1961) *J. Inst. Metals*, **90**, 22.

Poole, D. M. and Martin, P. M. (1969) *Metals Rev.*, **14**, 61.

Priestley, J. V. (1992) *Microsc. Analysis*, November, 15.

Rai-Choudhury, P. and Hower. P. L. (1973) *J. Elect. Chem. Soc.*, **120**, 1761.

Reed, S. J. B. (1975) *Electron Microprobe Analysis* Cambridge: Cambridge University Press.

Riviere, J. C. (1990) *Surface Analytical Techniques* Oxford: Clarendon Press.

Rogers, A. W. (1979) *Techniques in Autoradiography*, Amsterdam: Elsevier/North-Holland, p. 429.

Samuels, L. E. (1968) *Metallographic Polishing by Mechanical Methods*, London: Pitmans.

Schultz, W. W. (1959) ASTM Spec. Tech. Publ. No. 268.

Seager, A. and Schottky, G. (1959) *Acta Metall.*, **7**, 495.

Seah, M. P. (1980) *J. Vac. Sci. Technol.*, **17**, 16.

Seah, M. P. (1981) *Scr. Metall.*, **15**, 457.

Seah, M. P. (1977) *Acta Metall.*, **25**, 345.

Seah, M. P. (1975) *Surf. Sci.*, **53**, 168.

Seah, M. P. (1976) *Proc. R. Soc. (Lond.) A*, **349**, 535.

Seah, M. P. (1986) *Surface and Interface Analysis* **9**, 85.

Seger, L., Pinard, R. and Boulengrey, P. (1992) *Microsc. Analysis*, September, 1925.

Seto, J. W. Y. (1975) *J. Appl. Phys.* **46**, 5247.

Setti, C. (1987) *Mikrochim. Acta*, **1**, 437.

Sherman, R. (1984) *J. Mater. Sci. Lett.*, **3**, 711.

Slastry, C. R. and Judd, C. (1972) *Metall. Trans. A*, **3A**, 779.

Stephenson, T. A., Lorretto, M. H. and Jones, I. P. (1981) in *Quantitative Microanalysis. Gas with High Spatial Resolution*, London: Institute of Metals, Vol. 227.

Stobbs, W. M. (1975) *The Weak Beam Technique in Electron Microscopy in Materials Science*, eds E. Reudle and U. ValdrÉ, Brussels: European Commission.

Streir, L. L. (1970) *Werkst. Korros.*, **21**, 613.

Suzuki, S., Obata, M., Abiko, K. and Kimura, H. (1983) *Sci. Metall.* **17**, 1325.

Tatsumi, K., Okumura, N. and Yamamoto, M. (1986) *J. Phys. (France)*, **49**(C5), 699.

Taylor, W. E., Odell, N. H. and Fan, H. Y. (1952) *Phys. Rev.*, **88**, 867.

Tear, S. P. (1990) *Microsc. Analysis*, **19**, 7.

Thomas, W. R. and Chalmers, B. (1955) *Acta Metall.*, **3**, 17.

Thomas, W. R. and Winegard, W. C. (1952) *Can. J. Metals*, **15**, 26.

Tiller, W. A. and Winegard, W. C. (1955) *Acta Metall.*, **3**, 201.

Tixier, R. (1979) *Electron Probe Microanalysis of Thin Samples in Microbeam Biology*, eds C. Lechene and R. Warner, London: Academic Press.

Tsong, T. T. and Muller, E. W., (1966) *Appl. Phys. Lett.*, **9**, 7.

Turan, S. and Knowles, K. M. (1995) *J. Microsc.*, **177**, 287.

Twigg, M. E., Loretto, M. H. and Fraser, H. L. (1981) *Phil. Mag.* **43**, 1587.

Viefhaus, H., Moller, R., Erhart, H. and Grabke, H. J. (1983) *Scr. Metall.*, **17**, 165.

Walmsley, J., Spellward, P., Fisher, S. and Jenssen, A. (1995) Proc. Seventh Int. Symp. on Environmental Degradation of Materials in Nuclear Power Systems—Water Reactors, Breckenridge, Colorado, p. 985.

Was, G. S., Tischner, H. H. and Latanision, R. M. (1981) *Metall. Trans. A*, **12A**, 1397.

Wei, W. and Grabke, H. J. (1986) *Corros. Sci.*, **26**, 223.

Weinberg, F. (1963) *Trans. AIME*, **227**, 223.

Westbrook, J. H. (1964) *Metals Rev.*, **9**, 415.

Westbrook, J. H. and Aust, K. T. (1963) *Acta Metall.*, **11**, 1151.

Westbrook, J. H. (1970) *Interfaces*, ed. R. Gifkins, Butterworths, p. 283.

White, C. L. and Choudhury, A. (1987) *Mater. Res. Soc. Symp. Proc.*, **81**, 427.

White, C. L., Heatherly, L. and Padgett, R. A. (1983) *Acta Metall.*, **31**, 111.

Wiesendanger, R. (1994) *Scanning Probe Microscopy and Spectroscopy*, Cambridge: Cambridge University Press.

Wild, R. K. (1980) *Mater. Sci. Eng.*, **42**, 265.

Williams, D. B. (1988) *Analytical Electron Microscopy*, ed. G. W. Lorimer, London: Intstitute of Metals, p. 1.

Winnubst, A. J. A., Kroot, P. J. M. and Burggraaf, A. J. (1983) *J. Phys. Chem. Solids*, **44**, 955.

Wood, J. E., Williams, D. B. and Goldstein, J. I. (1981) *Quantitative Microanalysis with High Spatial Resolution* London: The Metals Society.

Yu, K. S., Joshi, A. and Nix, W. D. (1983) *Metall. Trans. A*, **14A**, 2447.

Yukawa, S. and Sinnot, M. J. (1955) *TAIMME*, **203**, 996.

Chapter 5

Mechanical Properties

5.1 ENGINEERING REQUIREMENTS

Materials including metals and their alloys, ceramics, composites and polymers are used for a vast range of engineering applications. As a consequence, there are different mechanical and physical properties, or combinations of these, that are important when selecting and assessing the appropriateness of a given material for a particular service application. As we have emphasised in the preceding chapters of this book, it is essential to recognise the role of the grain boundaries in polycrystalline materials since they are obvious candidates to be considered as a weak link in the material. In this respect there have been various attempts to engineer the types of grain boundaries to optimise overall properties. However, it has to be recognised that free surfaces and other interfaces such as interphase boundaries also have to be considered, and again it is the nanochemistry of these local regions within the overall microstructure that are so important to the final properties of materials: chemical, mechanical, electrical, etc. Moreover, if such factors are not correctly accommodated, then premature failure may arise. There are many examples of materials failure across a wide range of industries, but one particular example from the heavy engineering industry is that in 1969 of a low pressure steam turbine disc, 1.5 m in diameter, in an electricity generating power station in the UK (Figure 5.1). The cracks (Figure 5.1a) were the result of combined stress and corrosion, caused by the steam environment leading to stress corrosion cracking (SCC), propagating into the steel disc. The particular low-alloy ferritic chromium–molybdenum steel had reduced ductility and fracture toughness because it had been inadequately heat treated so that impurity phosphorus had segregated to grain boundaries; the phosphorus segregated to the prior austenite grain boundaries during the heat treatment. Figure 5.1b shows the consequential catastrophic effects. Fortunately, no lives were lost, but the cost ran into millions of pounds, made up from the cost to reinstate the plant and the lost electricity generation. The challenge here is to understand the nature of the change in cohesion, caused by the phosphorus and similar elements at the grain boundaries, and to be able to monitor the composition of these boundaries to provide a basis for determining continued safe operation.

Certainly, materials can fail in a variety of ways, and Figure 5.2 schematically illustrates some of the possible mechanisms. These mechanisms will be governed by several factors such as the mechanical properties of the material, the temperature, the

(a)

(b)

Figure 5.1 Catastrophic failure of a power generation turbine in 1969 at Hinkley Point Power Station caused by temper embrittlement of the ferritic steel low pressure stage disc showing (a) the reassembled failed disc and (b) the consequential damage to the turbine generator

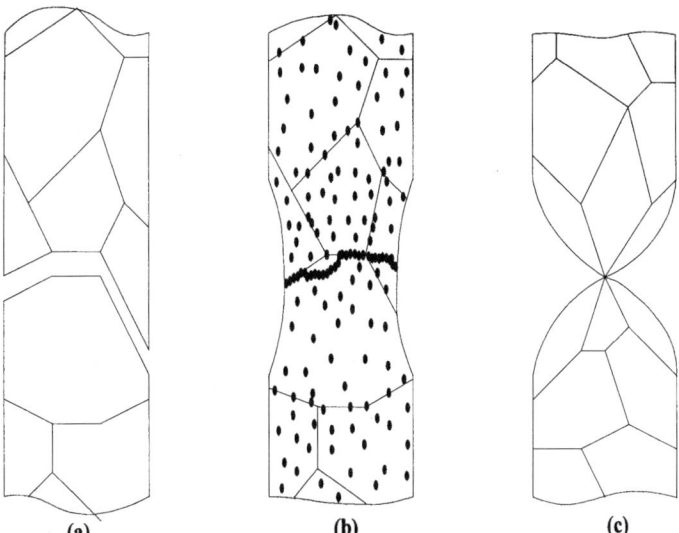

Figure 5.2   Failure mechanisms in metals (a) brittle fracture, cleavage and intergranular (b) ductile void interlinkage (c) ductile necking

environment and the stress. At low temperatures, $< 0.3T_{\mathrm{m}}$, where the ability of the atoms to diffuse is low, the material may fail by a low ductility process, such as cleavage fracture across specific low-index atom planes or intergranular fracture along a path following the grain boundaries (Figure 5.2a). At higher temperature, $> 0.3T_{\mathrm{m}}$, as materials become more ductile, some voids may grow within grains or cavities form at grain boundaries and the material will ultimately fail along an intra- or intergranular path by interlinkage of these voids and cavities (Figure 5.2b). In highly ductile materials the bulk may neck down to a point by plastic flow and fail by forming local necks or regions of shear (Figure 5.2c). At these higher temperatures, where the ability of the atoms to diffuse is much greater, failure by mechanisms similar to those shown in Figures 5.2b and c are likely to dominate. In this chapter we examine the effects of grain boundary, interface and surface chemistry on the mechanical properties of materials and the underlying mechanisms that lead to failure. Certainly, it is clear that there is no single mechanism leading to inter- granular failure in metals, but rather a variety of situations lead to cracks at grain boundaries.

   Ashby and his coworkers (Ashby *et al.*, 1979; and Gandhi and Ashby, 1979) in their classical work produced a series of maps for various pure metals and their alloys and ceramics that predict the predominant mode of fracture under a range of conditions. These maps are in two forms, one in which the tensile stress, at a fixed temperature, is plotted against time to fracture, and the other where the tensile stress is plotted against temperature for a given time to fracture. These plots are schematically of the form shown in Figure 5.3 for Nimonic 80A alloy and define

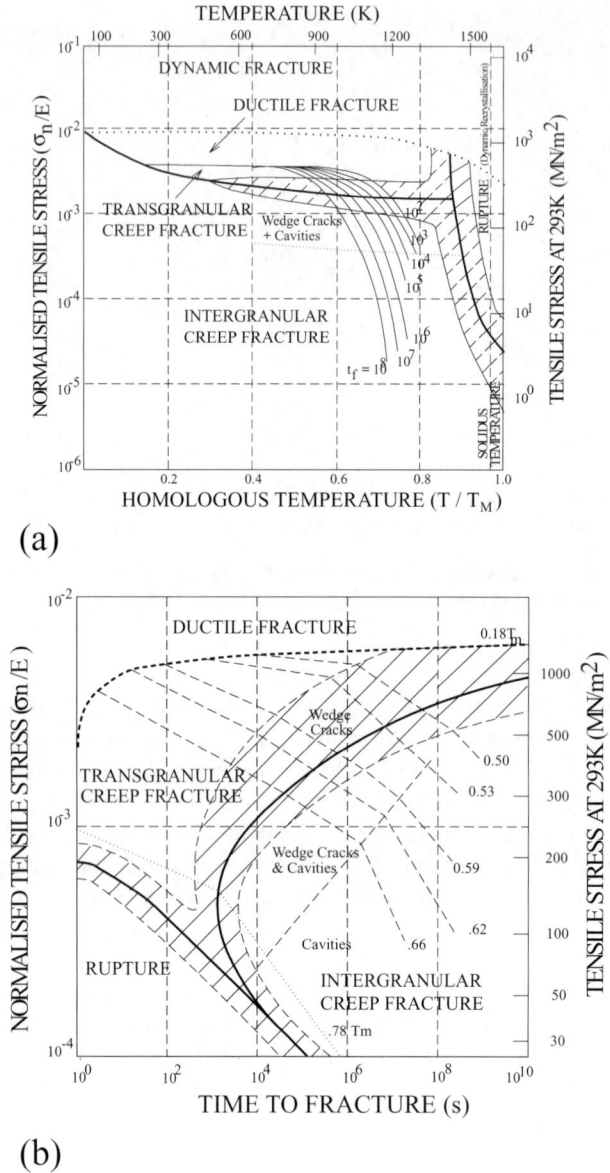

Figure 5.3 Ashby fracture map for nimonic 80 sharing (a) the effect of tensile stress with temperature for fixed time to fracture (b) for a given temperature the time to facture showing different failure regimes (Ghandi and Ashby, 1979)

regions where rupture, ductile fracture, transgranular creep fracture and intergranular creep fracture might be expected to occur. These maps are very useful in bounding the type of fracture that is to be expected for a homogeneous system subjected to known strains at given temperatures. However a range of other factors also play a significant role in determining the resulting fracture. For example, both the temperature at which fracture occurs and the thermal history are important factors to be taken into account. In this respect, local changes in the nanochemistry of the grain boundaries in metals and alloys arising from segregation of impurity or minor alloying elements have a significant effect by modifying localised layers of enrichment or, in the extreme, forming precipitates. Second phase precipitates or inclusions within the grains, where elemental redistribution can occur at the interface, may act as stress raisers or points for void nucleation and eventually crack initiation and growth. These local changes in chemistry can dramatically alter the mechanical behaviour, certainly of metals and alloys. Diagrams can be constructed that identify the competition between various fracture modes (Teirlinck *et al.*, 1988). These diagrams, constructed in stress space, show how brittle fracture is replaced by ductile failure as the pressure is increased. One example of this type of diagram is given in Figure 5.4 for a 4340 steel from data of Cox and Low (1974). This particular steel contains two types of particle, larger MnS type inclusions and smaller iron carbide,

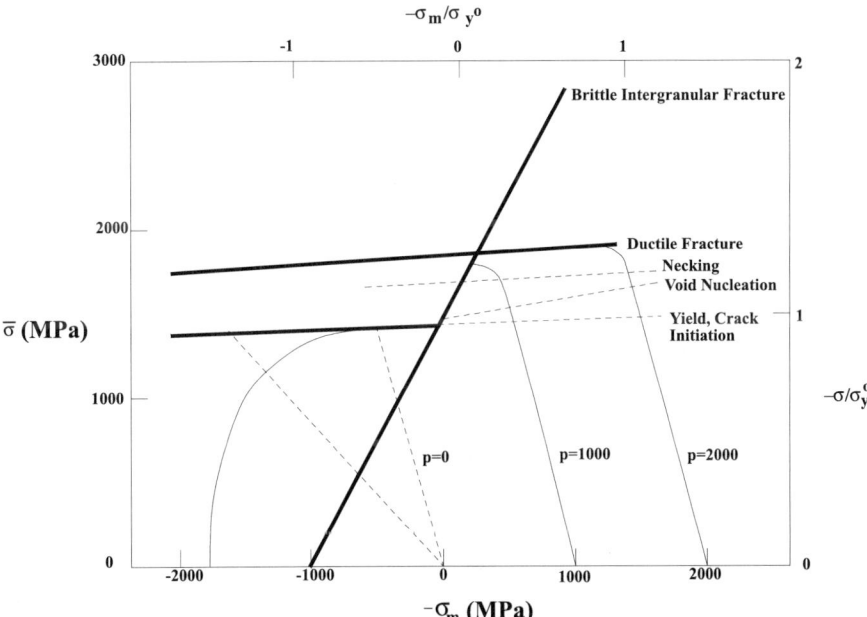

Figure 5.4 Fracture mechanism map in stress space for temper-embrittled 4340 steel in which the brittle fracture line intersects the line for ductile fracture, resulting in two fracture regimes. $\bar{\sigma}$ is the effective stress, σ_y^0 is the initial yield stress and σ_m is the mean stress (Cox and Low (1974))

Fe$_3$C-type precipitates. Temper embrittlement of this steel, where heat treatment causes segregation of trace elements to the grain boundaries, dramatically reduces the ductility. Voids form round the MnS particles when the matrix yields, and smaller voids form at the carbide precipitates which link with the MnS-induced voids. The act of particle growth may modify the composition of the matrix in the vicinity of the particle or there may be a redistribution of impurity or minor alloying elements to the interphase boundaries. These changes in composition can also modify the fracture characteristics, and in addition the environment in which the material is operating may determine the mode of fracture. In some cases, grain boundaries may be attacked preferentially, a typical example being the aqueous caustic attack that arises from chromium depletion at the boundaries in sensitised austenitic stainless steels. However, inclusions in ferritic steel such as manganese sulphides may be preferentially corroded where they meet the surface, and when this coincides with a grain boundary it forms a potential crack initiation site (see Section 5.9). In other environments, specific elements may diffuse preferentially along grain boundaries from the free surface and thereby progressively alter the grain boundary energy. This may occur when materials are exposed to liquid metal or hydrogen-containing environments (see Sections 5.9.5 and 5.9.6).

Irradiation of materials by fast and thermal neutrons, ions or γ-rays may also alter the mode of fracture by increasing the number of defects in the bulk to raise the yield strength and/or effect grain boundary segregation of certain elements (see Chapter 2). The latter may result in either weaker grain boundaries or boundaries that are more susceptible to chemical attack (see Section 5.10).

5.2   FRACTURE IN MATERIALS

5.2.1   Introduction

The redistribution of alloying or trace impurity elements to and from grain boundaries during heat treatment or elevated temperature service can lead to a reduction in intergranular cohesion and a concomitant loss of fracture toughness in a number of materials. The classical and probably the most complex example of this is the temper embrittlement encountered in a range of mild and low-alloy ferritic steels. The term temper embrittlement was first introduced by Dickenson in 1917. It describes the observation that intergranular fracture can intervene in fracture processes where the more usually encountered fracture mode varies from low-temperature cleavage to ductile crack extension at higher temperatures.

5.2.2   Continuum Models of Fracture

A perfect single crystal of a material containing no defects will fail at a stress determined by the forces between atoms. In such a theoretical solid, planes of atoms will separate by breaking atomic bonds at a stress, σ, given by

$$\sigma = (E\gamma_s/a)^{\frac{1}{2}} \tag{5.1}$$

where γ_s is the surface energy, E is the Young modulus of elasticity and $2a$ is the crack length.

In practice, polycrystalline materials fracture at stresses considerably lower than this because they contain defects. Griffith (1920) demonstrated that a crack with a length greater than a critical length, $2a_C$, could decrease its free energy by extending. The critical length is given by

$$a_C = 2E\gamma_s/\pi\sigma_a^2(1 - v^2) \tag{5.2}$$

where v is the Poisson ratio and $2\gamma_s$ is the energy expended in fracture equal to the surface energy. The crack will both extend and increase volume by surface and bulk diffusion until a critical volume is reached and the crack elongates indefinitely. For elliptically shaped cracks, in which the length is much greater than the diameter, the critical length is given by equation (5.2), but the major and minor axes a and b respectively have $a/b = 2.81$ and $a/a_C = 0.74$. Substituting in equation (5.2) values for a typical metal indicates that the stress to produce linear elastic fracture can be two orders of magnitude less than E for a crack 1 µm in length. In practice, plastic flow occurs near to the crack tip and increases the energy to form the two surfaces of the crack. Consequently, it is customary to replace γ with a term known as the strain energy release rate, G_E, which is given by

$$G_E = 2\gamma_s + \gamma_p \tag{5.3}$$

where γ_s and γ_p are the surface and plastic flow energies respectively.

In the case of a polycrystalline material, when fracture occurs in the region of the grain boundary the situation is further complicated. Firstly, the fracture energy of the grain boundary is not simply $2\gamma_s$ but is reduced by a contribution from the grain boundary energy which will be a function of its composition. Secondly, γ_s and γ_p are not independent variables; γ_p is assumed to be a large value but if γ_s reduces to zero then G_E will also become zero (Thomson and Knott, 1993). McMahon and coworkers (McMahon and Vitek, 1979; Jokl et al., 1980a, b) have described the relationship between γ_s and γ_p which helps in understanding the effect when γ_s decreases as a result of elements segregating to grain boundaries. They modelled the role of dislocations and grain boundary particles on brittle intergranular fracture. Here, dislocations are considered to pile-up at a grain boundary or boundary particle, leading to the formation of a crack along the path of the grain boundary. This can then be treated in the same manner as a Griffith crack by considering the grain boundary surface energy and plastic flow at the crack tip (Figure 5.5). Smith (1966) derived a similar equation for a crack nucleated at a brittle particle for the case where a dislocation pile-up and a crack are coplanar. If the crack length, $2a_c$, in the particle is much smaller than the length of the dislocation pile-up, L, this can be written as

$$\sigma_L^2 + \tau_{eff}^2 \left[\left(\frac{L}{2a_c} \right)^{\frac{1}{2}} + \frac{4}{\pi} \frac{\sigma_f}{\sigma_{eff}} \right]^2 > \frac{4E(2\gamma + \gamma_p)}{\pi(1 - \gamma^2)d_c} \tag{5.4}$$

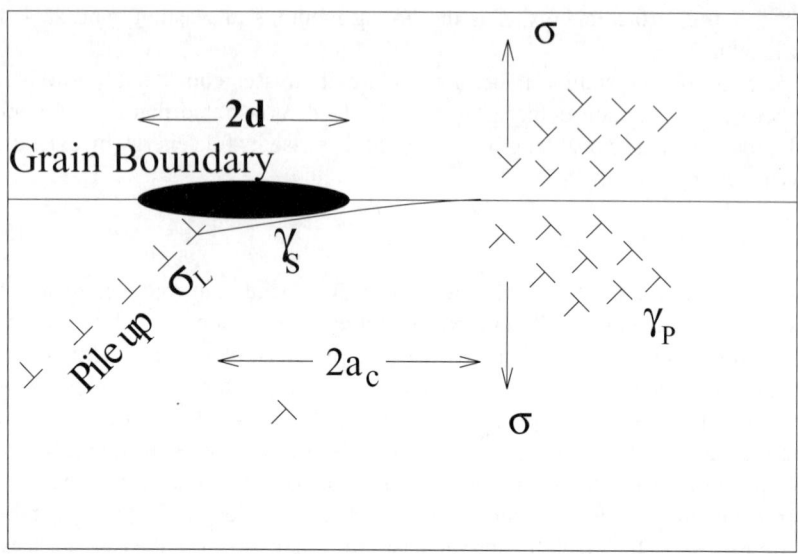

Figure 5.5   Schematic diagram of plastic flow at the tip of a crack, length $2a_C$, which has initiated at a grain boundary particle as a result of a dislocation pile up

where σ_f is the fracture stress, σ_L is the local stress ahead of the precrack in the region of the crack nucleation, σ_{eff} is the effective shear stress on the dislocation pile-up and γ_p is the plastic work associated with propagation of the crack in the matrix beyond the particle.

5.2.3   Temperature Dependence of Intergranular Fracture

In this section we examine the fracture of materials at high and low temperature. Simplistically, at very low temperatures atoms are not able to move between lattice sites and the material will fracture by breaking bonds between atoms. As the temperature increases, so the stability of atoms and vacancies decreases, with associated plasticity and elemental segregation, and the mode of fracture will change. For most materials there is a temperature at which the mode of fracture changes from brittle fracture, by breaking of bonds, to ductile fracture, where a considerable amount of plastic flow takes place. The temperature at which this occurs depends upon the geometry of the body and the crack it contains and varies between materials; for a specific material it depends on the microstructure arising from the thermal and mechanical history. The temperature at which the mode of fracture changes from brittle to ductile is known as the ductile-to-brittle transition temperature (DBTT), the fracture appearance transition temperature (FATT) or alternatively the onset of the upper shelf transition (OUST). This transition can be determined by fracturing a body, which has a given cross-section and notch or precrack, either by impact and measuring the amount of energy absorbed such as the Charpy test or by controlled displacement and measuring the resistance to crack extension—the fracture toughness. A plot of energy absorbed versus fracture temperature yields a plot of the form shown in Figure 5.6 where the DBTT is defined as the point of inflection. In the case of the variation in fracture toughness with temperature, this is defined by the onset of the upper shelf where ductile crack extension occurs.

Three factors can induce an upward shift in the transition temperature.

(a) an increase in yield stress arising from work or precipitation hardening,
(b) elemental segregation at precipitate to matrix interfaces and
(c) elemental segregation to grain boundaries.

The Davidenkov diagram (Davidenkov, 1938 and 1981) (Figure 5.7) shows that a vertical shift in the temperature dependence of the yield stress will lead to a shift in the DBTT. The exact form of the temperature dependence of the increased yield stress of the material will naturally be dependent upon the factors that lead to this change. When a crack propagates along a grain boundary, two free surfaces are created and the area of the grain boundary is removed. The energy, G_F, required to separate unit areas of interface between the two grains or indeed between two different phases by a thermodynamic process is given by

$$G_F = \gamma_1 + \gamma_2 - \gamma_{gb} \qquad (5.5)$$

where γ_1 and γ_2 are the specific energies of the two surfaces or planes, and γ_{gb} is the boundary or interfacial free energy per unit area. The work of fracture, G_F, is equivalent to the area under the stress–strain curve. For embrittlement to occur, the

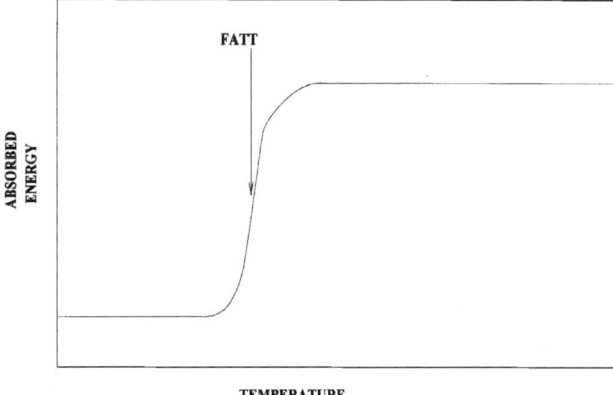

Figure 5.6 A schematic plot of energy absorbed versus temperature showing the fracture appearance transition temperature (FATT)

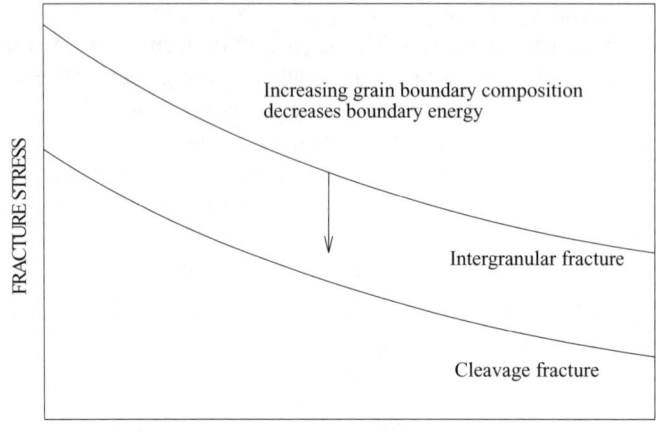

Figure 5.7   Schematic diagram showing the variation in cleavage and intergranular fracture stress with temperature. Increasing grain boundary composition by segregation lowers the intergranular fracture energy so that it ultimately dominates

value of G_F must be reduced by the presence of solute or impurity atoms at the grain boundary, and these reside subsequently on the fracture surfaces. In this model, γ_{gb} is $1/6(\gamma_1 + \gamma_2)$ so that for many segregated species changes in γ_{gb} make only a small contribution to G_F. Since many embrittling elements that segregate to boundaries of polycrystalline materials form strong, highly directional bonds, the grain boundary energy is likely to decrease. The most important contribution to embrittlement comes from the reduction in the surface free energy of each fracture surface, γ_1 and γ_2, owing to the presence of embrittling atoms (Hondros and McLean, 1974). We return to intergranular fracture in Section 5.3.

The fracture and its path may also be influenced by environmental factors. A surface microcrack will be modified by the local environment. Hydrogen may be released as the result of localised corrosion within a crack and diffusion to the tip where it embrittles the grain boundary ahead of the crack. Moreover, the corrosion may produce oxides that have a greater volume than the parent which, in turn, will force apart the two surfaces of the crack, thereby increasing the local stresses. Grain boundaries may be depleted in certain elements and this may increase their propensity to corrosion. For example, chromium-rich carbide precipitates may form on the grain boundaries of austenitic stainless steels, which depletes the volume around the grain boundary of chromium, an effect known as sensitisation. Chromium is responsible for the formation of a protective oxide on stainless steel and, in its absence, corrosive environments are able to penetrate into the bulk along grain boundaries (Lacombe and Parkins, 1977). Sensitisation of these steels can occur at high temperatures ($> 900\,\mathrm{K}$) in a few hours, but may also occur at temperatures as low as $600\,\mathrm{K}$ over a period of years (Povich, 1978). Neutron irradiation can also deplete the grain boundaries of chromium by a reverse Kirkendall effect (see Section 5.10). The embrittling tendency of the elements in a variety of solids has been collected by Finnis (1982). If a polycrystalline material is fractured while immersed in a fluid or liquid metal, the surfaces become covered and work is done such that

$$G_F = \gamma_{1s} + \gamma_{2s} - \gamma_{gb} \tag{5.6}$$

where γ_s is the interfacial free energy between the surface and the liquid. The energy values for wetted surfaces can be much less than for the corresponding non-wetted values, leading to a lower work of fracture. This results in glasses, rocks and ceramics fracturing much more readily when wet. This is known as the Rehbinder effect which is used to advantage in the petrochemical industry for a range of cutting and drilling processes. A similar effect exists for specific metal and environment combinations.

5.3 INTERGRANULAR FRACTURE

When a polycrystalline material is subjected to an increasing stress, it becomes strained until the elastic limit of the weakest point is reached. If this point or region is able to deform plastically, then the strain will continue, but if no flow is possible cleavage failure will occur across specific, low-index atom planes. Points of weakness in a material are frequently those areas containing stress raisers such as surface microcracks, particles, precipitates or inclusions which may fracture preferentially. If the grain boundaries are relatively strong and the material can flow plastically, the strain will increase, with dislocation density increasing, until a point is reached when the material fractures owing to a combination of loss of cross-section and work hardening. In materials that contain grain boundaries that are weak compared with the bulk properties, failure may occur by a combination of cleavage and/or ductile flow and intergranular fracture. The orientation of a grain boundary relative to the stress direction will determine the mode of fracture. If the grain boundary is oriented parallel to the plane of the advancing crack, then the component of stress is a maximum for intergranular fracture. As the orientation of the grain boundary deviates from the plane of the fracture path, so the component of stress for intergranular fracture decreases until it becomes zero for grain boundaries oriented normal to the crack plane.

In a specimen loaded in tension, fracture will frequently be initiated by microcracks or particles located close to a grain boundary oriented normal to the direction of the applied tensile stress. The fracture path then follows the grain boundary until the next grain is reached, when the failure will continue to be intergranular if the orientation and grain boundary strength are favourable; if unfavourable, ductile or cleavage failure will intervene. As the grain boundary strength decreases relative to the cleavage fracture energy, so the percentage of intergranular failure will increase.

Figure 1.6 shows examples of failure where there is almost total intergranular fracture, a mixture of intergranular and cleavage and ductile failure. In practice, most fracture surfaces contain a mixture of cleavage, intergranular fracture and ductile failure. The exact percentage of each is determined by the grain boundary strength, the strength of the matrix, the direction of stress and the orientation of the grain boundaries relative to the fracture path. It has been calculated that, in an ideal system, where the grain boundary fracture energy is equal to the cleavage energy, the percentage of intergranular fracture accompanying cleavage fracture can be as high as ∼ 25% (Smith *et al.*, 1997), but this is dependent upon a number of simplifying factors. When the material is able to deform plastically, then some of the accommodation at the grain boundary region will be by this mechanism rather than intragranular fracture. In the case of creep, cavities may form on grain boundaries. Figure 5.8 shows an optical micrograph of a polished surface of 1Cr–1/2Mo ferritic steel in the heat affected zone of a welded steel. These cavities will gradually weaken the material, resulting in grain boundary failure, and the failed boundary will have a cavitated appearance (see, for example, Figure 5.24).

The probability of an atom in a lattice site moving to an adjacent site is determined by the energy required for the atom to climb the potential barrier. If U is the activation energy needed to move the atoms and a stress σ is applied, the probability that the atom will jump in the direction of the applied stress is given by

$$P_\text{A} = K \exp - \left[\frac{U - v(a)\sigma}{kT} \right] \qquad (5.7)$$

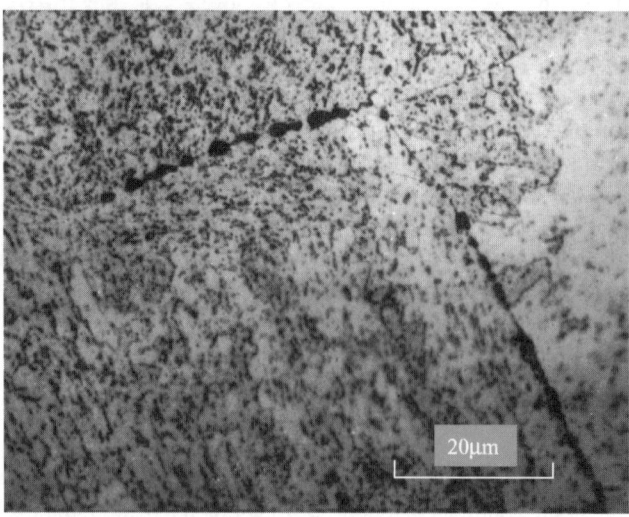

Figure 5.8 Optical micrograph of cavities found at grain boundaries in the heat affected zone of a 1Cr–$\frac{1}{2}$Mo welded steel

where $v(a)$ is an activation volume. The probability that the atom will jump in the direction opposed to the applied stress is given by

$$P_A = K \exp - \left[\frac{U + v(a)\sigma}{kT} \right] \tag{5.8}$$

giving a jump rate of

$$J = 2K \exp - \left(\frac{U}{kT} \right) \sinh \left[\frac{v(a)\sigma}{kT} \right] \tag{5.9}$$

At low temperatures, the height of this barrier is greater than the energy required to break the bonds between atoms and, when a high stress is applied, the bond preferentially shears and failure is by cleavage. As the temperature increases, so the energy required for atoms to climb the potential barrier decreases and a point is reached where this is less than the energy to break the atom bonds. At this point, the material flows plastically; the brittle-to-ductile transition temperature. Only rarely are perfect crystals encountered, since most materials contain many defects, such as vacancies, dislocations and grain boundaries. In these regions the activation energy for an atom to move is considerably reduced and, in practice, the brittle-to-ductile transition temperature has to be defined in the presence of these defects.

A typical polycrystalline material consists of grains randomly oriented, containing voids, particles, cracks and impurities, and failure will depend upon the nature of these defects. There have been many attempts to describe brittle fracture in materials, taking account of all these factors, and among these are papers by Thomson and Knott (1993), Cottrell (1989, 1990a, b) and Hirth (1980). A crack may form at a variety of sites in a material, and the particular location will be determined by the specific material, the thermal and mechanical history, the applied stresses and temperature. The defects may be internal particles formed during manufacture, such as oxides and sulphides, or internal particles formed during heat treatment and service, such as carbide precipitates. In addition, voids and cavities can form at higher temperatures around particles, or grain boundary intersections. Materials will frequently contain defects at, or near, to the surface as a result of deformation during machining. These defects may take the form of microcracks which in turn will be associated with near-surface particle defects. In some instances the segregation of trace elements to the interface between the particle and the matrix may promote the formation of a cavity. Certainly oxide particles in iron nucleate creep cavities in the presence of low levels of sulphur (George *et al.*, 1987).

5.4 STRENGTH OF GRAIN BOUNDARIES

Fracture along grain boundaries has been observed in many different materials spanning metals and alloys and refractory materials. The parameters that are important when analysing this process, which involves separation along interfaces, are the cohesive properties, the work to separate the interface and the maximum force that is necessary to separate a unit area of the interface (Rice and Thomas, 1974). The propensity to intergranular fracture is associated with the chemical composition of the grain boundaries (Lee *et al.*, 1984; Komeda and McMahon, 1981) which, in turn, can dramatically influence ductility and strength. As a consequence, it is important to be able to explain the processes leading to these changes in cohesion of the grain boundaries, and this has been strengthened by the ability to measure the composition and chemical state of atoms at the grain boundaries and interfaces (see Chapter 4). However, for the brittle fracture regime, when cleavage fracture predominates, a proportion of intergranular fracture will accompany the cleavage due simply to geometrical requirements. Certainly, when propagating from one grain to the next, the crack has a choice between either cleavage or brittle intergranular path. The mismatch between two misoriented cleavage planes can be accommodated in a number of different ways:

(a) it can propagate on several parallel planes, forming a large number of small cleavage steps,
(b) the mismatch can be bridged by ductile tearing and
(c) intergranular cracking may occur in the mismatch region.

The intergranular cracking generated between misoriented cleavage planes in adjacent grains can be regarded as geometrically necessary. Moreover, this type of intergranular cracking has to be distinguished from that arising from cohesive energy considerations. As shown by theoretical modelling (Crocker *et al.*, 1998), there is a minimum proportion of such geometrically necessary benign intergranular cracking which may vary with the temperature at which the brittle fracture occurs.

Two main approaches have been adopted to explain grain boundary cohesion, one based on thermodynamic arguments and the other invoking modern quantum theory at the atomic and electronic scale. The thermodynamic approach is attractive for applications involving adsorption-induced brittle fracture since it considers specifically the effect of interfacial cohesion, as a result of segregation of solute and impurity atoms to interfaces, determined by the kinetics of the transport process. Here, the essential relationship describing the ideal work of fracture per unit area, ϕ, is given by

$$\phi = 2\gamma_s - \gamma_{gb} \tag{5.10}$$

where γ_S is the energy per unit area of created fracture surface and γ_{gb} is the energy per unit area of the grain boundary. This has been the starting point for the theory developed by Seah (1976, 1980a), Rice and Thomas (1974), Hirth (1980), and Asaro (1980). For fast, low-temperature fracture, γ_S is assumed not to be an equilibrium value so that for a unit area of grain boundary

$$d\gamma_{gb} = V \, dP - S \, dT - \sum_i \Gamma_b^i \, d\mu_i \tag{5.11}$$

where V is the specific volume and S is the energy of the interface region, P is the pressure across the interface, T is the temperature and Γ^i is the quantity of species i

with chemical potential μ^i per unit area of interface. For a binary system A–B, where B is the solute, at constant T and P the Gibbs–Duhem relationship gives

$$d\gamma_b = \{[X^B/(1 - X^B)]\Gamma_{gb}^A - \Gamma_{gb}^b\} d\mu^B \tag{5.12}$$

where X^B is the solute molar fraction and, if Henry law is obeyed, $d\mu^B = RT \, d\ln a_a$, where a_a is the solute activity, so that $d\mu^B = RT \, d\ln X^B$. Therefore, for a dilute solution, substituting in equation (5.11) gives for a grain boundary

$$\gamma_{gb}^A = \gamma_{gb}^{A0} - RT\Gamma_{gb}^B \tag{5.13}$$

and for a surface

$$\gamma_s^A = \gamma_s^{A0} - RT\Gamma_s^B \tag{5.14}$$

where γ_{gb}^{A0} and γ_{gb}^{A0} are the grain boundary and surface energies for the pure A system. At fracture, $\Gamma_s = \frac{1}{2}\Gamma_{gb}$, irrespective of the amount of segregation, and then from equations (5.10), (5.13) and (5.14)

$$\phi = 2\gamma_g^{A0} - 2\gamma_{gb}^{A0} \tag{5.15}$$

However, as considered by Hirth (1980), an additional term has to be added to equation (5.15). In the Γ versus μ diagram (Figure 1.20), curve (i) shows the equilibrium excess of B atoms at the grain boundary, Γ_{gb}, as a function of chemical potential, whereas curve (ii) shows the excess for the fracture surface, $2\Gamma_s$.

In the case of slow fracture at higher temperatures, if equilibrium is maintained the system starts at position K on curve (i) and moves to position L on curve (ii). Hence, the level of segregation increases when fracture occurs so that, at constant μ, equation (5.10) gives

$$
\begin{aligned}
\phi &= 2\gamma_s^L - \gamma_{gb}^K \\
&= 2\gamma_s^{A0} - \gamma_{gb}^{A0} - RT(2\Gamma_s^L - \Gamma_{gb}^K)
\end{aligned}
\tag{5.16}
$$

Consequently, there is a decrease in ϕ as a result of solute segregation to grain boundaries in slow, high-temperature fracture. At lower temperatures, fast fracture starts at K and moves to M with the total amount of grain boundary segregation constant so that

$$
\left.
\begin{aligned}
\phi &= 2\gamma_s^M - \gamma_{gb}^K \\
&= 2\gamma_s^{A0} - \gamma_{gb}^{A0} - RT(2\Gamma_s^M - \Gamma_{gb}^K) \\
&= 2\gamma_s^{A0} - \gamma_{gb}^{A0}(\text{const. } \mu)
\end{aligned}
\right\}
\tag{5.17}
$$

However, in this case the segregated solute atoms release energy equal to

$$
\left.
\begin{aligned}
\phi &= 2\gamma_s^{A0} - \gamma_{gb}^{A0} - \int_{\mu^M}^{\mu^K} (2\Gamma_s^M - \Gamma_{gb}^K) \, d\mu \\
&= 2\gamma_s^{A0} - \gamma_{gb}^{A0}(\mu^K - \mu^M)\Gamma_{gb}
\end{aligned}
\right\}
\tag{5.18}
$$

This result, first observed by Rice (1976), shows that the maximum cohesive force for separating a boundary is directly related to the cohesive energy.

In the case of the atomic or electronic scale models, the calculations have been limited because of the complexity of the structure of grain boundaries. As a consequence, it has been more usual to consider polyhedral atomic cluster models where only the segregated atom and a limited number of grain boundary atoms are included in the local environment used in these calculations (Briant and Messmer, 1980, 1982 and 1984). The technique used for calculating the molecular orbitals is the self-consistent field scattered wave theory applied to many local bonding problems such as chemisorption and amorphous metals. Three types of information are obtained from this approach: the orbital energy level diagram, the one-electron molecular orbital wave functions and the valence electron charge density of atoms in the cluster. The results of these calculations indicate that a possible mechanism for embrittlement arising from segregation of an impurity atom is the formation of strong bonds between this atom and immediate neighbour atoms; the electrons participating in bonding come from surrounding metal–metal bonds which are correspondingly weakened. By quantifying the strength of embrittlement in terms of the amount of charge drawn to the impurity atom, Briant and Messmer were able to rank the relative embrittling potencies of certain impurity atoms at grain boundaries in a given host material to describe how the cohesive energy changes. Table 5.1 shows the results for three clusters in α-iron and two clusters in nickel. For impurity boron in polycrystalline Ni there is no change in the charge in the metal–metal bonds across the boundary, but a covalent-like bond is formed between the boron atom and the nickel atoms. However, the impurity atom–parent bonds are favoured at the expense of the Ni–Ni bonds when sulphur segregates to grain boundaries in nickel. Similar results have been obtained by Wang and Zhao (1994). Hashimoto *et al.* (1984) used cluster calculations with atom positions determined from atomic relaxation based upon large grain boundary cells with approximate pair potentials. For the case of boron and phosphorus in $\Sigma 5$ and $\Sigma 9$ grain boundaries in α-iron, the results indicate that a grain boundary is made up from an alternating structure of strong and weak bonds. The addition of impurity boron tends to strengthen the boundaries by replacing the weaker Fe–Fe bonds with stronger ones arising from the boron interactions. The addition of phosphorus atoms, on the other hand, rearranges

Table 5.1 Cluster composition for tetrahedral clusters (after Messmer and Briant, 1982)

Host metal	Impurity	Experimentally determined effect of impurity atoms
Fe	S	Strong embrittler
Fe	P	Moderate to weak embrittler
Fe	C	Cohesive enhancer
Ni	S	Strong embrittler
Ni	B	Cohesive enhancer

the atoms into clusters of Fe_9P which weaken the boundaries by removing charge from the stronger bonds responsible for cohesion. In the case of nominally FeAl (B2) alloy (Fe–37 wt% Al to Fe–48 wt% Al), the ductility decreases with increasing aluminium concentrations. However, an addition of boron shifts the ductile-to-brittle transition temperature to higher aluminium levels by segregating to grain boundaries and suppressing intergranular fracture. The ability of boron to suppress intergranular brittle fracture decreases with increasing aluminium concentration, although the grain boundary concentration remains the same, independent of the alloy stoichiometry. Consequently, even boron-doped Fe–Al alloys become embrittled as the stoichiometric composition is approached (Cohron *et al.*, 1998).

An alternative approach has been developed as a consequence of microanalysis of grain boundaries in α-iron (Figure 1.18), where the electron energy loss spectrum from a grain boundary that contains phosphorus is compared with a spectrum from the adjacent grain (Figure 1.18a) (see Chapter 1). The spectrum from a grain boundary that contains no phosphorus is compared with a corresponding spectrum from a nearby grain (Figure 1.18b). This analysis reveals that the d orbitals of the iron atoms in a boundary containing phosphorus have more electrons than the d orbitals of iron atoms within a grain. Hence, phosphorus atoms at grain boundaries donate electrons to the d bands of the iron. However, for boundaries where there is no phosphorus, the number of electrons in the d bands of the iron atoms is unchanged (Figure 1.18b). These results led Brown *et al.* (1997) to consider the contribution to the cohesive strength of the grain boundary (see chapter 1, section 1.8). They argued that transition metals such as iron have d bands that can contain 10 electrons. If these form covalent bonds in a linear combination of molecular orbitals, the d orbitals split into five states below the atomic reference level and five above, the so-called states of bonding and antibonding. For an isolated iron atom the energy levels are filled to the 3d and 4s states (Figure 1.19a) which contain six and two electrons respectively. In this notation, the first number ($n = 1$, 2, 3, 4, etc.) denotes the atomic shell (K, L, M, N, etc.) while the letter (s, p, d or f) denotes the subshell and refers to the angular momentum of the orbital state ($l = 0$, 1, 2, 3); the 3d state can contain up to 10 electrons. If an atom is excited by a fast incident electron, the six electrons in the full 2p state can move into the empty 3d states, causing the incident electrons to lose energy. The electron energy loss spectrum can be used to evaluate the number of empty 3d states. However, electrons in the 3d state have wave functions that are like standing waves with alternating positive and negative lobes (Figure 1.19b). If lobes of similar sign on neighbouring ions overlap, the electron waves interfere constructively, piling up charge in the region where the electron experiences the attractive potential of both ions, thereby lowering the total energy—bonding. If lobes of opposite sign overlap, they interfere destructively—antibonding. Each atomic d state splits into a bonding and antibonding state, and the difference in energy between these two states is controlled by the extent to which the wave functions overlap in space. The original atomic level remains roughly unchanged and marks the centre of the array of split levels—the reference level.

Maximum bonding occurs when all the bonding states are filled and all the antibonding states are empty, since the binding energy can be considered to be proportional to $x(10 - x)$ (Friedel, 1969), where x is the number of occupied states. Maximum bonding occurs when $x = 5$, but if the states are totally unoccupied $(x = 0)$ or totally full $(x = 10)$ then there is no bonding. This simple model gives equal weight to each of the 10 states, but it predicts the cohesive properties of transition metals and their alloys quite well. Since the electron energy loss spectrum of a grain boundary measures the density of unoccupied states, i.e. $(10 - x)$, the EELS spectra can be used to predict cohesion at a boundary. Muller (1996) developed a semiquantitative measure of the boundary cohesion on the basis of this approach. If an iron atom in a grain boundary containing an impurity phosphorus atom has Z nearest neighbours, then the binding energy of a single iron atom is proportional to $\beta Z^{1/2} x(10 - x)$, where β is the value of the overlap integral that controls the splitting of the state. If the grain boundary in pure iron is fractured near the newly created external surface, changes occur both to number of nearest neighbour atoms and to the electronic binding. According to experimental results, the presence of phosphorus atoms does not affect the surface binding: the phosphorus remains chemically unbound, held to the surface by the weak Van der Waals forces. However, at the boundary, phosphorus causes the iron to gain about half an electron, in addition to the six in a pure metal; x increases from 6 to 6.5. This weakens the boundary, so that, knowing the boundary energy per iron atom, the presence of phosphorus atoms in a grain boundary reduces the boundary cohesion by 0.04 eV per atom or about 10 %.

It is important in these calculations to know the bonding of phosphorus, both at the internal grain boundary and the free surface exposed when the material is fractured. Certainly, active radicals in the environment when a sample is fractured can alter the energy required to propagate the intergranular crack because of changes in the bond energies. However, it is clear that electron theory provides a link between the observed change in the electronic structure at a grain boundary, arising from changes in the nanochemistry, and susceptibility to intergranular fracture.

5.5 TEMPER EMBRITTLEMENT

As discussed in the introduction to this section, temper embrittlement is a term that was introduced to describe intergranular fracture in low-alloy ferritic steels that have been heated for long periods in the temperature range 620–820 K or slowly cooled through this range. Moreover, since this is considered to be associated with the equilibrium segregation of minor alloying or impurity elements to the grain boundaries, the process is classically considered as reversible and, depending upon the heat treatment cycle adopted, is referred to as either temper embrittlement or 620 K (350 °C) embrittlement. With each of these heat treatments the steels have reduced fracture toughness and enhanced intergranular fracture. As a consequence, the impact energy ductile-to-brittle transition temperature (DBTT) and the corresponding fracture toughness transition temperature (OUST) are displaced towards higher temperatures (Rellick and McMahon, 1974; Olefjord, 1978; Pugh, 1991). The distinction between these two terminologies is difficult to separate since, as suggested by Rellick and McMahon (1974), the underlying mechanisms are similar. However, as pointed out by Briant and Banerji (1978), this leads to the additional division of temper embrittlement into one-step or two-step embrittlement. Figure 5.9 schematically demonstrates the basis of these two classes of segregation-induced intergranular embrittlement in steels:

(a) One-step temper embrittlement describes that occurring in steels of high yield strength where the martensite or bainitic microstructure is tempered for short times at low temperatures.

(b) The two-step embrittlement describes steels of lower yield strength and results from tempering in the range 870–920 K which allows, for example, the martensite to ferrite and cementite reactions to proceed to near completion.

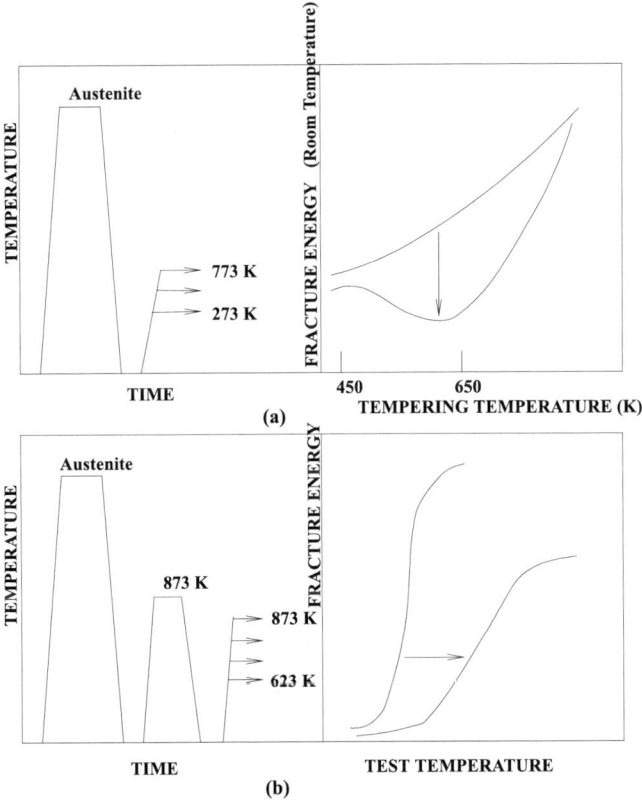

Figure 5.9 Schematic diagrams showing the heat treatment cycles involved in (a) one-step and (b) two-step temper embrittlement and the corresponding effect on the fracture energy (after Briant and Banerji, 1978)

Embrittlement occurs when the steel is subject to additional isothermal ageing at temperatures of the order of 770 K. This effect can be simulated by cooling the steel slowly through the embrittling temperature range following tempering.

Many steels used for rotors in the electrical power generating industry and pressure vessels in this industry and the petrochemical industry are subject, during manufacture, to slow cooling after tempering of such heavy-section components. Hence they are exposed to long times within the temper embrittlement temperature range; moreover, during service for extended periods of operation, the temperatures are again often within the embrittlement range. As a consequence, intergranular fracture resulting from temper-embrittling heat treatments producing thermally induced grain boundary segregation has been reported in a very wide range of low-alloy ferritic steels in the last 50 years (Figure 5.10). We will now examine a few examples.

5.5.1 $2\frac{1}{4}$Cr–1Mo Steels

Much of the work on the effects of thermal ageing on $2\frac{1}{4}$Cr–1Mo steel, which usually has a bainitic microstructure, arises from its use over the last 30–40 years as a material for thick-wall pressure vessels in oil refineries. In 1982, a major report on 25 years experience was published (Erwin and Kern, 1982) which examined transition temperature shifts, and this concluded that weld metal has a higher susceptibility to embrittlement than plate or forgings, with submerged metal arc weld metal being most susceptible; for example, transition temperature shifts of up

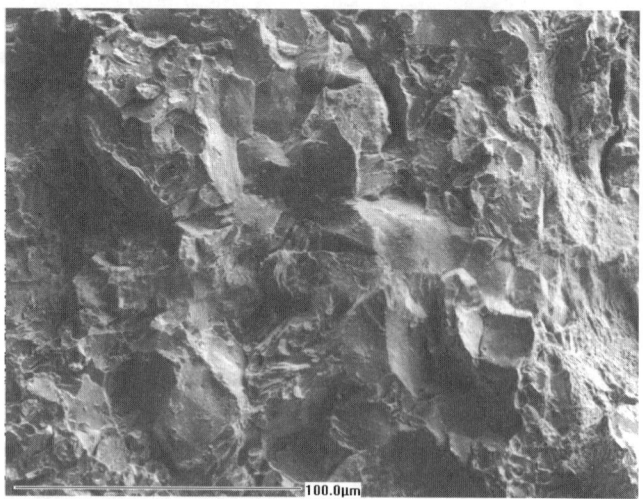

Figure 5.10 Scanning electron fractograph in the secondary electron imaging mode, showing intergranular fracture in a temper embrittled low-alloy ferritic steel

to 148 K have been recorded. Impurity phosphorus was recognised as a very significant element, and attempts to use bulk alloy composition to describe the behaviour has resulted in the development of various empirical relationships (Bruscato, 1970, and Brian and King, 1980). One is the *J* factor which attempts to provide a correlation with embrittlement susceptibility on the basis of the bulk composition (in wt%):

$$J = [(Si + Mn)(P + Sn)]10^4 \qquad (5.19)$$

It is noteworthy, however, that the relationship does not include any contribution from several important elements including carbon. Moreover, it has been shown by several workers, including Doig *et al.* (1982), that the distribution of phosphorus at the grain boundary can be influenced by the presence of carbide precipitates. Murza and McMahon (1980) and Jin Yu and MacMahon (1980) show that neither Mn nor Si if present on its own causes embrittlement, but both strongly promote P segregation and increase the fraction of grain boundaries which absorbs P, rather than increasing the maximum P level achieved. Nakamura *et al.* (1979) showed that additions of either Mo or C to a $2\frac{1}{4}$Cr–Fe alloy retard P diffusion to grain boundaries and lower the saturation concentration of the P at boundaries. It is well established that many components fabricated from $2\frac{1}{4}$Cr–1Mo steel, when exposed to long periods of service such as those used for electricity generation, have changed mechanical properties owing to impurity and solute atom segregation which has the

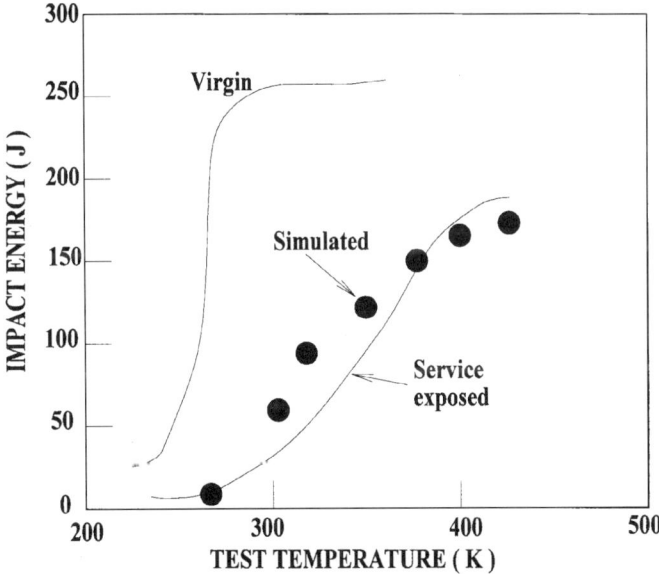

Figure 5.11 Charpy impact energy variation with temperature for $2\frac{1}{4}$Cr–1Mo steel after service exposure for 88 000 h at a nominal temperature of 813 K and accelerated simulation at a temperature of 873 K for 10,000 h (●) compared with the start of life property

opportunity to approach equilibrium conditions (Holdsworth and Thornton, 1993). Figure 5.11 shows the variation of Charpy impact energy for $2\frac{1}{4}$Cr–1Mo steel after service exposure for 88 000 h at a nominal temperature of 813 K. The data points represent samples of original material subject to an accelerated simulation heat treatment at 873 K for 10 000 h. The latter approach provides a tool to assess the contribution of thermal embrittlement (Wignarajah *et al.*, 1990).

5.5.2   3Cr–Mo, 1Cr–$\frac{1}{2}$Mo and Cr–Mo–V Steels

These low-alloy ferritic steels have been widely used as steam turbine components, at temperatures (up to 800 K), such as high-pressure and intermediate-stage rotors, so that in-service embrittlement is possible. Alternatively, these steels are used for low-pressure stage rotor discs which may enter service in a temper-embrittled condition owing to the slow cooling rate required a heat treatment to minimise the presence of high residual stresses in these thick-section components. Transition temperature shifts of over 100 K have been reported by Cheruvu (1989) in a Cr–Mo–V steel rotor after 200 000 h service, and shifts of up to 150 K have been reported for 3Cr–Mo steel turbine discs (King and Wigmore, 1973). The effect of Mn upon the extent of embrittlement, measured as a transition temperature shift, has been shown for these steels to vary with Mn concentration when compared with vacuum re-melted steel to remove manganese almost completely (Wigmore, 1973) (Figure 5.12). All the steels manufactured by the acid open hearth process had P levels of ~ 0.03 wt% whereas those manufactured by the basic electric arc process P ~ 0.01 wt%. A similar relationship between the manganese content and FATT is observed, but with the line displaced towards lower transition values for the higher-phosphorus steel.

Recently, Bulloch and Wild (1995) and Bulloch and Crowe (1998) have described an evaluation of 1Cr–$\frac{1}{2}$Mo steel turbine bolts after 209 500 h service at a temperature

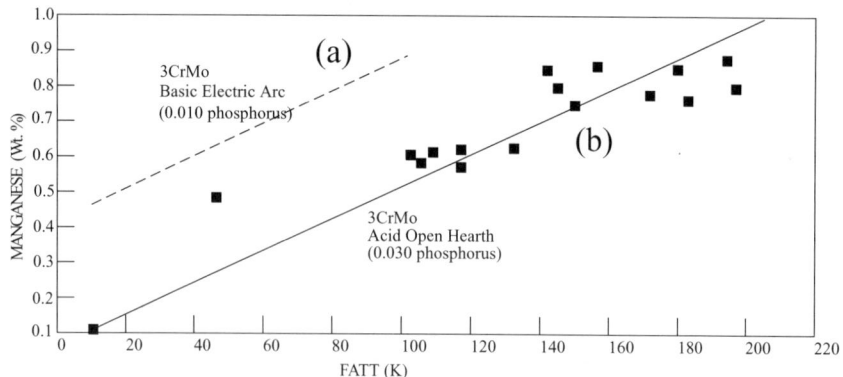

Figure 5.12   Effect of manganese concentration on the FATT for 3Cr–Mo steels: (a) a basic electric arc steel and (b) an acid open hearth steel (after Wigmore, 1973)

of 723 K. The bolts examined came from two populations, one containing 0.016 wt% and the other 0.032 wt% phosphorus with corresponding sulphur contents of 0.016 wt% and 0.04 wt% respectively. The Charpy impact energy data for these bolts after service are compared in Figure 5.13, and there is clearly an increase in the ductile-to-brittle transition temperature post-service, depending upon the bulk concentration of phosphorus. Indeed, measurement of the grain boundary concentration of phosphorus confirmed this to be enhanced by a factor of ~ 250 times that of the bulk. In an attempt to provide a diagnostic tool to evaluate fitness for continued service, a relationship between the average grain size of individual bolts and the bulk phosphorus was derived (Figure 5.14). In this figure, termed an embrittlement estimate diagram, distinct embrittled and non-embrittled regions are established, separated by a critical boundary defined by

$$Pd = 0.46 \qquad\qquad (5.20)$$

where P is the wt% phosphorus and d is the average linear grain size. Such empirically derived approaches can provide the operators of plant with a useful guide for assessing both the optimum time and the need to replace components during the operating life of such plant.

5.5.3 3Ni–Cr and Ni–Cr–Mo–V Steels

Much of the data for this family of steels comes from high-temperature turbine rotor materials (EPRI, 1983) where additions of Mn to a P-containing Ni–Cr–Mo–V steel

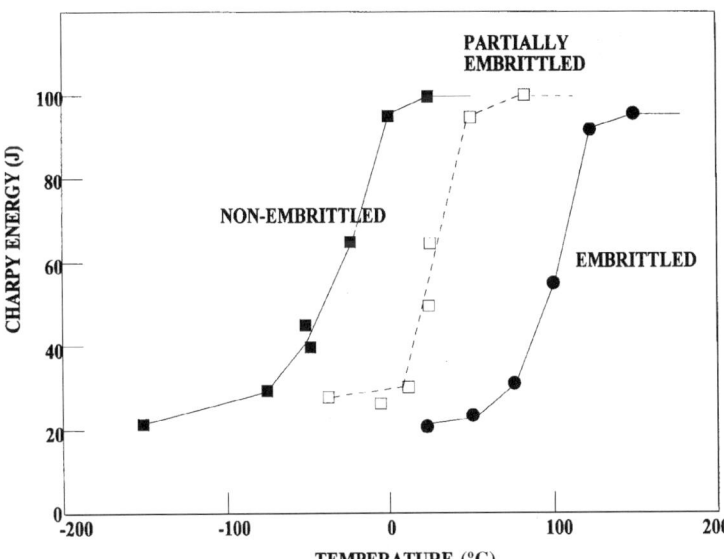

Figure 5.13 Charpy impact energy versus temperature trends for the various conditions of embrittlement of the bolts after 209500 h service at a temperature of 773 K (embrittled) (Bulloch and Crowe, 1998)

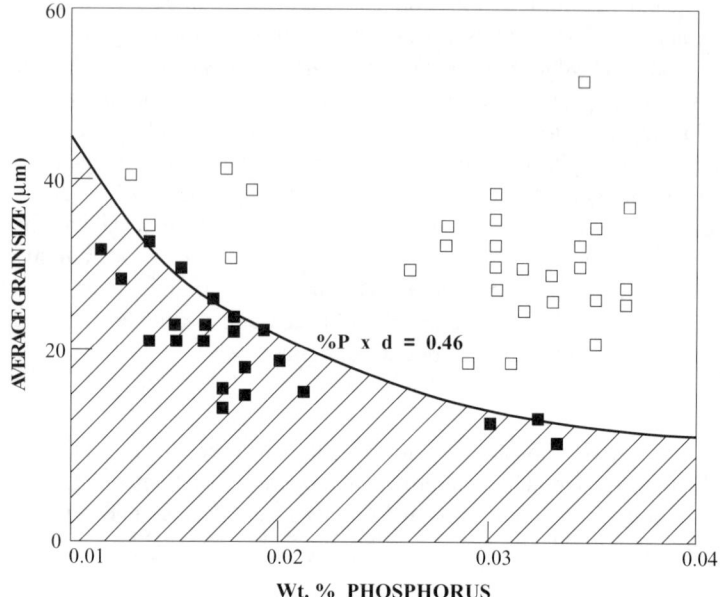

Figure 5.14 An embrittlement diagram showing grain size and bulk phosphorus trends for embrittled (■) and non-embrittled (□) bolts. The bonding line is given by Pd = 0.46 (Bulloch and Crowe, 1998)

produce a substantial shift in the FATT together with extensive intergranular fracture. This conclusion seems to have been equally applicable to Cr–Mo–V and Ni–Cr steels. In these cases, Mn is considered to promote intergranular fracture by increasing the fraction of grain boundaries that become embrittled by P. Smith *et al.* (1982) claim that Mn enhanced the kinetics of P segregation in $3\frac{1}{2}$Ni–Cr–Mo–V, whereas Cr and Mo additions retarded it. The effect of applied stress during thermal ageing of these steels promotes N as well as P or S segregation and hence intergranular fracture (Misra and Balasubramanian, 1990).

Ageing experiments have been undertaken on 4Ni–Cr steel (P 130 ppm, Sn 140 ppm, As 240 ppm) and Ni–Cr steels with added Sn (P 35 ppm, Sn 425 ppm, As ~ 50 ppm) have been subjected to prior temper embrittling heat treatments (Edwards *et al.*, 1980). Figure 5.15 shows that heat treatments for short periods are sufficient to effect grain boundary embrittlement, but there are subsequent further changes that divide broadly into three stages:

Stage 1. Here the DBTT for the 4Ni–Cr steel (Figure 5.15a) increases, with an attendant rise in the P, Ni and Sn concentrations at the grain boundaries. A similar trend occurs in the other Ni–Cr steel, but here the Ni and Sn levels are still increasing.

Stage 2. Within this stage for the 4Ni–Cr steel the P concentration at the grain boundaries remains essentially constant while the Ni and Sn levels fall and

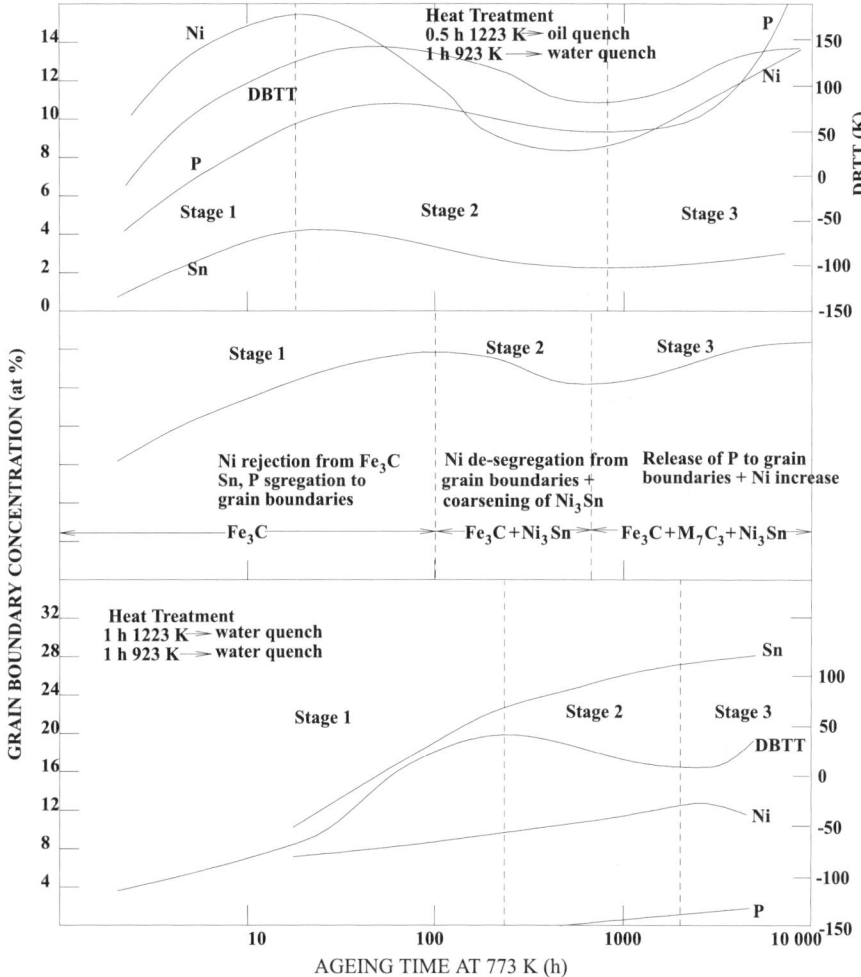

Figure 5.15 Effects of ageing at 773 K after quenching and tempering: (a) 4Ni–Cr steel, DBTT and segregate levels, (b) 4Ni–Cr steel, precipitate formation, (c) Ni–Cr steel with added Sn, DBTT and segregate levels (after Edwards *et al.*, 1980)

Ni$_3$Sn precipitates near to the grain boundaries. The levels of Ni and Sn on the grain boundaries at this stage correspond to the composition Ni$_3$Sn, leading to an additional contribution to the embrittlement. A decrease in the DBTT can be effected in the other NiCr steel by clustering of Ni and Sn, leading to a precipitation of Ni$_3$Sn$_2$ which is consistent with greater concentrations of Sn in this steel.

Stage 3. In this stage the DBTT again increases and, for the case of the 4Ni–Cr steel, there is a further rise in the P and Ni concentrations at these grain boundaries. For the Ni–Cr steel, which contains only 35 ppm P, there is a significant increase in this element at the grain boundaries which may be a consequence of site competition with Sn, cosegregation with Cr or rejection of P from the M_7C_3-type carbide precipitates. Clearly, the higher concentration of phosphorus at grain boundaries of both of these steels has a significant effect on the DBTT. This example shows the complexity of the tempering response of these steels and, indeed, many other steels to embrittling heat treatments.

5.6 HIGHER TEMPERATURE AND LIQUATION EMBRITTLEMENT

There are several phenomena considered in the heat treatment and fabrication of steels that are related to temper embrittlement, enhancing intergranular fracture and thereby reducing the mechanical properties of the material. Certainly, many ferritic alloy steels held in the temperature range 1473–1673 K and then quenched and tempered fail along prior austenite grain boundaries. This is often associated with segregation of sulphur to these boundaries at the higher temperatures, and the effect is removed if the S content is reduced to $\leqslant 0.002$ wt%. This embrittling effect can raise the DBTT by over 200 K, but is not observed with the presence of alloying elements such as Mn and Cr. It is argued that manganese lowers the solubility of S in both the austenite and ferrite phases and therefore, when S segregates to the prior austenite grain boundaries during a high-temperature heat treatment, MnS precipitate particles of $\sim 0.5\,\mu m$ in diameter are formed. Thus, grain boundary fracture arises from nucleation of cracks at these sulphide precipitates and the fracture mode is a function of their size and distribution which, in turn, is controlled by the rate of cooling from austenitising heat treatment temperature. At high austenitising temperatures, ~ 1673 K, extensive MnS precipitates grow, often in a dendritic form. In extreme cases, the partial formation of a liquid phase occurs (liquation) which, on subsequent heat treatment, greatly accentuates the intergranular fracture. In the absence of manganese in materials such as a simple wrought iron, liquid films of the iron–iron sulphide eutectic may cause embrittlement during hot working processes at temperatures down to 1273 K, and this is known as hot shortness. The phenomenon is important for high-temperature working processes such as forging and the likelihood is reduced by adequate temperature control. Again, in cast steels examination of the fracture surfaces reveals extensive sheets of manganese sulphide at grain boundaries, often only 0.2–0.5 μm thick, covering large areas. Marked embrittlement can occur in this 'as-cast' state or after subsequent heat treatment of the cast product in the range 773–923 K; this phenomenon is often referred to as cast brittleness. Liquation cracks can also occur during fusion welding of ferritic and austenitic steels either in the heat affected zone of the parent material or in previously deposited weld metal reheated by the subsequent weld run (Hansworth *et al.*, 1969; Robinson and Scott, 1980).

Equally, microfissuring in some nickel-base superalloys such as Inconel 718 is associated with constitutional liquation of grain boundary precipitates such as carbide, Laves and σ phase precipitates (Savage, 1980; Vincent, 1985). In addition, this phenomenon has also been attributed to the segregation to grain boundaries of boron and sulphur (Huang *et al.*, 1996; Thompson *et al.*, 1986). Certainly, Guo *et al.* (1999) have shown that there is a close relationship between intergranular liquation and boron concentration at grain boundaries in simulated heat affected zone microstructures of Inconel 718.

5.7 CREEP AND CREEP FATIGUE

5.7.1 Introduction

Creep deformation at high temperatures in engineering structures is of unquestionable technological importance, and there have been many attempts to explain this phenomenon both empirically and mechanistically. As a consequence, creep in metals and alloys has been reviewed by many authors and for a detailed description and the underlying mechanisms the reader is referred to the work of Nabarro and de Villier (1995), Pugh (1991), Evans and Wilshire (1985), Cocks and Ashby (1979) and Gittus (1978). In this section we briefly describe the effects of grain boundaries and their nanochemistry on creep deformation and overall life.

As the applied stress is increased on a material, so the strain increases linearly until the elastic limit is reached, after which the material either deforms by plastic flow or fails by fracture. The mode of deformation is determined by several factors, the temperature and the strain rate being among the most important, and the deformation maps produced by Ashby *et al.* (1979) and Gandhi and Ashby (1979) schematically describe the deformation regime for a series of materials under a wide range of conditions. At elevated temperatures, if the stress is maintained at a value below the elastic limit for a period of time, then the strain in the material will continue to increase, the phenomenon known as creep. The temperature that defines the onset of creep varies with the material, but in general significant creep occurs at temperatures above $\sim 0.3\ T_m$, where T_m is the melting point (K). Here, the steady state creep rate, $\dot{\varepsilon}$, is usually related to the applied stress, σ, by an equation of the form

$$\dot{\varepsilon} = K\sigma^n \tag{5.21}$$

where K is a constant incorporating the dependence on temperature and n is the stress exponent.

Creep can occur by deformation of the grains within a polycrystalline material as a result of diffusion of atoms in the matrix, and many methods have been devised to reduce the creep rate by, for example, solid solution strengthening, dispersion strengthening by inert particles or precipitation strengthening by coherent particles.

However, in a polycrystalline material, creep can also occur by the sliding of the grain boundaries, and methods have been developed to restrict this contribution. A common limit to creep life arises from failure of the grain boundaries under either net section loading or intergranular crack growth. Under these conditions, overall ductility is low and fracture results from the nucleation and growth of cavities at the grain boundaries and, in particular, those oriented normal to the axis of the maximum principal stress. Service failures of components and structures operating in a creep regime are often associated with weldments and, in particular, their heat affected zone where larger grain sizes, high residual tensile stresses in addition to the system stresses and higher residual impurity element concentrations on grain boundaries promote the intergranular fracture.

Creep in ceramics is similar to that in metals and alloys, again occurring at temperatures above $\sim 0.3\ T_m$ (Dokko *et al.*, 1977). The creep rate, $\dot{\varepsilon}$, of mullite, an Al_2O_3/SiO_2 ceramic, under steady state conditions is given by the complete form of equation (5.21) (Cannon and Langdon, 1983):

$$\dot{\varepsilon} = \left(\frac{KDGb}{kT}\right)\left(\frac{\mathbf{b}}{d}\right)^m \left(\frac{\sigma}{G}\right)^n \tag{5.22}$$

where K is a constant, D is the self-diffusion coefficient, G is the shear modulus, $\mathbf{b}$ is Burgers vector, k is the Boltzmann constant, T is the absolute temperature, d is the grain size, m is the inverse size exponent, σ is the stress and n is the stress exponent.

When the grain size has no effect on creep, m is 0, if grain boundary sliding is dominant then $m = 1$, if creep is controlled by intergranular Herring–Nabarro diffusion then $m = 2$ and for grain boundary diffusion $m = 3$. In the case of mullite, m has values in the range 1.25–2.5 (Schneider *et al.*, 1994) where creep is controlled by grain boundary sliding combined with diffusion and cavitation (Ashizuka *et al.*, 1989; Nixon *et al.*, 1990). Mullite can be strengthened by the incorporation of ZrO_2 in the form of small particulates in the bulk and on grain boundaries (Figure 3.35) (Shiga *et al.*, 1990; Rundgren *et al.*, 1990).

It is important in creep to evaluate the overall response of the material to establish creep life and, as a consequence, it is necessary to understand the role of the grain boundaries and their interrelationship with intergranular crack growth at elevated temperatures. Certainly, the characteristic features of grain boundary failure at elevated temperatures have been known for nearly 100 years (Huntington, 1912). For example, copper deformed in tension has a minimum reduction in area at temperatures between 573 and 773 K which is associated with a change from transgranular to intergranular failure. However, this embrittlement of the grain boundaries is influenced by the presence of impurity atoms within the material. Another example is the embrittlement of gold by small additions of bismuth (Arnold and Jefferson, 1896). Intergranular fracture at high temperatures is not restricted simply to creep but embraces a range of closely related phenomena such as creep fatigue, stress relief and reheat cracking. In this section we will deal with creep and creep fatigue and related mechanisms, all of which have a major technological importance for metals, alloys and ceramics.

5.7.2 Mechanisms

CREEP DEFORMATION

The increased ability of the atoms to diffuse facilitates plastic deformation at temperatures above $\sim 0.3\ T_m$. As shown schematically by the strain versus time curve obtained for constant uniaxial loading (Figure 5.16), a material will initially extend elastically but at stresses below the yield point will continue to deform. Immediately after the elastic region, creep is relatively fast. This is known as primary creep and is the result of the relative ease with which unpinned dislocations may move before becoming pinned by obstacles or other dislocations. There then follows a stage where creep is relatively uniform but slowly and can represent a considerable fraction of the total life. During this stage, recovery occurs at the same rate as hardening and this balance usually leads to the formation of stable dislocation subcells of a size inversely proportional to the magnitude of the applied stress. Over a period of time, the hardening processes may become less effective, and this combined with the loss of net section due to strain accumulation leads to an increase in the creep rate. In addition, cavities may form on grain boundaries which progressively interlink, and, again, when they are present in sufficient numbers the creep strain accelerates. These factors produce the third stage of creep (tertiary creep) where the strain rate increases rapidly with time until the material fails. Essentially, within polycrystalline materials there are at least six separate mechanisms that can effect deformation, and these mechanisms involve the motion of lattice defects such as vacancies or dislocations either within the grains or around the grain

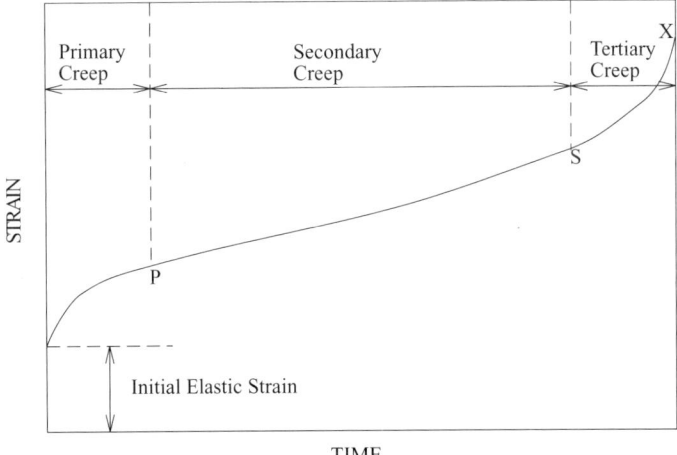

Figure 5.16 A schematic diagram showing strain versus time curve obtained for a uniaxial loaded test specimen subject to a constant loading; P is onset of secondary creep, S is onset of tertiary creep and X is point of failure

boundaries. Ashby has devised maps (see Chapter 1 and Section 5.1) that show the range of stress and temperature where one of the steady-state creep deformation processes dominates. Figures 5.17a and b are typical deformation maps, in this case for iron (Figure 5.17a) and copper (Figure 5.17b), which show the main mechanisms copper of polycrystalline deformation.

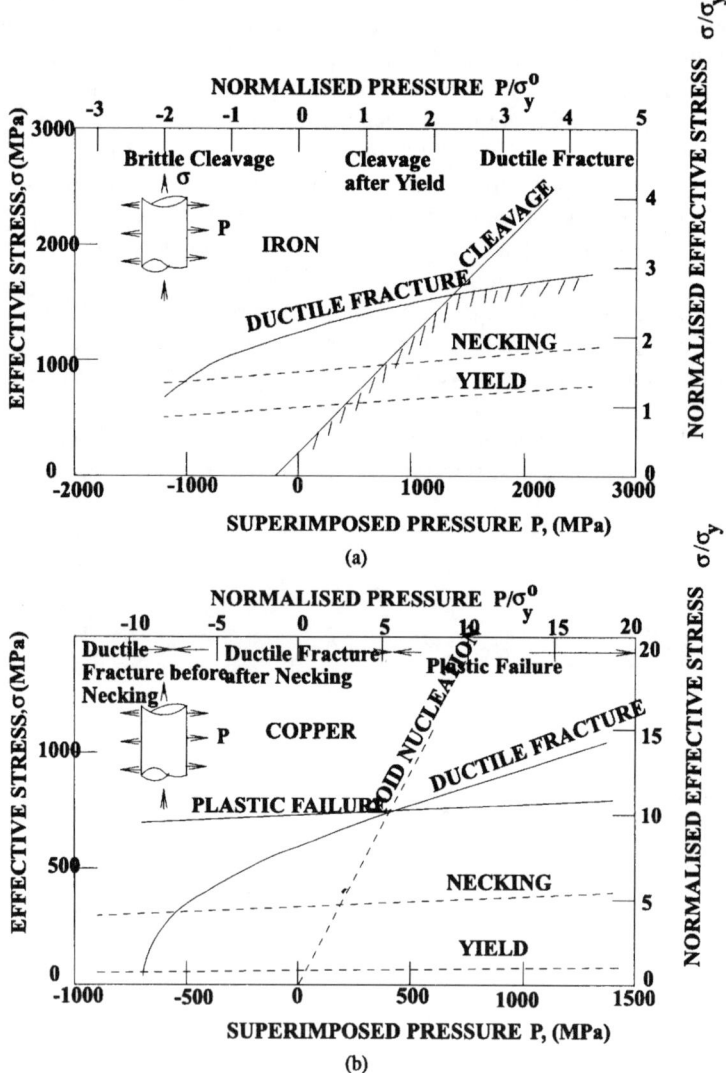

Figure 5.17 Maps for uniaxial loaded specimens showing the principle deformation mechanisms in polycrystalline (a) iron and (b) copper

For plastic flow to occur at stresses below the theoretical shear stress, it is necessary for the stress to be sufficient to allow climb of planes of atoms in a crystal. By contrast, dislocation glide is an easy process, particularly in fcc and hcp metals, that occurs at very low stresses, but, in the presence of obstacles such as polycrystalline grain boundaries, other dislocations and solute or impurity atoms, this becomes more difficult. An external stress applied to a polycrystal can also direct the diffusion of point defects either through a crystal or around a grain boundary path. Nabarro and Herring independently described creep by the diffusion through grains. However, in a pure metal, vacancies can also flow around grain boundaries, causing them to elongate in the direction of the applied stresses. Ashby addressed the problem of sliding at non-planar grain boundaries where diffusional flow of matter occurs from point to point on a boundary and showed that the creep rate, $\dot{\varepsilon}$, is given by

$$\dot{\varepsilon} = 14 \left(\frac{\sigma \Omega}{kT} \right) \left(\frac{D_V}{d_g^2} \right) \left(1 + \frac{\pi d_{gb} D_{gb}}{d_g \, D_V} \right) \tag{5.23}$$

where Ω is the atomic volume, d_g is the grain size, D_V is the bulk diffusion coefficient, D_{gb} is the grain boundary coefficient and d_{gb} is the effective cross-section of a grain boundary for diffusional transport. For a large grain size material, the term $\pi d_{gb} D_{gb}/d_g D_V$ is very much less than unity so that diffusion around a grain boundary path is small and, therefore, Herring–Nabarro creep dominates. Higher temperatures reduce the ratio of the grain boundary to bulk diffusion, D_{gb}/D_V, again favouring this creep process. Hence, there is a separation of creep due to grain boundary diffusion (Coble creep) from that due to lattice diffusion (Herring–Nabarro creep). Certainly, for the Coble process local changes in grain boundary composition become important by modifying the grain boundary diffusion coefficient. Coble creep has the activation energy of grain boundary diffusion which may be as low as half that for the bulk so that, as a mechanism, it becomes more important at lower temperatures. The range of behaviour is summarised in the deformation mechanism maps by considering grain size and applied stress as variables at a given temperature. As a result of the large inverse grain size dependence of diffusional creep, it takes over from dislocation creep at smaller grain sizes, $\leqslant 10 \, \mu m$. The feature that distinguishes diffusional creep is that the strain rate is directly proportional to the applied stress.

In addition to the mechanisms for creep deformation of grains at elevated temperatures there are contributions from grain boundary shear, boundary migration and high rates of diffusion along the grain boundaries. Strain accommodation between grains and, in particular, at grain corners is normally achieved by a combination of grain boundary sliding, vacancy diffusion and dislocation climb (Figure 5.18). Grain boundary sliding and accommodation occur on an atomic scale, where the creep deformation tends to be confined to grain boundaries, thereby promoting the sliding mechanism (Murr, 1995). Equation (5.23) applies when considering the relative contributions to this process from volume and grain

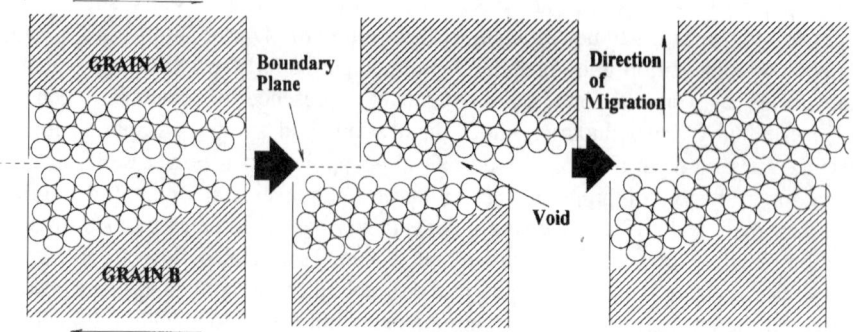

Figure 5.18   Atomic model of grain boundary sliding, showing schematically how accommodation may occur (after Murr, 1975) as sliding between grain A and grain B develops

boundary diffusion which are a function of the stress, grain size and temperature. In materials that fail in a creep brittle manner, the fracture path follows the grain boundaries where damage is initiated from second-phase particles or precipitates and then propagates under the control of the cohesive energy, which is a function of the impurity concentration at the boundary, or, if sliding occurs, the damage can nucleate at discontinuities in the grain boundary such as ledges or triple point junctions (Figure 5.18). At stresses lower than within the grains, plastic flow is able to occur in regions close to the grain boundaries as a result of local stress concentration. The relationship between sliding creep and stress in polycrystals has been described by Hart (1967) and extended by Crossman and Ashby (1975). Essentially, the proportion of the strain resulting from grain boundary sliding decreases as the stress increases. At higher temperatures, specimens with a larger grain size have a lower creep rate than specimens with a smaller grain size. In addition to grain size, other parameters that influence the creep properties (Wilcox and Clauer, 1972) include the grain geometry, and here the ratio of the grain size parallel to and normal to the tensile stress direction is important. Elongated grains, aligned parallel to the direction of the applied stress, will fracture normal to their axis, while shorter-aspect grains will pull out. As a result, polycrystalline materials with grains that have a high aspect ratio have higher creep strength than those with a low aspect ratio (Figure 5.19).

Grain boundary sliding can be inhibited if the boundary can be strengthened. This can be achieved by heat treating the material either to form particles on, or to change the local composition of, the boundary. In particular, in steel and nickel-base superalloys the formation of carbide precipitates containing chromium, molybdenum, titanium, etc. significantly reduces grain boundary sliding. The creep properties of non-metals such as ceramics and glasses remain of considerable importance. Figure 5.20 summarises the deformation maps for a range of non-metals (McLean, 1968), and these show that a small change in temperature produces a large change in the stress needed to generate a flow rate of 10^{-7} s^{-1}. By contrast, the temperature

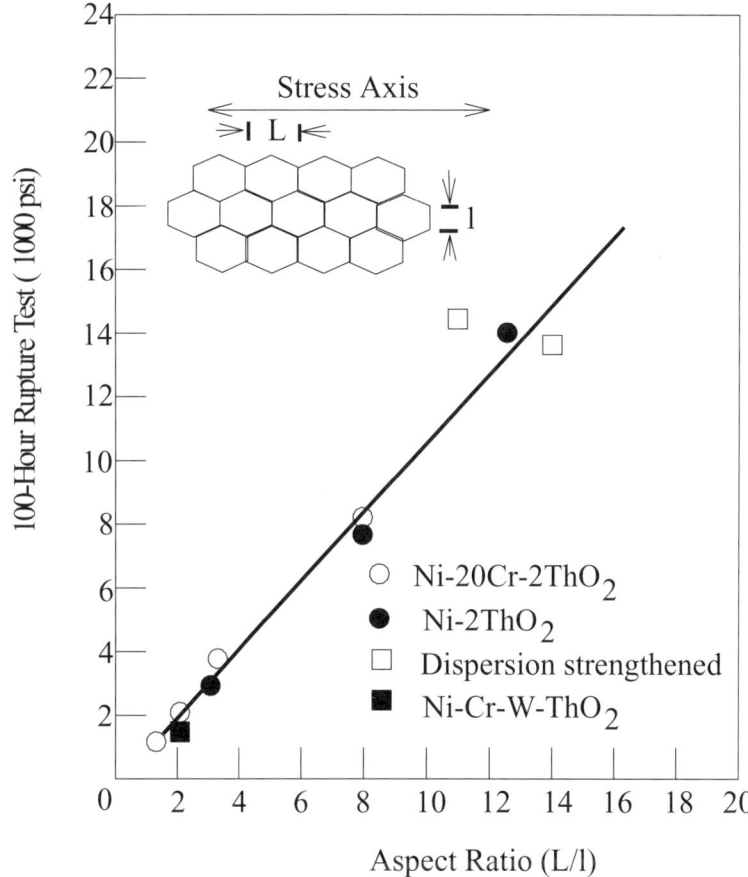

Figure 5.19 Effect of the grain aspect ratio L/l in dispersion strengthened nickel alloys on the strength properties at 1373 K in terms of the 100 h rupture strength (after Wilcox and Clauer, 1972)

dependence of the flow stress in metals is comparatively small since the activation volume is significantly different from that for non-metals (Cottrell, 1964).

CREEP CAVITATION AND STRESS RUPTURE

During creep at higher stresses materials have generally good ductility, whereas at lower applied stresses failure occurs with poor ductility. For many years it has been recognised that the composition of the grain boundary can significantly influence creep ductility. In addition to the wedge-type cracking described in the previous section, brittle fracture at high temperatures is frequently associated with the

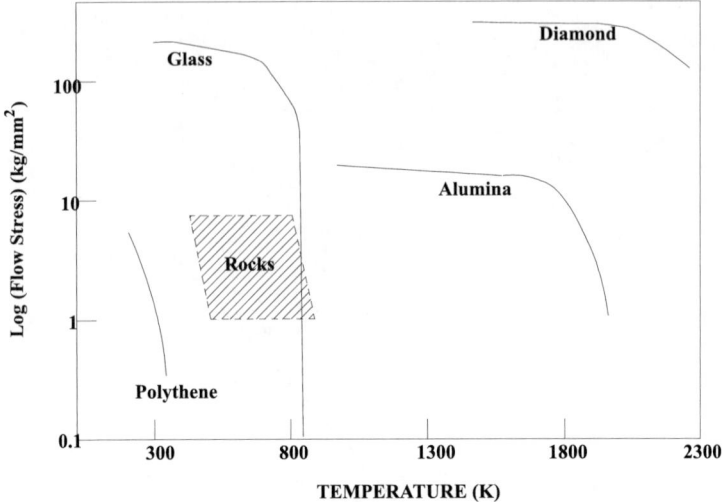

Figure 5.20 Summary of deformation map for a range of materials (after McLean, 1968) excluding metals

nucleation and growth of small cavities on the grain boundaries oriented normal to the maximum principle stress. In their classic paper, Hull and Rimmer (1959) described the growth of cavities at grain boundaries (Figure 5.21), making a number of assumptions:

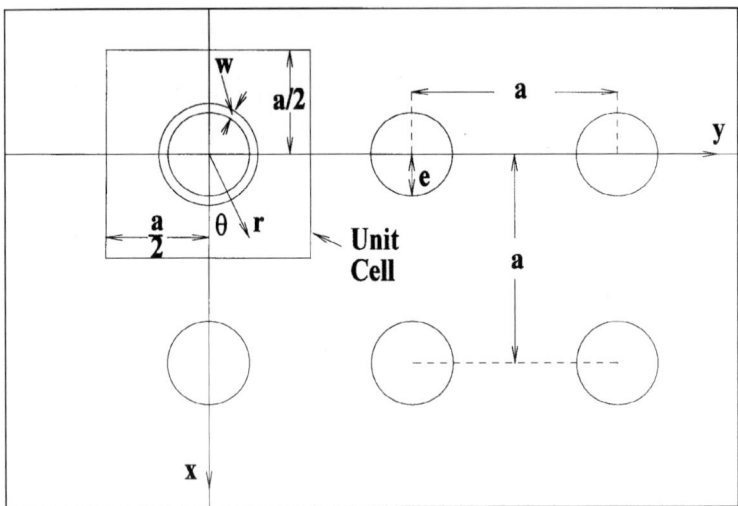

Figure 5.21 Growth of cavities at grain boundaries. Square array of spherical voids with a repeat distance, c, having grown to a radius r, lying on a grain boundary normal to the direction of uniaxial stress σ (Hull and Rimmer, 1959) [stress axis is normal to plane of page]

(a) Spherical cavities are maintained by surface diffusion.
(b) The grains are regarded as blocks of material.
(c) Diffusion is slower in the grain than within the grain boundary.
(d) The grain boundary is a perfect source of vacancies.
(e) Vacancy diffusion is described by classical diffusion theory.
(f) The boundary stress at the cavity is equal to $2\gamma/r$, where r is the radius and γ is the surface energy.
(g) There is no restraint affecting grains moving apart.

Since then, the basic model has been refined to take account of a range of additional factors, summarised in Figure 5.22. However, the simplified calculation based upon the original Hull and Rimmer approach but assuming circular geometry, gives the rate of change of cavity volume, v_c, as

$$\frac{\mathrm{d}V_c}{\mathrm{d}t} = \frac{8\pi D_g \delta \Omega}{kT}\left[\frac{(\sigma - 2\,\gamma/r)}{4\ln(c/r) - (1 - r^2/c^2)(3 - r^2/dc^2)}\right] \tag{5.24}$$

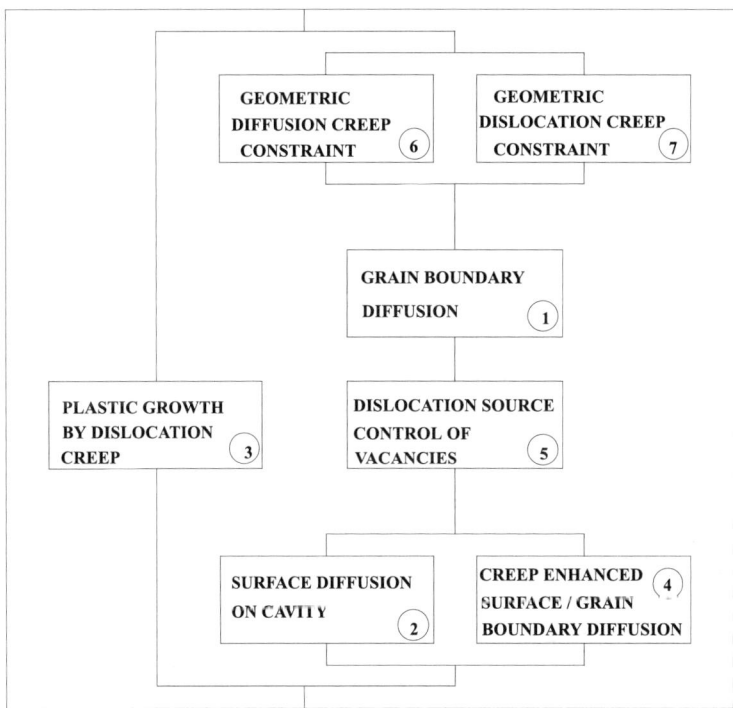

Figure 5.22 Sequential and independent relationships between growth mechanisms. Independent mechanisms act in parallel, sequential mechanisms act in series. The numbers refer to equations (Beere, 1981). [Numbers refer to equations in Table 5.2]

where t is the time, σ is the applied uniaxial stress, r is the cavity radius and $2d_c$ is the cavity spacing; the remaining terms have been defined earlier. The rate of growth of cavities is often many orders smaller than given by this simple equation, mainly because one or more of the basic assumptions have been invalidated. Figure 5.22 and Table 5.2 summarise the basic equations for the various assumptions and modifications to relate this process of cavity growth to practical applications in metals and alloys (Beere, 1981). Here, independent mechanisms are represented by parallel circuits and sequential mechanisms by series circuits. When a mechanism is regarded relative to the other mechanisms, it has low impedance. Thus, the diagram

Table 5.2   Cavity growth equations

(1)   Diffusion control

$$\dot{v}(1) = \frac{B(S-2)(1+2cD_v/\pi\delta D_g)}{2v^3 q\varDelta}$$

(2)   Surface/grain boundary diffusion

$$\dot{v}(2) = \frac{2\sqrt{2}B}{v^3}\left\{\frac{(S-1-v^2/c^2}{(1-v^2/c^2)\left[1+\sqrt{\frac{1+4\sqrt{2q\varDelta(S-1-v^2/c^2)}}{(1-v^2/c^2)^3}}\right]}\right\}^3$$

(3)   Plastic growth of spheres by dislocation creep

$$\dot{v}(3) = \frac{B}{v^3}\frac{\alpha^2}{2}$$

(4)   Crack-like cavities growing by creep enhanced diffusion

$$\frac{v^3\dot{v}(4)}{B} + \left[\frac{v^3\dot{v}(4)}{B}\right]^{1/3}\left(\frac{3}{4}\alpha\varDelta\right) - \left(\frac{3}{4\sqrt{2}}\alpha S\varDelta\right) = 0$$

(5)   Vacancy source control

$$\dot{v}(5) \geqslant \frac{\pi}{8}\frac{c^2}{v^2}\lambda\dot{\varepsilon}_1$$

(6)   Geometric diffusion creep constant

$$\dot{v}(6) = \frac{10}{\pi}\frac{c^2}{v^2}\frac{\sigma\Omega D_v}{dkT}\left(1+\frac{150D_g\sigma}{\pi\,dD_v}\right)$$

(7)   Geometric dislocation creep constraint

$$\dot{v}(7) = \frac{1}{2}\frac{c^2}{v^2}\frac{d\dot{\varepsilon}}{1}\exp\left[\frac{(n-1)\ln 10}{5}\right]$$

v = cavity radius; $B = D_s\gamma\Omega^{4/3}/kT$; $2c$ = cavity spacing; d = grain size; D_g = grain boundary self-diffusion coefficient; D_s = surface self-diffusion coefficient; D_v = volume self-diffusion coefficient; k = Boltzmann constant; n = creep exponent; $q = \ln(c/v) - (3 - v^2/c^2)(1 - v^2/c^2)$; R = gas constant; $S = \sigma v/\gamma$; T = temperature (K); $\alpha = v^2\sqrt{\dot{\varepsilon}/B}$; γ = surface tension; δ = grain boundary width; $\varDelta = D_s\Omega^{1/3}/D_g\delta$ = uniaxial macroscopic creep rate; $\dot{\varepsilon}$ = creep rate local to cavity; λ = precipitate spacing; σ = applied uniaxial stress; Ω = atomic volume.

is analogous to an electrical current; one mechanism controls when the impedance is a factor in the impedance of the matrix. The overall growth rate can be found approximately by treating the reciprocal of cavity growth rate as the impedance.

The effect of grain boundaries on the overall creep rate and life varies significantly between materials. Pure metals show little effect of grain size on creep rate, which leads to the conclusion that grain boundaries contribute little under these conditions (Parker and Wilshire, 1980). However, in precipitation strengthened alloys, the effect of grain boundaries on creep can be very marked (Wilcox and Clauer, 1972) because, while dislocations are pinned effectively within the grain, it is easier for rearrangement and annihilation of dislocations to occur at the boundaries. This view is supported by the fact that there is an increase in creep resistance with particle size, where the larger particles act as more efficient obstacles to sliding. Failure during creep will ultimately occur as a result of the increase in the number and size of cavities at grain boundaries. Chen and Argon (1981) plotted the variation in cavity concentration and cavity diameter at various points on the creep curve for a Type 304 austenitic stainless steel. Here, cavities increase by approximately 50% in diameter and an increase in concentration by a factor of 15 from the start of the linear steady state condition to the onset of tertiary creep (Figure 5.23). Cavities may be nucleated at grain boundaries by the presence of inclusions (Svoboda and Sklenicka, 1990). The critical stress for thermally activated cavity nucleation, σ_n, has been estimated to be (Riedel, 1987)

$$\sigma_n = \left(\frac{\gamma_s^3 f_v}{10kT} \right)^{1/2} \tag{5.25}$$

where γ_s is the surface energy of the cavity, f_v is a shape factor comparing the cavity volume with the cavity radius, k is the Boltzmann constant and T is the absolute temperature. However, cavity nucleation is not readily observed in pure polycrystalline materials with 'clean' grain boundaries. For cavities to form there has to be some form of elemental segregation so that cavity nucleation and composition are linked by the process that drives both vacancies and impurity elements to the grain boundaries.

5.7.3 Role of Local Composition

It is acknowledged that the creep life of materials can be changed by the presence of different impurity elements in a range of ferrous and non-ferrous metals and alloys (Lonsdale and Flewitt, 1979) and ceramics. In general, creep life is extended by reducing the overall concentration of the impurity elements and/or minor alloying elements that segregate to the grain boundaries. The local composition modifies the surface energy in the relationship between stress σ and the critical size for the cavities, σ_c, which defines the condition for cavity nucleation $\sigma_c = 2\gamma/\sigma_c$ [see equation (5.14)]. Thus, segregation to the surface of grain boundary cavities also becomes important since under these circumstances there is both an interface, the

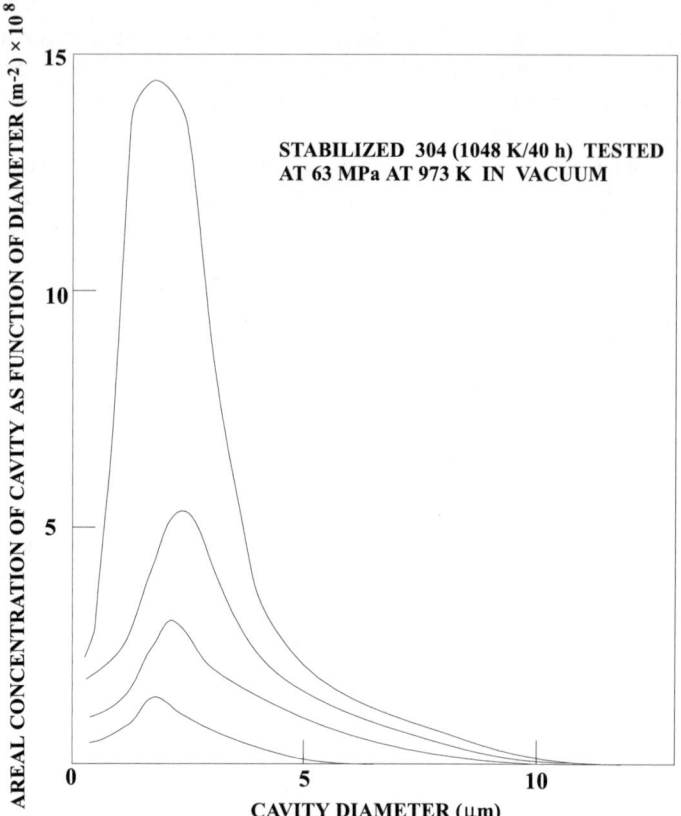

Figure 5.23 Variation in cavity concentration and cavity diameter at various points on the creep curve for Type 304 stainless steel (Chen and Argon, 1981)

grain boundary, and a free surface, the internal surface of the cavity. The build-up of segregated atoms to the surface of a spherical cavity is faster than to a flat surface because of the larger volume of material available to supply the segregating atoms. Seah (1976), extensing an existing analysis, gives the segregation level in mol/m^2, Γ_S, to be

$$\Gamma_S = C_i[(4Dt/\pi)^{1/2} + (Dt)/r] \tag{5.26}$$

where C_i is the impurity concentration, D is the bulk diffusion coefficient, r is the cavity radius and t is the elapsed time.

Indeed, if the kinetics of surface segregation are sufficiently rapid, the decrease in the surface energy of the cavity can be more rapid than the reduction in the cavity radius by sintering at a given stress. Hence, any cavity initially greater than the critical size will stabilise and grow and contribute to creep failure. Therefore, provided the segregating elements are sufficiently surface active, surface kinetics dominates. There have been several observations of grain boundary cavities having a different surface composition to that of the adjacent grain boundary (Wild, 1997).

Furthermore, for a grain boundary precipitate, surface segregation can affect the rate of cavity nucleation when nucleation occurs at the interface between the precipitate and matrix phase. Under the applied stress, σ, this interphase boundary decoheres and a cavity with a large initial radius starts to grow and, in the process, the radius, and hence the surface energy, decreases. Again, the cavity will achieve a stable radius for the given applied stress and thus will be influenced by the impurity atom concentration originally at the interphase boundary, but subsequently by any additional segregation within the surface of the cavity. Certainly, for minor alloying or impurity elements in materials generally, there is sufficient time to enable elements to redistribute. Indeed, it has been shown on several occasions that the interface local to carbide precipitates on grain boundaries in ferritic steels has a different composition to the adjacent grain boundary on which they are distributed (Lonsdale and Flewitt, 1979; Seah, 1980a). An example of this mechanism was provided by Rellick and McMahon (1974) who showed that phosphorus enrichment weakens the ferrite–carbide interface. Other mechanisms have been proposed that also fall into the category of reducing the cohesive energy. For example, changes in grain boundary composition arising from segregation can alter the cavity density by modifying the density of second-phase MnS particles (Figure 5.24) (Middleton, 1980) where a gain in the grain boundary energy promotes the nucleation of MnS particles. Cavities can nucleate by sliding on the boundaries between grains which induces stress concentrations at any grain boundary second-phase precipitate particles. The more slowly the stresses are relaxed by either diffusion or plasticity, the greater the propensity to cavity nucleation. Since the relaxation near to precipitate particles will be controlled by interfacial diffusion, a decrease in the grain boundary self-diffusion will produce high nucleation rates for the associated cavities or voids (Angus *et al.*, 1980).

Changes to the nanochemistry of grain boundaries have a significant effect on the self-diffusivity for this localised region of the microstructure, and this can thereby modify the growth rates of creep cavities (Wilkinson *et al.*, 1980). Certainly, an impurity atom effect on grain boundary diffusion in a gold–tantalum alloy has been observed (Gupta, 1977). Grain boundary segregation in polycrystalline materials can lead to a decrease in the binding energy of atoms at the boundaries, γ_{gb} (Seah, 1977).

$$\gamma_{gb} = \frac{kT}{\alpha a^2 N_m} \ln\left(\frac{D_{gb} d_{gb}}{a D_L N_m}\right) \tag{5.27}$$

where D_L is the lattice diffusion, N_m is the number of atomic layers of the grain boundary, a is the interatomic spacing and α is a parameter equal to 1 for interstitial diffusion and 2 for vacancy diffusion. For vacancy diffusion this reduces to

$$\frac{D_{gb}}{D_L} = \frac{a N_m}{\delta} \exp\left(\frac{2\gamma_{gb} a^2 N_m}{kT}\right) \tag{5.28}$$

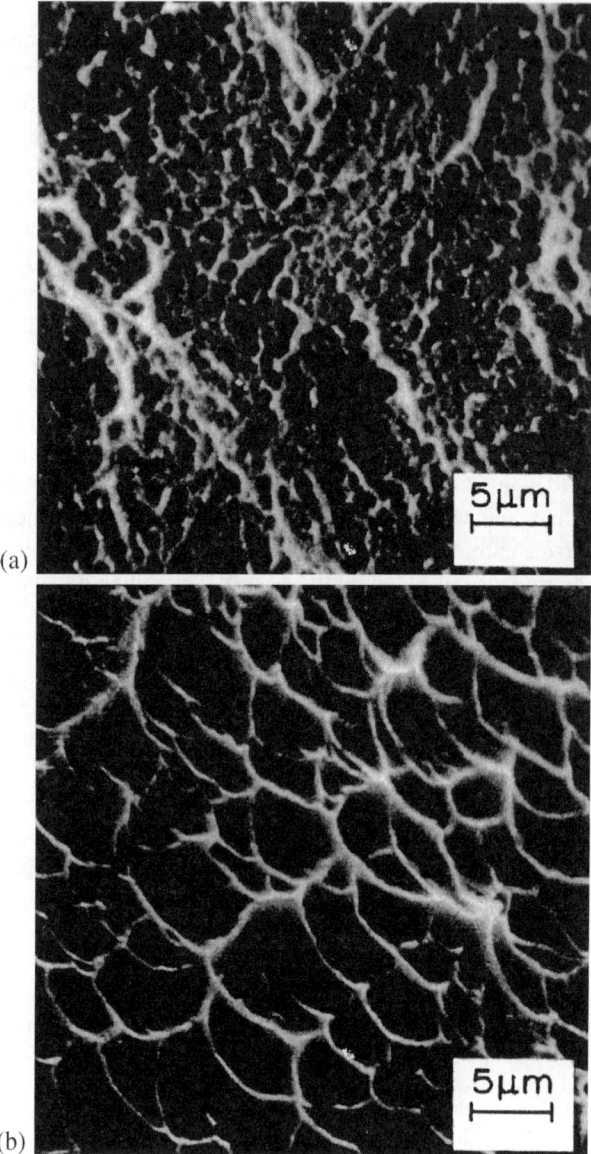

(a)

(b)

Figure 5.24 Influence of surface segregation on the rate of cavity nucleation when nucleation occurs at the interface between the precipitate and the matrix phase. Reproduced by permission of the Institute of Materials from Middleton, C. J. (1981) *Metal Sci.*, **15**, 154

If the segregated elements improve the grain boundary order, reducing γ_{gb}, they will also reduce the self-diffusivity for the grain boundary, and this prediction is supported by experimental observation (Gupta, 1976). Hence, the effect of grain boundary composition on the time to creep rupture can be derived from the relationship (Skelton, 1975)

$$t_r = (aG^{-1})^{2/5} N^{-3/5} \tag{5.29}$$

where G is given by $2\pi D_{gb}d_{gb}d_g\Omega/kT$, D_{gb} is the grain boundary self-diffusion, d_{gb} is the width of the boundary, Ω is the atomic volume and d_g is the grain size.

In the early stages, the grain boundary cavities are usually of a crystallographic shape so that any segregation of impurity or solute atoms will be distributed over the surface in relation to specific crystallographic planes of each facet. Indeed, the presence of the particular atom species will modify and influence the formation of the cavity facets by local changes to the surface energy. When creep cavitation is extensive, since material is transferred between boundaries, extra layers of atoms will form transverse to the stress axis and additional segregated atoms are distributed within these atoms. However, as pointed out by Yu *et al.* (1983), only one of the layers will contain these segregated atoms, since those remaining will achieve a regular crystal structure related to the adjacent grains.

5.7.4 Creep Crack Growth

Not all materials readily form cavities at high temperature and in such materials a crack may form and grow by the motion of vacancies. Impurity and minor element segregation can modify the rate of growth of the cracks which extend by a diffusion-controlled mechanism (McMahon, 1975). For an intergranular crack subjected to a normal stress growing by a diffusion mechanism (Figure 5.25) the atoms to the left of the tip diffuse over the crack surface and then along the grain boundary, driven by the stress normal to the boundary. The net driving force is $(\sigma - 2\gamma_{SV})\Omega/r$, where γ_{SV} is the solid to vapour surface free energy, r is the crack tip radius, Ω is the atomic volume and θ is the angle between the crack surfaces at the crack tip. When local equilibrium is established by surface diffusion, then r and θ at the crack tip will be governed by the local composition. Certainly, Hondros (1965) has shown that $\gamma_{gb}/2\gamma_{SV}$ increases with solute atom segregation to grain boundaries, leading to a lowering of the angle θ. As a consequence, the crack tip becomes sharper and the crack grows further for a given amount of atom transport (Figure 5.25). The resulting sharper crack will grow faster by the diffusion-controlled process.

In materials that do form cavities, a crack may nucleate and then extend by interlinking cavities or bypassing between cavities. Cavities are more likely to form on grain boundaries that are aligned normal to the applied stress, whereas shear stresses may allow cracks to form on grain boundaries inclined at a low angle to the applied stress. The deformation will therefore continue by shear on grains aligned along the stress axis but by passing through cavities on those aligned normal to the

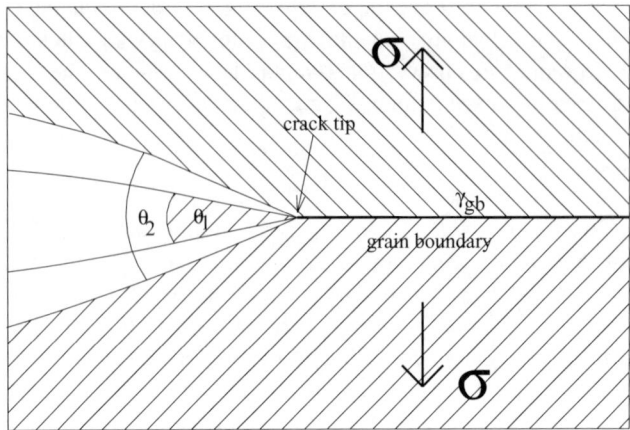

Figure 5.25 Formation of an intergranular crack growing by a diffusion mechanism. Increasing with the stress axis vertical solute atom segregation to the grain boundary lowers θ i.e. $\theta_1 \rightarrow \theta_2$

stress axis. For this reason, many observations of intergranular fracture surfaces show some grain boundary facets having almost no cavities, while others will have a high concentration. Moreover, some austenitic stainless steels form internal cracks under the influence of an applied stress at relatively high temperatures. These cracks can open at triple points and extend across grain boundaries. Cavities, particularly gas-filled cavities, just ahead of the crack can cause sharp cracks to propagate.

In general, the local composition of a grain boundary or interphase boundary arising from segregation of impurity or minor alloying elements will modify the cavity nucleation rate as described previously.

The cavity growth rate is affected by the rate of grain boundary self-diffusion and hence the vacancy concentration within the cavities. Both, therefore, control the overall creep life of materials.

5.7.5   Creep–Fatigue

Engineering structures and components are rarely operated at constant stress and constant temperature for long periods of time. For example, engines, particularly aircraft engines, experience considerable thermal and stress cycles during operation. During these cycles, the components, particularly those operating at high temperature, can be subject to intense local stresses. These high local stresses, at temperatures greater than $0.3\,T_m$, can result in an increase in the nucleation of cavities, grain boundary cracks, grain boundary segregation and the growth of precipitates, and all of these can modify the overall operating life. If the effects of creep and fatigue are purely additive so that time to failure in a static test is t_f and the number of cycles to failure in fatigue is N_f for a given temperature, stress and stress range, then at time t and after N cycles

$$(N/N_f) + (t/t_f) = 1 \qquad (5.30)$$

This provides a simple framework for more sophisticated models to accommodate interaction between the fatigue and creep mechanisms. Creep cavitation occurs mainly on grain boundaries oriented normal to the maximum principle stress, whereas at higher temperatures fatigue damage accumulates on boundaries subject to high alternating shear stresses. Hence, the two effects may be more deleterious in combination than separately (Koterazawa and Mori, 1980). For example, in a martensitic stainless steel tested at 898 K, cavitational damage was observed to be completed during the first dwell period in the overall cycle, with no further effects in subsequent dwells (Min and Raj, 1978). Hence, during further cycles it is the interaction of this damage with the alternating shear stresses that controls overall life. Clearly, during such periods of dwell the role of impurity and solute elements at grain boundaries will be similar to those encountered during static creep. However, the creep–fatigue mechanism interaction is important since in many cases crack growth rates and failure times are different from predictions based upon linear addition of short-period fatigue and creep data (Evans and Wilshire, 1985; Beere and Roberts, 1982).

Once a crack has nucleated either at the surface or grain boundary, growth will be controlled by the plant operating cycle which could be, for example, alternately tensile and compressive stresses. Some cavities will be present on grain boundaries subject to tensile stress at high temperature and will therefore grow, while others will be subject to tensile stresses at low temperature and will remain unaltered. In a typical system, the cavities form on the grain boundaries inclined at a low angle to the direction of stress, while fatigue cracks would appear on those inclined at high angles. Failure occurs by a combination of these two mechanisms. Segregation of alloying or impurity elements to grain boundaries during the creep–fatigue cycle will modify the growth of the crack by changing the surface energy of the grain boundary which will influence the rate of growth and size of the cavities. However, predicting the type and degree of segregation of the various atom types to grain boundaries is complicated by the cycling process. In the simplest form, the thermal history needs to be known to predict the resulting grain boundary composition. At intermediate temperatures, elements may move towards the grain boundaries to form segregated layers or even precipitate particles, but at higher temperatures reverse diffusion may result. The experience of the influence of trace impurities in nickel-base superalloys used for aero-engines shows that first-stage turbine blades have lives significantly influenced by a range of impurity elements at trace levels. Here, the creep–fatigue lives were modified by the presence of Sb, As, Bi, Ga, In, Th, Si, Pb and S which ultimately have to be closely controlled to achieve consistent and reliable performance (Wilcox and Clauer, 1972).

5.8   STRESS RELIEF AND REHEAT CRACKING

5.8.1   Introduction

Almost all large engineering structures are constructed with extensive use of welding processes. The mass and the volume of material being welded vary widely and, although each weld is heated to above the alloy melting temperature, it then cools over a period of time to ambient temperature. The time taken to cool depends upon the mass of material welded, the cross-sectional area of the conduction paths and the availability of convective cooling, and as a result cooling times vary considerably (Rosenthal, 1946). When cooling, the weld pool material will attempt to reduce size and this will result in considerable strains and residual stresses within the final weldment. At the same time, alloying elements and impurities may segregate to grain boundaries to change the local composition or form precipitate particles. To reduce the residual stresses within weldments and improve mechanical properties such as the fracture toughness, the weldments are frequently subject to a post-weld heat treatment. In some cases, however, this heat treatment may cause the weldment to crack, usually within the heat affected zone (Figure 5.26). This is known as stress relief cracking. The heat treatment undertaken to relieve the residual stresses varies with material, but typically in ferritic steels the weldment is heated to temperatures in excess of 900 K with controlled heating and cooling. It is under these conditions that intergranular cracking may occur, usually along coarse prior austenite grain boundaries in susceptible steels. There have been numerous studies of stress relief cracking in ferritic steels (Shin and McMahon, 1984; Bowen *et al.*, 1990; Rauh *et al.*, 1989; Hippsley, 1987; Bowen and Hippsley, 1988; Hippsley *et al.*, 1988) and in stainless steels (Ortner and Hippsley, 1992). In other instances it is not possible to relieve the stresses by post-weld heat treatment because of the size, complexity or position of the component. In such cases the residual stresses remain when the component is taken into service. During service it may operate at temperatures within the creep range, and this will tend to relieve the stresses and again may cause cracking. This is known as reheat cracking and this type of cracking in ferritic steels has been reviewed by Dhooge and Vinckier (1987). Weld metal reheat cracking is mostly confined to low-alloy ferritic steel weldments such as $2\frac{1}{4}$Cr–1Mo and Cr–Mo–V, although it has been observed in austenitic stainless steels (Thomas, 1984) and high-alloy, ferritic 9 % Cr–Mo (Hippsley *et al.*, 1986).

In either stress relief or reheat cracking, the factors that contribute to formation of the cracks may be summarised by a Venn diagram (Figure 5.27) where the three main contributions are stress, temperature, and microstructure; removing one of these will eliminate the cracking. Welding is carried out by melting and laying down successive beads of weld metal. The size of the bead, speed of travel, bead spacing

Figure 5.26 Coarse-grained heat affected zone (HAZ) of low alloy ferritic steel weldment, showing intergranular cracking

and current and voltage conditions used determine the temperature characteristics both at the bead and as a function of distance from the bead which, in turn, determine the microstructure of the overall weldment. This heat affected zone (HAZ) in the parent plate contains a region of coarse-grained material close to the bead with a fine-grain region extending away from the bead. Overlap of weld beads will result in overlap of these heat affected regions within parent plate and the weld metal, and some regions will experience multiple heat cycles, while others will only experience a single thermal transient.

Figure 5.28 schematically shows the microstructure around a series of weld beads. Alberry *et al.* (1977) have shown that, for small initial grain size material, the final grain size, d_F, is given by

$$d_F = \sum_{t_{Ts}}^{t^*_{Ts}} K \exp\left(\frac{-Q_A}{RT}\right) \Delta t \tag{5.31}$$

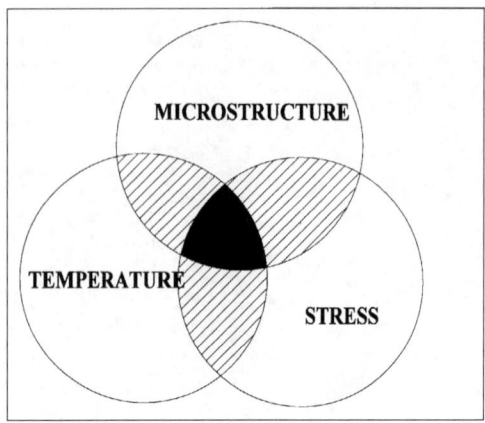

Figure 5.27 Venn diagram showing the factors that contribute to the formation of the cracks in stress relief or reheat cracking which may be summarised by three main contributions: stress, temperature, and microstructure

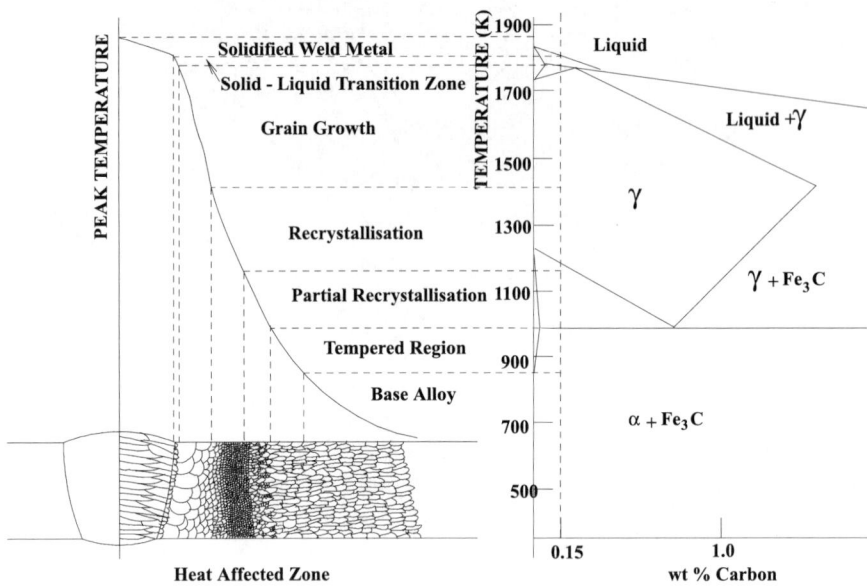

Figure 5.28 Schematic of microstructure around weld beads in relation to the the iron carbon phase diagram (Alberry, 1977). All relative to 0.15 wt.% C

where t_{Ts} is the time at which transformation to austenite is complete, t_{Tf} is the time at which transformation from austenite to ferrite is completed and all grain growth ceases, Q_A is the apparent activation energy for grain growth, K is a constant and Δt is the time interval of integration.

It is convenient now to address these two modes of cracking, recognising that the mechanisms involved are common to both. Both stress relief and reheat cracking occur when the relaxation strains, arising from the residual stresses in the weldment, cannot be accommodated owing to reduced creep ductility resulting in cavitation or wedge cracking at the grain boundaries (Emmer *et al.*, 1973; Ritter and McPherson, 1974). As a consequence, the composition of the grain boundaries will have similar effects to those described for creep failure and crack growth.

Stress relief cracking and reheat cracking occur predominantly within the coarse-grained HAZ immediate to the weld metal. At higher temperatures and usually higher stresses, brittle intergranular fracture occurs (Kussmaul *et al.*, 1977), while at lower temperatures and stresses the failure is by creep cavitation (Hippsley *et al.*, 1986). Dhooge and Vinckier (1982, 1987) have reviewed this type of cracking in low-alloy steels, ferritic creep-resisting steels, nickel-base alloys and austenitic stainless steels. This form of failure has caused significant problems for the power generation industries, in particular with austenitic stainless steels of Types 310, 316 and 347 and ferritic steels ($2\frac{1}{4}$Cr–1Mo and Cr–Mo–V) used in steam pipework, valves and headers. Both reheat and stress related cracks were found to form in the HAZ of low-alloy steels (Myers, 1977, 1980; Exworthy *et al.*, 1999; Batte and Murphy, 1979), and alloy additions (Nakamura *et al.*, 1970) can increase the susceptibility of the metal to reheat cracking.

Following welding, the weld metal solidifies and cools to room temperature at a rate determined by the size and shape of the component. During this period, recrystallisation, recovery and grain growth occur, depending on temperature, time at temperature, the amount of prior cold work and the impurity content in the parent alloy (Ashby and Easterling, 1982). Carbide precipitates may form at grain boundaries with bulk element depletion, and other precipitates, such as sulphides, may form on grain boundaries or within the material bulk. Segregation to grain boundaries of impurity and alloy elements such as phosphorus, tin and carbon can occur, resulting in temper embrittlement in the HAZ of the weldment. The strains set up as a result of the temperatures and thermal expansion differences will give rise to creep deformation in the bulk and at grain boundaries damage may promote wedge cracks or the initiation and growth of cavities.

The predominant mode of failure experienced during the post-weld heat treatment of welded structures and components is creep cavitation (Middleton, 1981; Hippsley, 1981). Most ferritic steels contain chromium and manganese which combine with the trace element sulphur to form sulphides. Manganese sulphide is thermodynamically the more stable and, during a high-temperature welding cycle, will partially or fully dissolve and reform on the prior austenite grain boundaries (Schultz and McMahon, 1973). These become the initiation particles for cavity formation under the influence of stress and temperature and ultimately lead to cracking through interlinkage of these cavities. As an extension to this model McMahon (1984) and Shin *et al.* (1982) argue that some steels initially containing higher levels of chromium and hence chromium sulphides will be more susceptible to stress relief cracking by cavitation. These steels will dissolve more sulphide during the high-temperature phase which reprecipitates on cooling as predominantly MnS precipitates on grain boundaries. Thus, steels with higher chromium levels would generally be expected to be more susceptible to stress relief cracking. Copper has a similar deleterious effect in $2\frac{1}{4}$Cr–Mo submerged arc weld metals (O'Brian and Wolstenholm, 1984). However, in this case copper precipitation in the form of copper-rich phases at prior austenite grain boundaries promotes cracking. The copper becomes included in the weld metal as a result of using copper-coated welding consumables and increases the susceptibility to cracking during post-weld heat treatment. Unfortunately, the use of uncoated submerged arc wires can introduce other welding difficulties.

Stress relief cracking and reheat cracking are enhanced by the segregation of trace elements such as P, Sn, Sb and S to grain boundaries but the thermal history of the weldment is important in determining the likelihood of significant grain boundary segregation. If, for example, the weldment is within a massive structure, it is likely that the heat affected zone will remain in the temperature region of 850 K for some considerable time, and during this time significant trace element segregation to grain boundaries may occur. However, if the structure has a low mass and is well ventilated, cooling will be relatively rapid, thereby minimising trace element diffusion and segregation to grain boundaries. In commercial ferritic steels of the Cr–Mo–V and Cr–Mo type weldments with stress relief at ~ 973 K, the effect of grain boundary composition within the heat affected zone is of equal importance compared with the other parameters that may influence susceptibility to cracking. To accommodate this, various empirical relationships have been developed to account for the contribution of different impurity elements (Seah, 1980b and Boniszewski, 1982). For example, levels of impurities in the weld metal have been correlated with embrittlement (Boniszewski, 1982), and this resulted in a rating for the contribution of each impurity element to give a metal composition factor (MCF) that indicates a total embrittling potency:

$$MCF = Si + 2Cu + 2P + 10As + 15Sn + 20Sb$$

where the composition for each element is in wt%. However, such approaches can be used only as a guide since, for example, they will be conditional upon the heat input: lower heat inputs consistently result in lower measured embrittlement.

A mechanism to explain the less common brittle fracture mode observed in ferritic steels that is known as high-temperature brittle intergranular fracture (HTBIGF) has been given by Hippsley *et al.* (1984), Hippsley (1981, 1985) and McMahon (1984). Cr–Mo and Cr–Mo–V steels used in steam pipework and valve assemblies have been found to crack intergranularly, transversely across the weld, following post-weld heat treatments or after long periods in service. High-temperature brittle intergranular fracture has been used to describe this cracking which occurs in these steels at temperatures between 573 and 873 K, in the presence of local and high hydrostatic stresses, and impurity elements such as sulphur, phosphorus and boron, when austenitised at temperatures above 1373 K. Hippsley and others conclude for a $2\frac{1}{4}$Cr–1Mo steel that sulphur promotes fracture in the presence of stress (Hippsley,

1987; Rauh *et al.*, 1989; Bowen and Hippsley, 1988; Hippsley *et al.*, 1988; Bowen *et al.*, 1990; Bika and McMahon, 1993, 1995). At low stresses, $< 55\,\mathrm{MPa\,m}^{1/2}$, the crack advances in discrete steps which occur when a critical concentration of sulphur is reached. The sulphur is driven to diffuse to a distance ahead of the crack by high tensile stresses that are produced a short distance from the crack tip. At higher stresses, $> 55\,\mathrm{MPa\,m}^{1/2}$, the crack advances by intergranular microvoid coalescence and is controlled by a distribution of fine sulphide precipitates. The transition between these two modes is controlled by a balance between the rate of HTBIGF crack growth, and hence the dynamic crack tip sulphur segregation, and the rate of sulphide-nucleated microvoid growth in front of the crack tip in the period of time between high temperature brittle intergranular fracture steps (Bowen and Hippsley, 1988).

High-temperature brittle intergranular failure has been observed in types 310, 316 and 347 austenitic stainless steels (Asbury *et al.*, 1960; Christoffel, 1962; Ortner and Hippsley, 1992; Hour and Stubbins, 1990). Cracking frequently occurs in the heat affected region of the weldments, associated with low creep ductility at temperatures between 673 and 1023 K, where there are high local stresses and hydrostatic stresses. In particular, it is related to steels with higher carbon contents in the presence of carbide-forming elements, where the grain size is large, and in the presence of impurities, particularly phosphorus, sulphur and tin. However, for materials with enhanced levels of phosphorus, sulphur and tin at temperatures between 823 and 1023 K, increasing the levels of these elements above the residual levels has little effect on brittle intergranular fracture (Ortner and Hippsley, 1992). The cracking results from stress-assisted diffusion of impurities, principally sulphur but also oxygen, to the grain boundaries ahead of the crack tip, provided residual element levels are sufficient to allow cracking to proceed (Bika *et al.*, 1995, and Lewandowski *et al.*, 1987). An extension of this process is dynamic embrittlement, which arises from brittle fracture along grain boundaries caused by stress-driven diffusion of a surface-absorbed embrittling element. It usually arises when a stress is applied to a material containing a surface-active, low melting-point element so that it has high mobility and low binding energy. However, since this is a diffusion-controlled process, the crack growth rates, which are proportional to the grain boundary diffusivity of the embrittling element, are typically ten orders of magnitude slower than for conventional brittle fracture. This mechanism has been used to describe dynamic embrittlement in a range of non-ferrous materials including tin in bronze and copper–tin alloys (Bika and McMahon, 1993; Barrera *et al.*, 1992; Muthiah *et al.*, 1999), copper beryllium (Misra *et al.*, 1994), and nickel-base intermetallic and superalloys (Chang *et al.*, 1990; Lui and White, 1987; Hippsley and De Van, 1989).

5.8.2 Irradiation Creep

When an elastic collision occurs between a neutron and an atom, energy is transferred and the neutron may displace an atom from its site. The displaced atom is a primary knock-on and dissipates energy by displacing other atoms from lattice sites. To displace an atom, the energy of a neutron must exceed a critical value; for α-iron this is of the order of 25 eV. Each primary knock-on can produce many displacements to dissipate all the associated energy, thereby creating Frenkel pairs which are a combination of an atom occupying an interstitial site and a vacancy. A supersaturation of vacancies is produced which may diffuse, aggregate and form voids or cavities. Voids and cavities, the latter on grain boundaries, may ultimately form throughout a material subjected to severe neutron irradiation. An applied stress will produce vacancy flow, and the cavities will grow on grain boundaries at a rate that is enhanced by the presence of intersecting dislocations. Although second phase particles may be necessary for grain boundary cavity nucleation, they may also restrict the growth of pre-existing cavities. Particles on grain boundaries that do not easily deform will prevent the boundary moving, and, when the stress across the boundary is relaxed by diffusion of atoms from the cavity, the particle will absorb the stress and the creation of vacancies will stop (Beere, 1980, and Burton, 1976). Neutron irradiation will also introduce helium by boron transmutation. When thermal neutrons react with the ^{10}B isotope, alpha particles are produced which become helium atoms by acquiring electrons. These gas atoms diffuse to grain boundaries where they agglomerate to form helium gas bubbles with an equilibrium radius, r_0, of

$$r_0 = \frac{2\gamma}{P_0} \tag{5.32}$$

where P_0 is the internal gas pressure and γ is the surface tension (Lovell, 1975). As a consequence, it is possible to develop deformation maps to describe the interrelationship between irradiation and thermally controlled creep (Figure 5.29). At temperatures below about $0.5T_M$, irradiation creep is the most important deformation mechanism if the material is subject to severe neutron irradiation. Here, the mechanism is controlled by climb and glide of dislocations with the attendant processes of void or cavity nucleation and growth.

Under neutron irradiation, minor alloying and impurity elements can segregate to grain boundaries in materials. As a consequence, for irradiation creep it would be anticipated that the creep rupture process would be controlled by grain boundary composition as described for thermal creep. The effect of irradiation of austenitic stainless steels can have a significant effect upon creep rupture. For example, if there is prior thermally induced segregation of boron to grain boundaries, the transmutation of the boron to form helium bubbles could lead to creep rupture at low values of elongation (Figure 5.30). Specimens of annealed type 316 stainless steel irradiated in the temperature range 723–1033 K to neutron doses from $\sim 10^{21}$ to 10^{23} n/cm^2 at energies of > 1 MeV show generally a reduction in rupture life owing to irradiation. In addition, once formed, the free surface of voids or grain boundary cavities provide energetically favourable sites for segregant elements which modify swelling rates under neutron irradiation. Such effects have been observed in a range of materials: vanadium (Okamoto et al., 1974), types 316L and 321 austenitic stainless steels and nickel (Gittus, 1978). Ashby and Brown (1963) used strain contrast to consider the

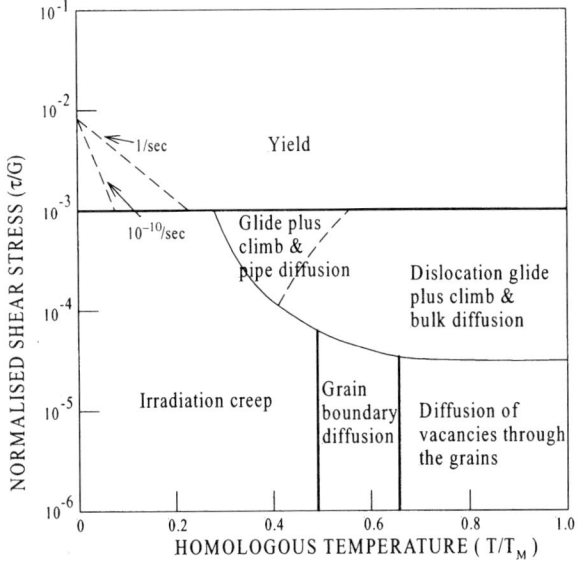

Figure 5.29 Notional deformation map for a bcc metal under stress in a fast breeder reactor. Irradiation creep gives way to thermal creep at about one-half the absolute melting point. τ is shear stress, G is shear modulus, T is absolute temperature; T_m is melting point. Contours enclose dominant creep mechanisms (Lovell, 1975)

role of segregation to the surface of cavities or bubbles in 18Cr–8Ni–1Si stainless steel and nickel irradiated at 723 K with 1 MeV electrons. Figure 5.31 shows the concentration of impurity atoms, $C_S(r_0)$, in the matrix surrounding a bubble, void or cavity of radius r_0 which is given by

$$C_S(r_0) = C_S^0 \exp\left\{\frac{K_\alpha n d_R}{3 D_S}\left[d_R/d + 1/2(d/d_R)^2 - 3/2\right]\right\} \qquad (5.32)$$

where C_S^0 is the impurity concentration at $d = d_R$, n is the net point defect production rate, $2d_R$ is the mean distance between voids, K_α is a proportionality constant ($\sim 10^{-2}$) and D_S is the diffusion coefficient of the segregating atoms. Hence, this equation predicts that the impurity concentration at the void surface will increase exponentially with dose rate, consistent with observations in vanadium. Certainly, these concepts are supported by observations of thermally induced segregation in an Fe–3 % Ni alloy containing low concentrations of phosphorus and tin, heat treated to maximise the segregation of these elements to grain boundary; phosphorus segregated to the grain boundary and tin to the cavity surface (Wild, 1997).

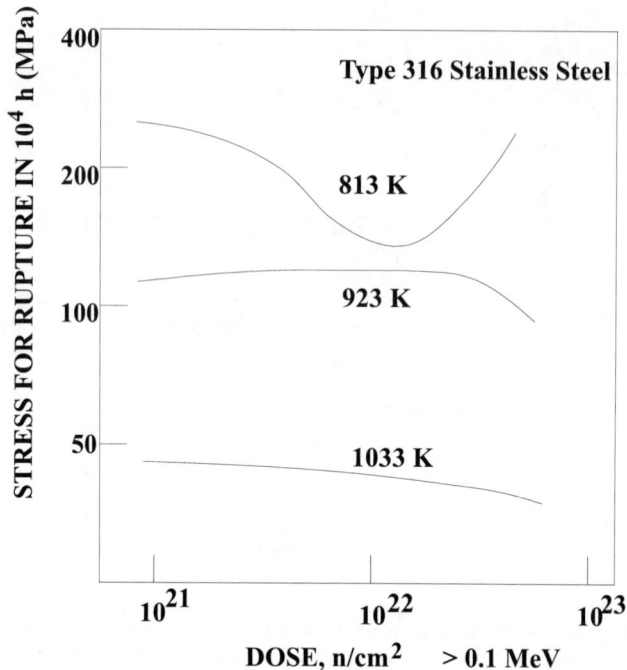

Figure 5.30 Effect of fast reactor irradiation on the creep rupture life of stainless steel irradiated at, or near, the test temperature. As calculated by Lovell (1975)

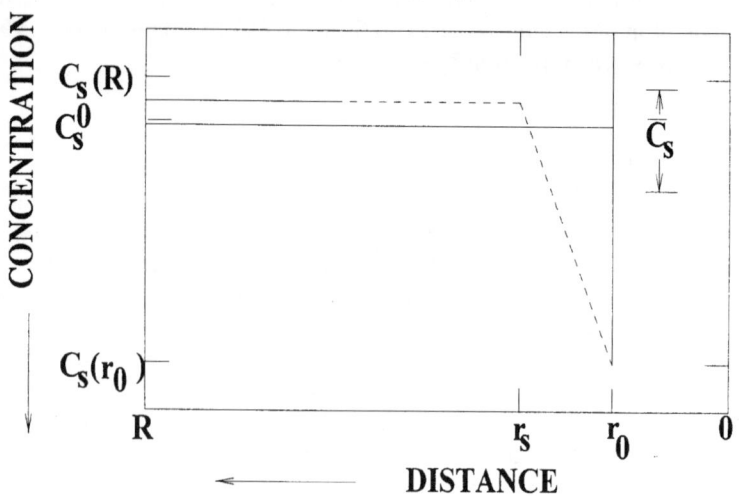

Figure 5.31 Concentration of impurity atoms in the matrix surrounding a bubble in vanadium (Okamoto, 1974)

5.9 ENVIRONMENTALLY ASSISTED CRACKING

5.9.1 Introduction

Until now we have considered the material, the temperature and the stresses and not, apart from irradiation, included the environment in which a body is operating. Almost all materials operate in an environment that will influence performance and ultimately failure. For example, materials may operate in oxidising environments where there may be surface corrosion and, in some cases, corrosion into the bulk of the material via grain boundaries. The presence of a stress may modify this process since, when combined with corrosion, this leads to commonly encountered failure mechanisms, both at low and elevated temperature. Certain austenitic stainless steels can crack by this mechanism at room temperature in the chloride environment near swimming pools and in the higher-temperature water environments of power plants. The possible types of environmentally assisted cracking can be visualised by reference to a Venn diagram made up of three parts:

(a) the specific material and its microstructure and hence mechanical properties
(b) tensile stress either externally applied or residual and
(c) the environment (Figure 5.32).

Where the environment and the microstructure of the material, particularly changes in the grain boundary composition, interact in the absence of a microstress, attack

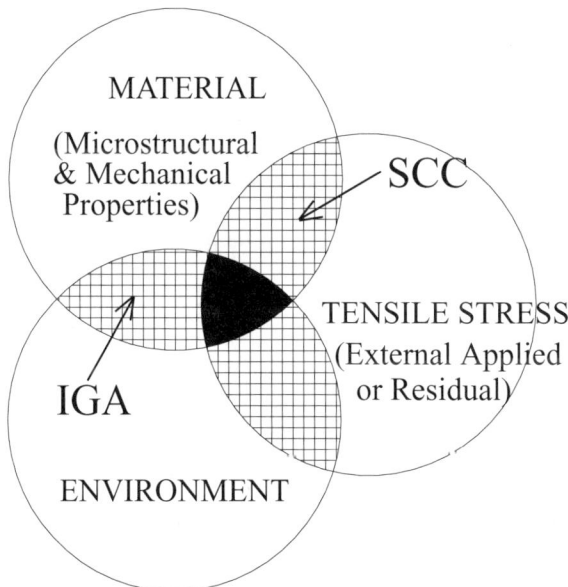

Figure 5.32 Venn diagram showing factors contributing to environmentally assisted cracking, both intergranular attack and stress corrosion cracking

via the grain boundaries can occur, intergranular attack (IGA). When the third factor, stress, is introduced (Figure 5.32), this leads to stress corrosion cracking (SCC). This can be intergranular, which will be considered in this chapter, but the cracking can also be transgranular. All of these processes of failure are termed environmentally assisted cracking (EAC).

5.9.2   Mechanisms

Grain boundary composition has been identified as influencing intergranular fracture which arises in the environmentally assisted cracking regime in a wide range of material systems. However, despite the availability of a range of microanalytical techniques for a considerable time (Doig and Flewitt, 1978; Bruemmer, 1998), there remains a need to improve the quantitative relationship between interfacial composition and these failure mechanisms. Certainly, quantitative measurements of grain boundary composition are essential to enable reasonable assessment of the variables controlling either intergranular attack or intergranular stress corrosion cracking.

As described above, environmentally assisted cracking can proceed in the absence of an applied tensile stress, by preferential attack at the grain boundaries, frequently initiated either at a defect in the protective surface oxide in the vicinity of a grain boundary or a particle such as a manganese sulphide inclusion. The latter in a ferritic steel reacts with the surface oxide and the environment, producing a pit where the local chemistry can develop to allow corrosion to proceed along an associated grain boundary. Once initiated, the crack can propagate, under the control of one of the electrochemical processes (Figure 5.33a), such as anodic dissolution if exposed to moist air or water. Here, the crack tip becomes anodic and the environment within develops a local acidic pH. However, under these circumstances, since oxidising species such as oxygen, in solution, are not readily replenished within the crack, passive film formation is limited and a flow of current continues ($M \rightarrow M^{2+} + 2e$). Metal ions in solution are transported along the potential gradient to the mouth of the crack where oxides or hydroxides form. As oxidising agents in solution are reduced, the anodic region becomes more oxidising by the absorbing electrons, which drives the metal to a more positive potential. When the aqueous environment within a deep crack does not contain species that form insoluble compounds with the metal ions, the 'bare' surface is covered by an electrical double layer which regulates movement of ions to the crack tip and hence the rate of crack propagation. However, it has to be recognised that even the simplest models for stress corrosion cracking are complex because of the need to describe the distribution of the passivating oxide film formed at the crack tip and the interaction between the electrochemistry and deformation at that tip. Important factors controlling crack advance are summarised in Figure 5.33b, but, in nearly all cases of intergranular cracking, the grain boundary composition is dominant. Local equilibrium and non-equilibrium changes in grain boundary composition influence crack propagation by conjoint changes to the electrochemical and mechanical behaviour. For example, carbon can diffuse to grain boundaries of austenitic stainless steel, where it may interact with chromium-rich carbide preci-

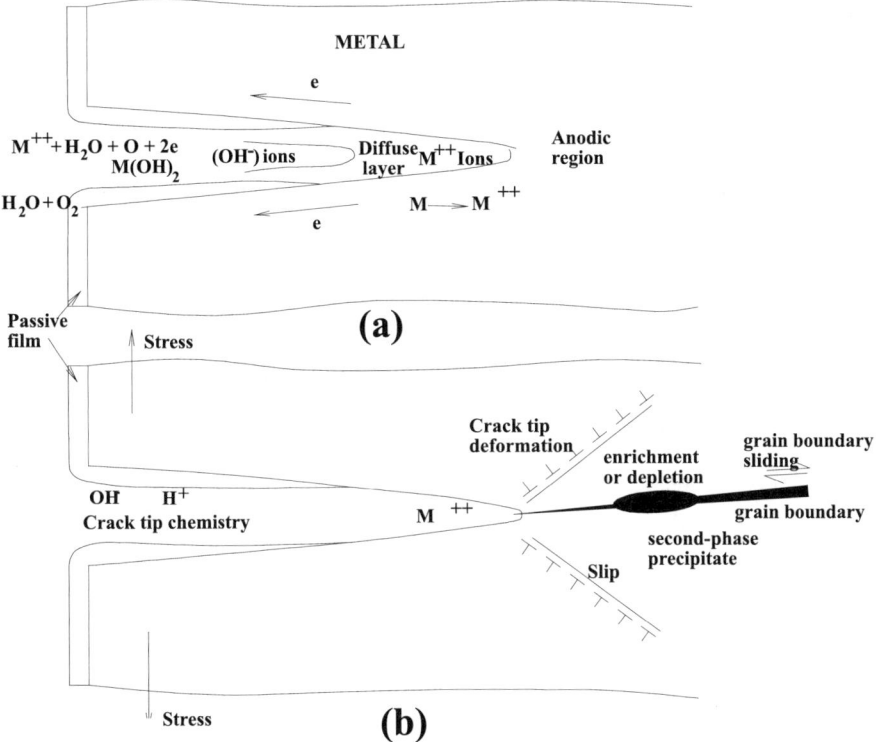

Figure 5.33 Schematic diagram of anodic stress corrosion of a metal crack exposed to water (a) showing electrochemical features and (b) the interaction with a grain boundary

pitates. These precipitates deplete the grain boundary and the region adjacent to the boundary in chromium, so that the iron-enriched grain boundary is more susceptible to a corrosive environment. Again, segregation of other atom species may also alter the local electrochemical potential, thereby modifying the rate of propagation of the crack.

The different (Doig and Flewitt, 1984) models proposed to describe stress corrosion cracking in an aqueous environment divide into:

(a) metal loss by anodic dissolution at a crack tip to give dissolution rates controlled by the kinetics of the various electrochemical reactions and

(b) adsorption phenomena at a crack tip and associated diffusion into the strained volume of material ahead of the crack.

Adsorption phenomena include concepts of the reduction in surface energy brought about by chemisorption of specific environmental species that, in the case of hydrogen atoms, can be absorbed into the strained material ahead of the crack tip. Both adsorption and dissolution processes have been used in mechanistic descriptions of crack initiation and growth, with each contributing at different stages to the crack development. The dominating mechanism gives the faster crack growth, with the overall kinetics conforming to the specific underlying process. Removal or reduction in one essential component may stop cracking by this mechanism, but this does not preclude the possibility of cracking continuing by another. A simple example is the influence of electrochemical polarisation, where impressed cathodic polarisation decreases cracking susceptibility resulting from anodic dissolution but increases that due to hydrogen embrittlement (Doig and Flewitt, 1984). In view of the generally accepted dependence of environmentally assisted cracking on both the detailed microstructural and local chemical environment, it is surprising that correlations between bulk material properties or bulk environments and crack growth rates continue to be attempted.

5.9.3 Intergranular Attack

A frequently observed example of intergranular attack is the grain boundary corrosion of austenitic stainless steels in chloride environments when the steel is sensitised. Austenitic stainless steels typically have a bulk composition that contains 18–25 wt% chromium. When this steel is exposed to a corrosive environment, the chromium forms a protective layer of rhombohedral chromium oxide. If the chromium content falls below about 12 wt%, then the oxide that forms is a mixed iron–chromium spinel which is not protective. Heat treatments at temperatures of about 900 K for some hours or of 650 K for many thousands of hours allow the carbon, in solution, to segregate to the grain boundaries where it is able to form chromium-rich $M_{23}C_6$-type carbides. This precipitation process has the effect of depleting the region in the vicinity of the grain boundary in alloying elements, such that the chromium content decreases to below the critical value of ~ 12 wt%. A chloride environment is then able to attack this chromium-depleted boundary and, locally, spinel oxide forms at the grain boundary intersecting a crack tip and chloride ions are free to diffuse through this spinel.

5.9.4 Stress Corrosion Cracking

Under certain environmental conditions and in the presence of stress, many materials may fracture with reduced ductility—stress corrosion cracking (SCC). When fracture occurs along grain boundaries it is referred to as intergranular stress corrosion cracking (IGSCC). Certainly, many structures and components fabricated from metals, alloys and other materials are designed for operation in corrosive environments while under applied stress or when subject to a prestress.

NON-FERROUS MATERIALS

Yamashita *et al.* (1991) studied IGSCC in Cu–Al alloys by producing bicrystals with different orientations, and these were exposed to ammonium hydroxide and sodium hydroxide solutions of various pH under a range of stresses. A strong dependence of

IGSCC on the misorientation of the grain boundaries was established and obviously the nanochemistry of these boundaries can dramatically influence the rate and depth of cracking. Age hardening of another non-ferrous alloy, in this case the high-strength 7075 Al–Zn–Mg–Cu series of aluminium alloys, provides a good example of the influence of heat treatment on intergranular stress corrosion susceptibility. Overageing of these alloys beyond the maximum hardness recommended for commercial applications is known to reduce susceptibility to stress corrosion cracking. Reduced susceptibility to IGSC of the 7075 aluminium alloy occurs when overaged for 1 h at 738 K and then water quenched, followed by 24 h at 393 K and 24 h at 433 K. This can be shown to be a consequence of the changes in electrochemical properties of the grain boundaries that result from a decrease in the local copper concentration at these microstructural features (Doig *et al.*, 1977).

FERRITIC STEELS

As described in previous chapters and sections of this chapter, there is a base of information on the segregation of impurity and alloying elements to grain boundaries in ferritic steels, arising from considerations of temper embrittlement. In the simplest case, pure iron alloyed with carbon, nitrogen or phosphorus is rendered susceptible to intergranular stress corrosion cracking in a variety of environments, whereas ultrapure iron is resistant. For fracture in air, both carbon and nitrogen are recognised grain boundary strengtheners, whereas phosphorus effects grain boundary embrittlement. The influence of a temper embrittling heat treatments on the intergranular stress corrosion susceptibility in a caustic environment for an Fe–3 wt% Ni alloy containing additions of impurity tin or phosphorus has been described by Doig and Flewitt (1978). The mean stress corrosion crack lengths were measured after various exposure times following either an embrittling or annealing heat treatment. Crack growth rates in the embrittled specimens, where segregation of either tin or phosphorus to the grain boundaries occurs, increases by a factor of three compared with unembrittled specimens (Figure 5.34); crack initiation times were similar for all conditions. These changes in crack growth rate, brought about by such highly localised changes in grain boundary nanochemistry, were interpreted either as demonstrating changes in the grain boundary 'oxide-free' anode dissolution characteristics or a hydrogen embrittlement mechanism (Latanision and Staehle, 1969, and Doig and Flewitt, 1977). The latter arises from enhanced segregation of the impurity tin or phosphorus to the grain boundaries which, in turn, acts as a hydrogen recombination poison (Latanision and Oppenhauser, 1974). These results are consistent with observations for Fe–P alloys heat treated to cause grain boundary segregation, where the rate of intergranular crack propagation in an aqueous 55 % $Ca(NO_3)_2$ solution at a temperature of 333 K increased with enrichment in phosphorus at the grain boundaries. This approach supports the prevention of local passivation arising from the change in grain boundary composition (Kupper *et al.*, 1981).

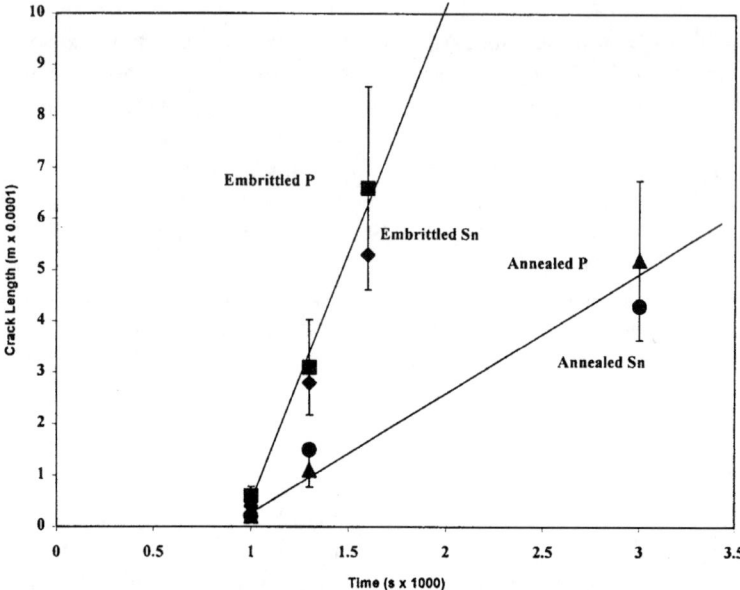

Figure 5.34 Mean length of stress corrosion cracks as a function of the testing time for embrittled and annealed Fe–3 wt% Ni-base alloys containing additions of impurity tin and phosphorus (after Doig and Flewitt, 1978)

Ni–Cr–Mo–V steels are used for applications where a component or structure experiences service stresses in mildly corrosive environments. For example, they are frequently used for steam turbine plant, but infrequently have been found to fail by intergranular stress corrosion cracking in both dilute caustic environments (Kupper *et al.*, 1981; Bandyopadhyay and Briant, 1985; McMinn *et al.*, 1985) and in pure condensing steam (Denk, 1994). Further failures have been detected during service of low-pressure (LP) rotors fabricated from low-alloy ferritic steels as a result of IGSCC caused by a combination of high local stress, an aggressive environment and a susceptible material (Lyle and Burghard, 1982; Jaffee, 1986). The time to failure as a function of applied stress increases for a phosphorus containing low carbon NiCr steel in a caustic environment when compared with a similar but higher purity steel (Figure 5.35) (Bandyopadhyay and Briant, 1982). The uncertainty of the service behaviour of these steels arises because they can frequently contain P, S, Sn, Sb, As, Mn and Si, either as impurities or minor alloying elements, all of which may either segregate or influence segregation to grain boundaries. In addition, carbon present in the steels may form carbide precipitates preferentially at these locations in the microstructure. Again, carbide precipitates can be chromium rich and result in a chromium-depletion in the boundary region. Heat treatments that enhance grain boundary impurity segregation have been shown to cause increased IGSCC in chromium–molybdenum ($1Cr–\frac{1}{2}Mo$) ferritic steels exposed to caustic environments

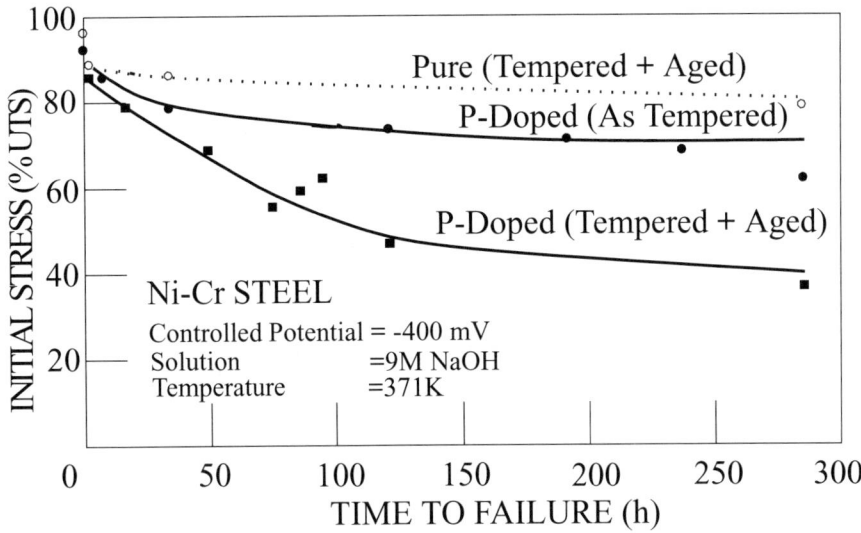

Figure 5.35 The time to failure plotted as a function of the applied stress (normalised by the ultimate tensile strength) for a high purity–low carbon NiCr steel and a phosphorus doped low carbon NiCr steel. The phosphorus-doped steel was tested in both the As-tempered (948 K for 100 h) and tempered and aged (948 K for 100 h, quenched, 753 K for 100 h) conditions (Bandyopadhyay and Briant (1982))

(Harrison *et al.*, 1977). In particular, phosphorus segregation enhances IGSCC in deareated pure water (McMinn *et al.*, 1985) and aqueous caustic (Bandyopadhyay and Briant, 1982, 1983), and nitrate solutions (Molonznik *et al.*, 1979; Kupper *et al.*, 1981) in a range of chromium–molybdenum steels. By reducing the level of impurities in these steels, both temper embrittlement and resistance to IGSCC have been greatly improved, particularly by reducing the levels of P and Sn, which segregate to the grain boundaries, and Si and Mn, which enhance impurity segregation (Jaffee, 1986; Jaffee *et al.*, 1986). Manganese sulphide inclusions are often active sites for pitting and nucleation of voids that can be the origin of a stress corrosion crack path in these steels (Lindley *et al.*, 1982), reducing manganese sulphide inclusions decreases IGSCC (Jaffee *et al.*, 1986).

Overall, however, there remains little correlation between the effects of grain boundary composition on fracture and the contribution of an aqueous or aggressive environment leading to intergranular stress corrosion cracking. Certainly, it is recognised that chemical effects can be dominant so that small local changes in electrode potential can modify susceptibility to intergranular stress corrosion cracking. The effects can modify crack tip events and crack sidewall passivation rates or, more indirectly, change the composition of grain boundary precipitates and colloids formed in the liquid phase within the crack or even at its mouth.

AUSTENITIC STAINLESS STEEL

Austenitic stainless steels are used commercially because they generally have good corrosion resistance because of the protective surface oxide formed as a result of the higher levels of chromium in the bulk; they are used for applications that experience severely aggressive environments (Wyatt, 1971). However, these steels have been observed to be susceptible to intergranular attack and, as a result of tensile stress, will also be susceptible to IGSCC. This is particularly the case following heat treatments that produce carbides, usually the chromium-rich $M_{23}C_6$ type, that carry attendant chromium-depletions (Weiss and Stickler, 1972). Susceptibility to IGSCC can be reduced by lowering the phosphorus and sulphur levels in the bulk (McIntyre *et al.*, 1996; Younes *et al.*, 1997). Typical levels of sulphur and phosphorus in Type 316 austenitic stainless steel are 0.008 % and 0.025 % respectively, whereas in the high-purity version of the alloy these levels can be reduced to less than 0.002 %. The effect of reducing the level of impurities on the IGSCC has been studied by Younes *et al.* (1994), where a high-purity type 316H austenitic stainless steel was compared with a conventional commercially available steel of similar specification but higher impurity level. Most element compositions were maintained constant, including carbon, but P and S were reduced to levels as low as could practicably be achieved. These were then given standard heat treatments before being subjected to exposure corrosion tests in aerated deionised water at a temperature of 353 K. Figure 5.36 shows that the depth of attack is at least an order of magnitude less in the high-purity austenitic steel.

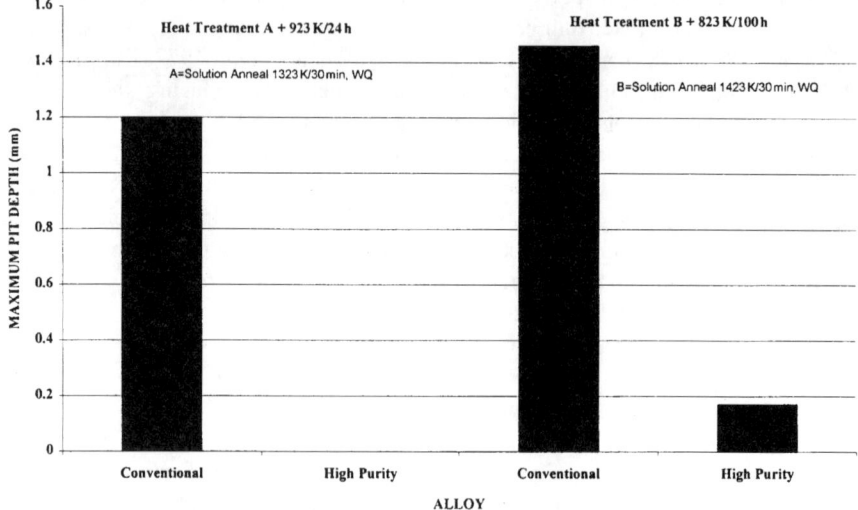

Figure 5.36   Pit depth as a function of time for stainless steel of different purity (after Younes *et al.*, 1994)

The high-purity alloy performed much better when subjected to slow strain rate tests in dilute sulphate solutions. The effect of reducing the phosphorus and sulphur contents to below 0.002% modified the $M_{23}C_6$-type carbide precipitate size and distribution at the grain boundaries that developed during sensitisation heat treatments at 873 and 923 K, with a corresponding less pronounced chromium depletion profile. It has been suggested that phosphorus segregation to grain boundaries in the conventional steel may influence the $M_{23}C_6$-type carbide precipitation process by modifying the equilibrium between carbon and chromium at the grain boundary (Weiss and Stickler, 1971). Phosphorus is known to decrease the activity of carbon in Ni–20% Cr alloys which leads to a change in carbide precipitate morphology (Grabke, 1989). This effect may be more pronounced in austenitic stainless steels containing molybdenum because phosphorus and molybdenum interact to delay the development of carbide precipitates. It has been shown by Muraleedham *et al.* (1999) that, for AISI Type 304L austenitic stainless steel, the effect of phosphorus segregation to grain boundaries will not cause IGSCC in a non-susceptible heat treatment condition in hot water environments. Rather, the effect of any phosphorus enrichment at the grain boundaries is to enhance the IGSCC susceptibility of this type of austenitic stainless steel provided chromium depletion is already present at these grain boundaries. Certainly there are many observations where impurities such as P, Si, and S have been implicated in intergranular stress corrosion cracking in austenitic stainless steels. However, these assertions have often been made without detailed grain boundary composition measurements. Even with these discrepancies in the contribution of impurity segregation the results support the important contribution that phosphorus segregation can make to grain boundaries in these steels (Briant, 1980; Was and Latanision, 1981).

In the case of nickel, phosphorus segregated to grain boundaries has been observed to accelerate intergranular stress-corrosion cracking at passive potentials in $1NH_2SO_4$ solutions. By contrast, sulphur has a negligible effect (Jones *et al.*, 1984). The resistance to corrosion of nickel-based alloys 600 and 690 depends on the grain boundary composition (Younes *et al.* (1997). Grain boundaries that contained a high concentration of segregated impurities, such as phosphorus, without carbide precipitates have the highest susceptibility to corrosion. Typical grain boundary composition profiles across a grain boundary measured on thin foil specimens using FEG-STEM in combination with energy dispersive X-ray analysis (EDX) for the alloys 600 and 690 (Figure 5.37a). In Figure 5.37a alloy 690 has a lower phosphorus segregation but also significant chromium depletion, whereas Figure 5.37b the alloy 600 has little or no chromium depletion.

5.9.5 Hydrogen Cracking

Hydrogen is a common cause of environmentally assisted cracking in many materials. This element may be present in structures and components as a result of the processes used in their manufacture, which might include quenching, heat

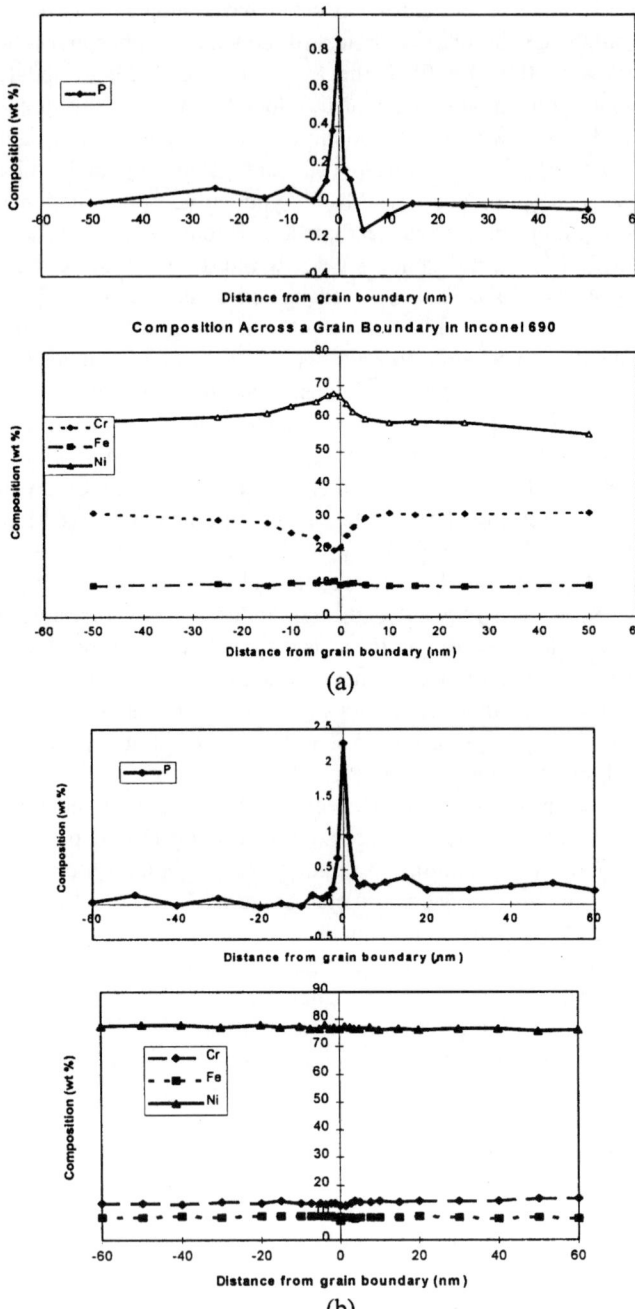

Figure 5.37 FEG-STEM EDX determination of phosphorus, chromium, iron and nickel distributions across a grain boundary in (a) alloy 600 and (b) alloy 690 (Younes *et al.*, 1997)

treatment in atmospheres containing hydrogen or welding. However, even if the material starts life relatively free from hydrogen, it may be absorbed in several ways, as gaseous hydrogen, by hydrolysis or chemical reaction. The mechanisms by which hydrogen may be absorbed into the metal are shown schematically in Figure 5.38. Gaseous hydrogen molecules impinging on a surface of the material dissociate and are absorbed. The rate of hydrogen absorption is determined by the rates of dissociation and recombination. As the temperature increases, so will the likelihood that a hydrogen molecule will dissociate and the rates of diffusion of hydrogen ions will increase leading to greater hydrogen penetration. Hydrogen ions in solution can acquire an electron to form a hydrogen atom which can either combine to form a hydrogen molecule, which would bubble off, or be absorbed into the material. Hydrolysis can occur when a material is immersed in a solution containing hydrated H^+ ions such as an acidic environment. Such an environment may develop in pits and cracks when the overall solution has a pH value in the range 1–3.5 (Marek and Hochman, 1970; Brown *et al.*, 1969; Sedriks *et al.*, 1971; Ateya and Pickering, 1974). Chemical reactions arise when a molecule containing hydrogen reacts with the material to release hydrogen which can then enter the metal. This type of reaction frequently occurs in solutions containing an alcohol (Blackburn *et al.*, 1973) and

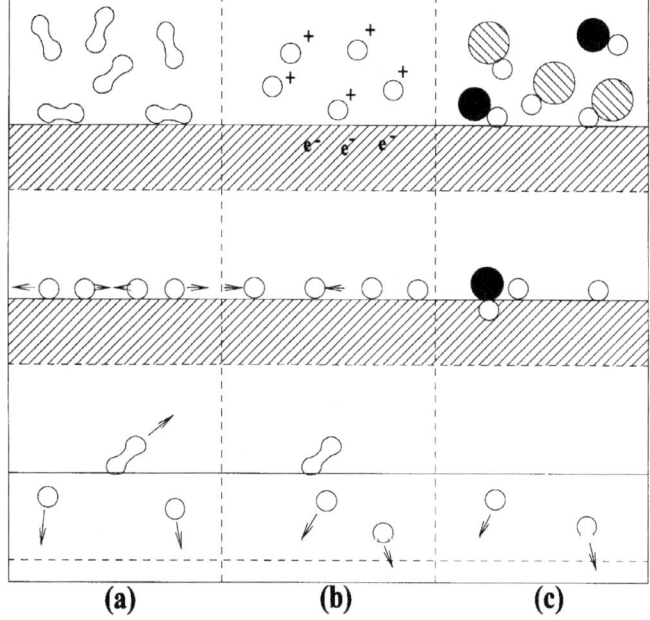

(a)   (b)   (c)

Figure 5.38 Schematic diagram of mechanisms showing the stages by which hydrogen may be absorbed into a metal: (a) hydrogen gas molecules adsorbing on the surface, dissociating and entering as nascent [H]; (b) hydrogen ions combining with electrons at the surface to form [H]; (c) hydrogen-bearing molecules reacting at the surface to provide [H]

leads to the formation of hydride phases (Thompson, 1973). The natural dissociation of water molecules may also be a source for this type of reaction:

$$2H_2O \Leftrightarrow H_3O^+ + OH^- \tag{5.34}$$

Certain elements such as sulphur, arsenic, antimony and others present in either the environment or the material can inhibit the recombination of hydrogen and hence assist in the dissolution of hydrogen into the material. These elements act as poisons that block hydrogen recombination by preventing two hydrogen atoms being chemisorbed at adjacent surface sites. The poison and hydrogen molecules compete for sites on the surface, with the result that the total hydrogen concentration is reduced, although the ratio of hydrogen atoms to hydrogen molecules is increased, releasing hydrogen atoms capable of diffusing into the material.

Poison atoms can also exist within the bulk of a metal or alloy where they may segregate to grain boundaries or interphase boundaries, such as those between carbide precipitates. Poison atoms behave similarly at grain boundaries to a free surface. Indeed, a grain boundary intersecting a surface may produce local absorption of poison atoms at that site. Latanision and Opperhauser (1974) proposed that poison atoms present at grain boundaries and other sites should promote the absorption of hydrogen into the metal bulk. Since hydrogen diffuses faster along grain boundaries than through the bulk, suitable poison atoms segregating to grain boundaries may have a dramatic effect on the embrittlement by hydrogen; this provides a mechanism for hydrogen embrittlement in nickel-base alloys.

The locations for hydrogen within a metal are shown schematically in Figure 5.39: the lattice, grain boundaries, incoherent and coherent precipitates, voids and

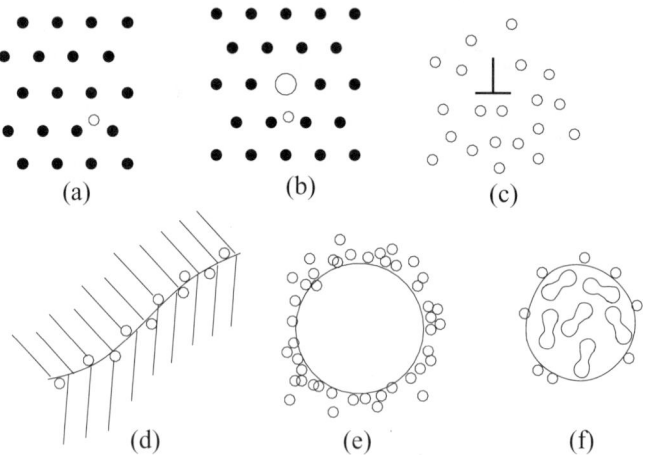

Figure 5.39 Locations for hydrogen within a metal (after Trioano, 1960) (a) Interstitial sites, (b) interaction with solute atom, (c) interaction with dislocation, (d) interaction with grain boundary (e) interaction with interphase boundary (f) cavity

dislocations. Once within the metal, the hydrogen can either combine with other elements or remain uncombined. In particular, hydrogen can diffuse to carbide precipitates at high temperatures, and hydrogen molecules may combine to form gas bubbles. Gas accumulating at voids, vacancies and dislocations is considered to be responsible for fracture (Vitovec, 1964). Hydrogen may react with oxides present within the metal bulk, forming water vapour, and this has been shown to promote failure in nickel (Beevers and Arrendu-Edmonds, 1968). The hydrogen may also form hydrides with elements such as nickel or titanium (Beevers and Arrendu-Edmunds, 1968). Hydrogen may remain in the metal as free protons (Plusquellec *et al.*, 1957; Plusquellec, 1960), atomic hydrogen, which will segregate to areas of high stress, for example local areas containing interstitials, atoms or vacancies or carbide precipitate interfaces or sulphide precipitates. Atomic and in some cases molecular hydrogen may diffuse to defects such as grain boundaries (Bodenstein, 1922; Berg, 1958; Petch, 1956; Besnard, 1952; Besnard and Talbot, 1959).

Many authors have considered the diffusion of hydrogen in metals and alloys (Plusquellec *et al.*, 1957; Plusquellec, 1960; Darken and Smith, 1949; Foster *et al.*, 1965; Azou, 1967; De Kazincky, 1959; Keeler and Davis, 1953; Johnson and Hill, 1960; Werner and Davis, 1961; Coehn and Sperling, 1933; Edwards, 1924; Baukloh and Retziaff, 1937; Kelemen, 1955; Speidel, 1974; Steigerwald *et al.*, 1959; Johnson, 1969). Hydrogen ions (protons) will move to regions of negative potential by diffusing through the lattice. Diffusion rates are rapid in most metals, particularly in those that have a bcc structure such as ferritic steels and β-titanium alloys. However, the rate of diffusion through the lattice is not sufficient to account for the observed rate of cracking (Williams and Nelson, 1971). Other mechanisms are required to allow hydrogen to diffuse at faster rates. Bastein and Azou (1951a and b) and Bastein *et al.* (1960) proposed that hydrogen moved with dislocations, and Blanchard and Troiano (1960) demonstrated that the energy of dislocation arrays increased in the presence of hydrogen. Other workers have concluded that dislocation transportation of hydrogen occurs in nickel and nickel alloys (Donovan, 1976; Frandsen *et al.*, 1973; Thompson, 1974; Thompson and Wilcox, 1972) and titanium alloys (Beck, 1971). Tien *et al.*, 1975) have calculated that hydrogen transport can be accelerated by factors from 10^2 to 10^4 if grain boundaries act either as paths for short-circuit diffusion or as traps to enhance or reduce the rate of hydrogen diffusion respectively. The mechanisms by which hydrogen can embrittle a metal or alloy are summarised in Figure 5.40 (Thompson and Bernstein, 1980).

Several theories have been put forward to explain the role of hydrogen in embrittlement of metals and alloys. Pfeil (1926) proposed that hydrogen reduced the cohesive forces in the metal lattice, particularly near the crack tip. This idea was developed by Troiano and Oriani and has become known as the Pfeil–Troiano–Oriani cohesive strength mechanism such that hydrogen interacts with the 3d electron shell of the matrix to lower the binding energy. The cohesive energy between adjacent metal atoms is reduced by an increase in the atomic distances caused by the hydrogen atom donating an electron to the d band of the metal (3d orbitals) (Blanchard and Troiano, 1960; Troiano, 1960). Petch and Stables (1952)

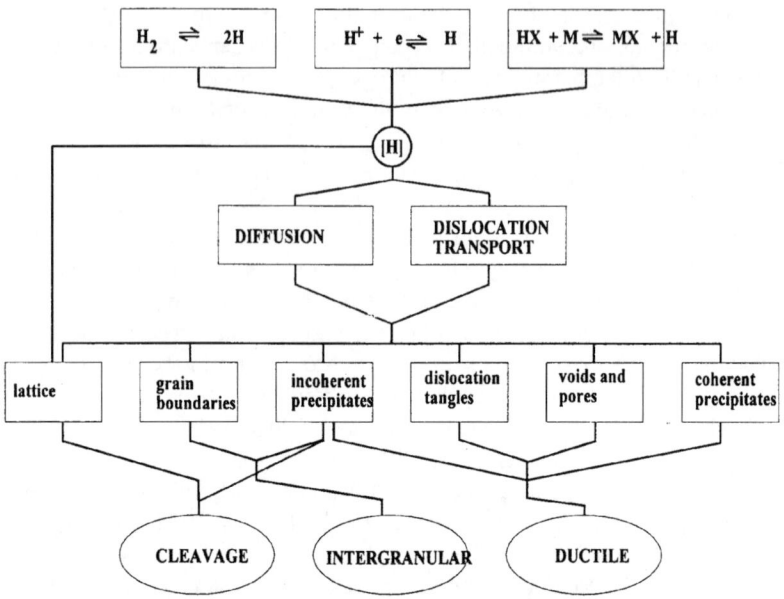

Figure 5.40 Mechanisms by which hydrogen can embrittle a metal or alloy (after Thompson and Bernstein, 1980) and interact to produce different forms of cracking

proposed a mechanism in which a dislocation pile-up at a grain boundary or inclusion interface effects microcrack formation. Hydrogen transfer to the micro-crack by dislocations, becomes adsorbed on the crack surface, thereby lowering the surface energy of the crack which is able to advance more readily. Bastein and Azou (1950, 1951a,b) proposed that hydrogen moved with dislocations which formed pile-ups under the influence of an applied stress. These pile-ups then develop into microcracks which grow by the increase in pressure caused by hydrogen recombina-tion, and this embrittles the material (Bastein *et al.*, 1960). Grain boundaries are clearly an important microstructural location for hydrogen because they feature both in hydrogen embrittlement alone and in fracture involving intergranular stress corrosion. The grain boundaries have an important role in increasing the diffusivity of hydrogen. Figure 5.41 compares the diffusivity for three nickel-base alloys, Inconel 600, 690 and 800, as a function of temperature (Uhlemann and Pound, 1998). The diffusivity increases with increasing nickel content as a result of a decrease in the apparent activation energy. This agrees well with the diffusion coefficient determined from the depth of intergranular fracture described previously by Younes *et al.*, (1997). Thus, although microcracks may form both at grain boundaries and within the matrix, because hydrogen can diffuse preferentially along the grain boundaries the resultant fracture path is likely to be intergranular. In addition, grain boundaries will frequently have both segregated impurities and precipitates present, this can further enhance the hydrogen segregation (Kimura

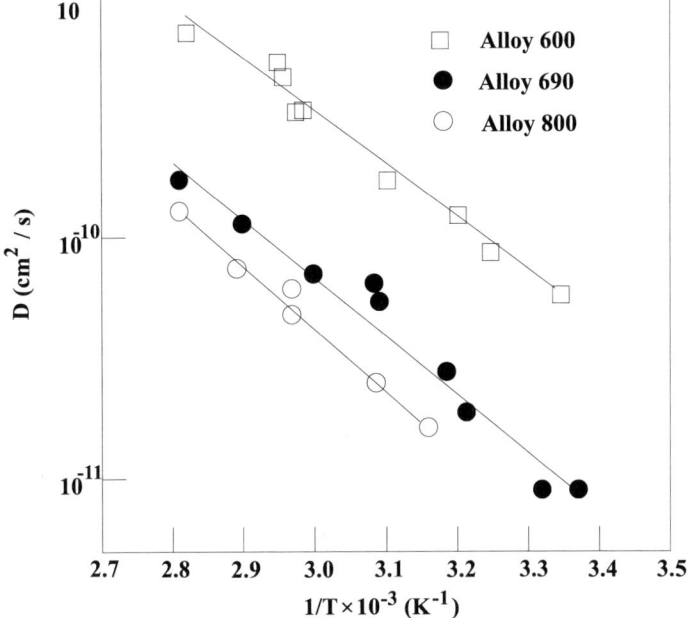

Figure 5.41 Grain boundary diffusivity of hydrogen in three nickel-base alloys, Inconel 600, 690 and 800 (after Uhlemann and Pound, 1998)

and Birnbaum, 1988) while also reducing the grain boundary energy. Hydrogen-assisted cracking requires a microcrack to form, either at a precipitate or grain boundary. As the crack opens, hydrogen diffuses, by one or more of the above mechanisms, to the plastic zone at the crack tip where it embrittles the region immediately ahead of the crack. In the presence of a stress the crack will continue to open.

5.9.6 Liquid Metal Embrittlement

When certain ductile metals and alloys are exposed to specific liquid metal environments they can become embrittled. It is not strictly necessary for the metal being embrittled to be immersed in the liquid metal; embrittlement will occur if the metal is exposed to a relatively high vapour pressure of the embrittling metal. Moreover, this behaviour occurs when the body is externally strained, but the phenomenon can also occur in the absence of applied mechanical stress. The classic example of the latter is the intergranular attack of β-brass by mercury at room temperature. However, there are now many examples of liquid metal embrittlement, for example, aluminium embrittled by liquid gallium and copper embrittled by liquid bismuth (Table 5.3).

Table 5.3   Metals susceptible to liquid
metal embrittlement

Metal	Embrittling metals
Fe	Bi, Pb, Sn, Zn
Zn	Bi, Cd, Hg, Sn, In, Ga
Cu	Bi, Cd, Pb, In, Hg
Al	Cd, In, Na, Sn, Hg, Zn, Ga
Ag	Ga, Hg
Cd	Ga, In, Sn
β-Brass	Hg

However, a recent review undertaken by Joseph *et al.* (1999) concluded that these mechanisms are not well established. One proposed model is based upon crack growth similar to diffusional cavity growth on boundaries. Under the influence of a tensile stress, the atoms in the vicinity of the crack tip move apart sufficiently to allow an atom from the embrittling liquid to occupy a lattice site that has been enlarged. This causes the stresses in this local region near the crack tip to become compressive, and it is this stress that is relieved by the formation of crack and the process repeats. A thermodynamic approach for the copper–liquid bismuth system is based upon the formation of a disordered metastable phase at the copper grain boundaries, which is a transitory state leading to the presence of a bismuth liquid film (Joseph, 1998). Irrespective of the mechanism, polycrystalline copper is embrittled by liquid bismuth when tested under a tensile load at a temperature of 573 K (Figure 5.42). There is a reduction in elongation and stress at fracture when tested in liquid bismuth compared with tests conducted in argon gas. The elastic behaviour of the copper is not changed by the presence of the liquid bismuth, whereas plastic deformation is necessary prior to rupture. In these tests, the strain rate did not affect elongation to rupture but, rather, increasing strain rate decreased rupture life.

The segregation of elements present in the bulk alloy can have a significant influence on the effectiveness of the liquid metal in embrittling the metal. In some cases embrittlement is increased while in others it is decreased. Dinda and Wark (1976) demonstrated that phosphorus and arsenic segregated to grain boundaries in a steel reduced the effectiveness of liquid metal embrittlement by lead, whereas segregation of tin or antimony to the grain boundaries increased the embrittling effect of the lead environment.

5.10   IRRADIATION DAMAGE

As described previously, irradiation can dramatically alter the grain boundary composition of a material, and the principles involved in radiation-induced segregation have been reviewed by English *et al.* (1990). A high flux of neutrons will

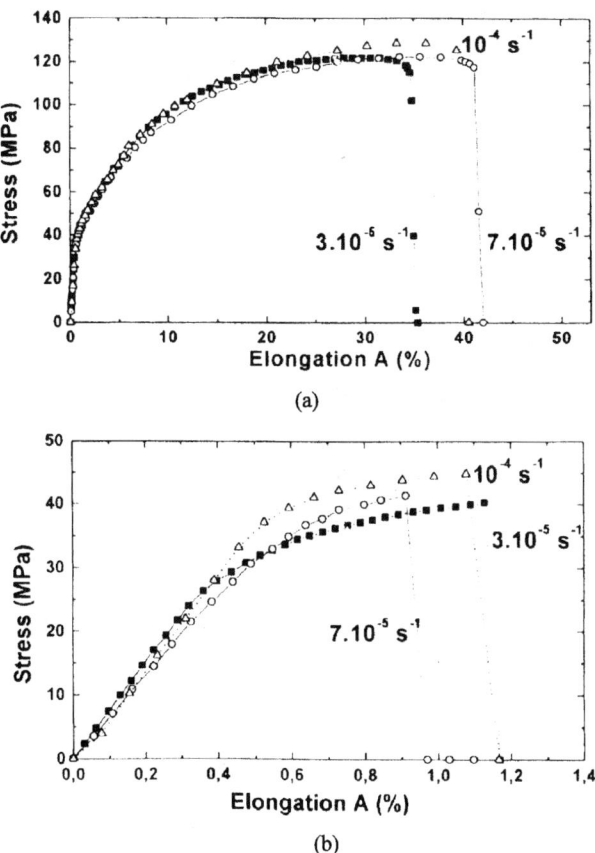

Figure 5.42 Embrittlement of copper by liquid bismuth when tested under a tensile load at 573 K (a) under argon gas (b) under liquid bismuth. Reproduced by permission of Pergamon Press from Joseph *et al.* (1999) *Eur. Phys. J. Appl. Phys.*, **5**, 4341

introduce a high density of defects into a material and create both vacancy and interstitial defects in the crystal lattice. In addition, interaction with the isotope [10]B produces an alpha particle which becomes a helium atom on acquiring an electron. At low temperatures the damage introduced by the neutron flux remains locked in original sites, but as the temperature increases so the vacancies, interstitials and helium atoms are free to diffuse. The vacancies diffuse towards grain boundaries. A flux of vacancies in one direction is equivalent to a flux of atoms in the opposite direction. Different atomic species will diffuse at different rates because of their different jump frequencies, and this will result in a change in atom concentrations in the direction of the vacancy flux; this is referred to as an inverse Kirkendall effect.

Nickel-base superalloys such as PE16, an alloy with 44 wt% Ni, 34 wt% Fe and 16 wt% Cr, have been examined following an irradiation of 18.9 GW days/t at a temperature of 631 K. It was found that the grain boundary composition of the nickel had increased to 65 at.%, whereas both chromium and iron had been depleted to 12 and 16 at.% respectively (Nettleship and Wild, 1992) (Figure 5.43). Similar effects have been observed in 20Cr–25Ni–Nb stabilised stainless steel (Norris *et al.*, 1992)

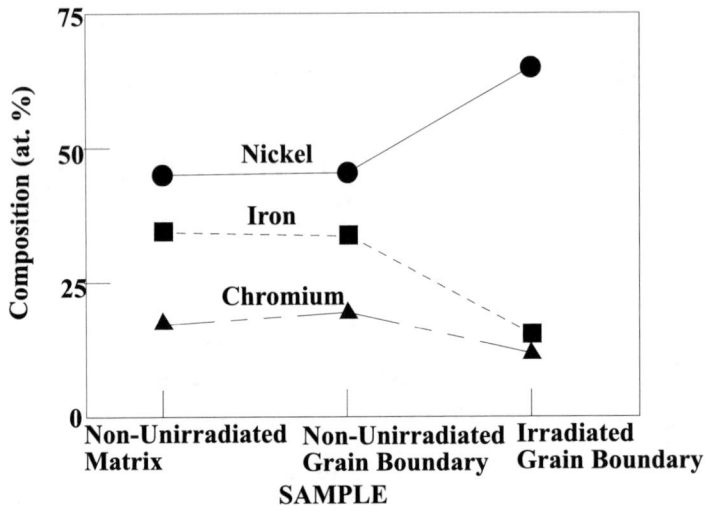

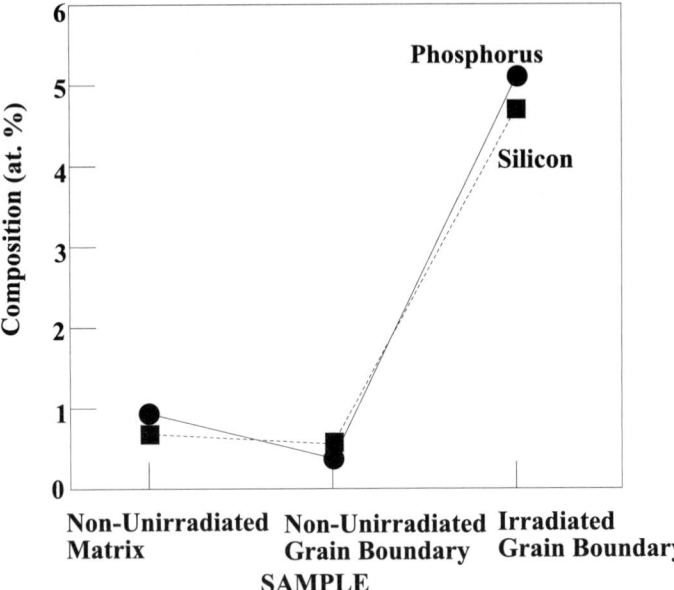

Figure 5.43   The effect of irradiation on grain boundary composition in alloy PE16 showing enrichment of trace elements P and Si and bulk element Ni and depletion of Cr and Fe (after Nettleship and Wild, 1992)

and in 12 % chromium martensitic steel (Morgan *et al.*, 1992), while the segregation and precipitation of tin in Zircaloy has been observed following irradiation damage (Woo and Carpenter, 1988; Motta *et al.*, 1992). Enhanced levels of trace elements such as silicon and phosphorus have also been detected on grain boundaries. These elements segregate to the boundaries in the absence of a flux of neutrons and, although it might be expected that the increase in defects would enhance the rate of diffusion, this has not been confirmed.

Neutron irradiation can also promote stress corrosion cracking of stainless steels in high-pressure water environments (Andersen *et al.*, 1990). Simonen *et al.* (1992) have modelled the effect of depletion of grain boundaries of chromium with neutron flux and related this to irradiation-induced IGSCC. Grain boundary composition changes and the effect on the intergranular stress corrosion susceptibility of austenitic stainless steels have been reviewed by several workers (Simonen and Bruemmer, 1998; Bruemmer *et al.*, 1992; Jenssen *et al.*, 1998). Certainly, chromium depletion at the grain boundaries is recognised as the major change promoting IGSCC in most service environments. Here, cracking susceptibility is a direct function of the local chromium concentration irrespective of the exact form and spatial extent of the profile. For specific alloys of given heat treatment, mechanical loading and environment threshold grain boundary compositions for the onset of IGSCC have been established. Indeed, in some materials very small depletions of chromium, ~ 1–2 wt% below the bulk concentration, are sufficient to promote IGSCC in oxygenated, high-temperature water typical of that encountered, for example, in a boiling water reactor. Both commercial and high-purity Type 304 and commercial purity Type 316 austenitic stainless steels can be rendered susceptible at doses of 1×10^{21} n cm^{-2} ($E > 1$ MeV). The preirradiation condition of the material is of major importance in regulating the segregation occurring at grain boundaries during irradiation. Here a prior heat treatment, such as quenching from a high temperature, can lead to non-equilibrium segregation of chromium and molybdenum to the grain boundaries, (Doig and Flewitt, 1985) but subsequent irradiation leads to phosphorus, silicon and nickel segregation to the boundaries with an associated depletion in chromium and molybdenum (Figure 5.44); these processes are again non-equilibrium. However, molybdenum-containing stainless steels may retain their chromium levels at low radiation fluences which then are depleted at higher fluences. Certainly, chromium depletion is important in the response to IGSCC of these steels, and the molybdenum exerts an influence on the segregation of chromium during irradiation. It is recognised, however, that other segregants such as Ni, Fe, Si and P can also influence susceptibility to IGSCC for this class of steels. Since the measured enrichment of nickel and the depletion of iron are nearly always consistent with the depletion of chromium, similar relationships to IGSCC result. There is, to date, little evidence that the minor elements or impurities influence the IGSCC susceptibility under neutron irradiation conditions, since high-purity steels show cracking similar to commercial materials where significant boundary enrichment of minor alloying or impurity elements occurs.

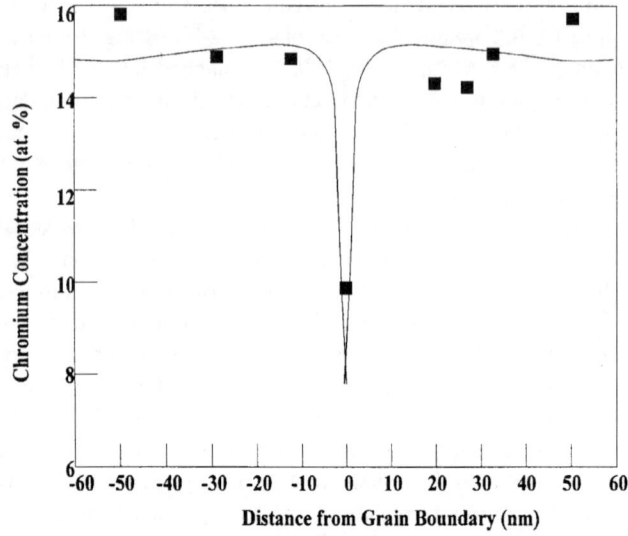

Figure 5.44 Comparison of predicted and measured chromium concentration profiles for ion irradiation in Type 316 stainless steel at 723 K (after Simonen *et al.*, 1992). Grain boundary at 0 nm position

5.11 OTHERS

5.11.1 Superplasticity

Most metals and alloys will elongate a few % before necking and then fracture. There is, however, a class of materials that extend several orders of magnitude more than these conventional materials without fracturing or necking. These are known as superplastic materials and their properties have been reviewed by Sherby and Wadsworth (1989). Extreme examples of these alloys include Pb–62 wt% Sn (Ahmed and Langdon, 1977) and copper–10 wt% aluminium (Higashi *et al.*, 1985), and they can all extend many thousands of % at strain rates of 10^{-1}–10^0, compared with strain rates of 10^{-3}–10^{-4} for conventional alloys (Cui *et al.*, 1994). Aluminium, titanium, zinc and iron alloys can all be made superplastic, as can certain ceramics. In general, most superplastic materials will elongate from 300 to 1000 %. All superplastic materials have a small grain size, typically between 1 and 5 μm, because the strain rate for superplasticity increases with decreasing grain size when grain boundary sliding is the rate-controlling mechanism. Also, the flow stress decreases with decreasing grain size, reducing the energy required for superplastic forming.

Superplasticity is rarely observed in single-phase materials because, where grain boundary sliding occurs, grain growth is too fast at superplastic forming temperatures. In almost all cases a second phase is present at the grain boundaries to pin the boundary and maintain the small grain size. The relative strengths of the second-phase particles is important. In materials where there is a large difference in strength, the alloys tend to be susceptible to the formation of cavities during superplastic flow, whereas alloys with second-phase strengths that are similar to the matrix do not exhibit this form of damage. It has been shown that, if hard second-phase particles are distributed as small evenly dispersed particles on the grain boundaries, then the tendency to cavitation is reduced. Chung and Cahoon (1979) have shown small silicon particles reduce cavitation in a fine-grained Al–Si eutectic. Grain boundaries in superplastic materials should generally be high-angle, mobile, readily slide for deformation, and not easily separate under a tensile load. Ceramic materials have a high surface energy and can readily separate, and this may account for the relatively low number of superplastic materials in this category. Polycrystalline zirconia can be made superplastic by producing material with very fine grains ($< 0.3 \, \mu m$) which reduces the stress needed for plastic flow below that for tensile fracture of the grain boundaries (Wakai and Kato, 1988). Aluminium alloys can readily be made to deform in a superplastic manner by the addition of Zr, Cr and Mn (Gandhi and Raj, 1991). They have shown that, to model superplastic flow a balance must be set between the arrival and emission of dislocations at grain boundaries. The emission rate depends on the angle of the boundary, while the arrival rate is determined by the mobility of the dislocations and this is controlled by the size and spacing of coherent intermetallic dispersoids in the lattice. The addition of these elements to conventional aluminium alloys leads to a fine-grained microstructure because the intermetallic phases formed by the additions pin the grain boundaries (Mishra *et al.*, 1995).

5.11.2 Nanocrystalline Materials

This is a class of polycrystalline materials that are made up of grains with a crystal size of only a few nanometres ($< 10 \, nm$). At this size, each grain will only contain a few thousand atoms, while the grain boundary surrounding the grain will consist of a few hundred atoms and the distinction between the grain and the grain boundary becomes less well defined. Figure 5.45a illustrates the atomic structure of a two-dimensional nanocrystal. Open or closed circles represent atoms of two chemically different crystals, and the dashed lines mark the boundary between adjacent crystals (Herr *et al.*, 1990). The properties of nanocrystalline materials have been reviewed by Gleiter (1989).

Nanocrystalline materials are formed in a number of ways, but all involve the production of small clusters of atoms. This can involve sputtering or laser ablation in a vacuum, condensation of volatilised clusters in inert atmospheres or condensing solutions of particles from the liquid phase. Figure 5.45b is an example of a transmission electron image of a gold nanowire, only a few atom layers thick, which exhibits an hexagonal structure in the centre (Takayanagi *et al.*, 1999). Most of these nanocrystalline materials contain impurities that may form interphase boundaries. In a nanocrystalline material consisting of silver and iron, the resulting composite was found to consist of iron and silver crystallites with solid solutions of iron in silver at

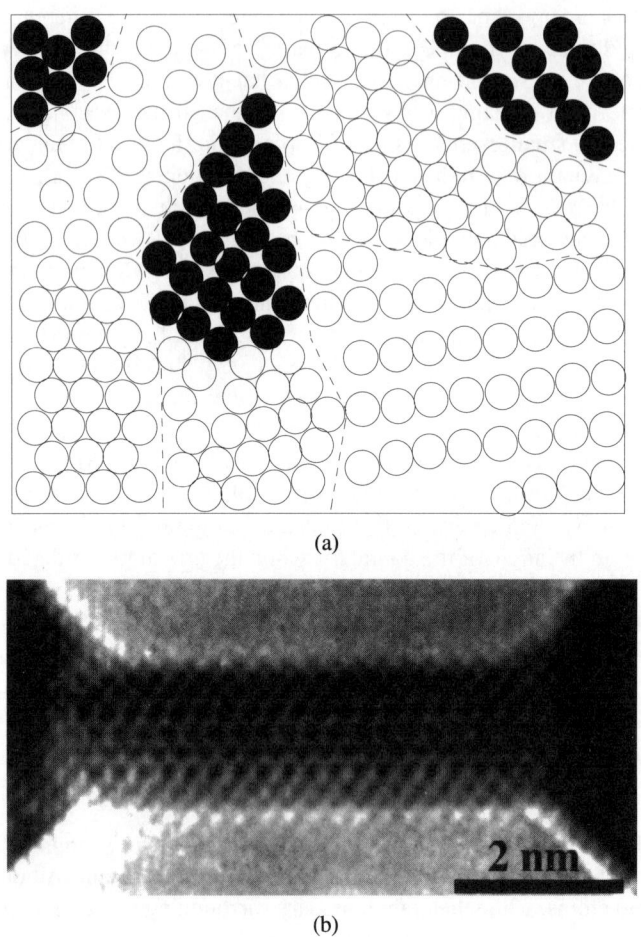

(a)

(b)

Figure 5.45 (a) Schematic cross-section through a mixture of nanometre-sized crystals. Open or closed circles represent atoms of two chemically different crystals and the dashed lines mark the boundary between adjacent crystals (after Herr *et al.*, 1990). (b) transmission electon image of a gold nanowire 2 nm in width, exhibiting an hexagonal lattice at the centre. Reproduced from Tagayanagi *et al.* (1999) *JEOL News*, **34E**, 20

the interfaces (Herr *et al.*, 1990). The mechanical properties of nanocrystalline materials show many similarities to superplastic materials. Conventionally brittle materials such as ceramics can exhibit high ductility at low temperatures (Karch *et al.*, 1987).

5.12 REFERENCES

Ahmed, M. M. I. and Langdon, T. G. (1977) *Metall. Trans. A*, **8A**, 1832.
Alberry, P. J., Chew, B. and Jones, W. K. C. (1977) *Metals Technol.*, June, 317.

Andersen, L. U. A., Bottiger, J. and Dyrbye, K. (1990) *Nucl. Instrum. Meth. B*, **51**, 125.

Angus, A. S., Chen, D. W. and Lau, C. W. (1980) *Intergranular Cavities in Creep, Theory and Experiments*, eds R. M. Pilloux and N. S. Stoloff, New York: The Metallurgical Society of AIME, p. 46.

Arnold, J. O. and Jefferson, J. (1896) *Engineering*, February.

Asaro, R. J. (1980) *Phil. Trans. R. Soc. (Lond.) A*, **295**, 151.

Asbury, F. E., Mitchell, B. and Toft, L. H. (1960) *Br. Weld. J.*, **7**, 598.

Ashby, M. F. and Brown, L. M. (1963) *Phil. Mag.*, **8**, 1083.

Ashby, M. F. and Easterling, K. E. (1982) *Acta Metall.*, **30**, 1969.

Ashby, M. F., Gandhi, C. and Taplin, D. M. R. (1979) *Acta Metall.*, **27**, 699.

Ashizuka, M., Okuno, T. and Kubota, Y. (1989) *J. Ceram. Soc. Jap.*, **97**, 662.

Ateya, B. G. and Pickering, H. W. (1974) in *Hydrogen in Metals*, eds Bernstein, I. M. and Thompson, A. W., Metals Park, OH: ASM, p. 207.

Azou, P. (1967) *Mécanique*, **52**, 7.

Bandyopadhyay, N. and Briant, C. L. (1982) *Corrosion*, **38**, 125.

Bandyopadhyay, N. and Briant, C. L. (1983) *Metall. Trans. A*, **14A**, 2005.

Bandyopadhyay, N. and Briant, C. L. (1985) *Corrosion*, **41**, 274.

Barrera, E. V., Menyhard, M., Bika, D., Rothman, B. and McMahon, C. J. (1992) *Scr. Metall.*, **27**, 205.

Bastein, P. and Azou, P. (1950) *C. R. Acad. Sci.*, **231**, 147.

Bastein, P. and Azou, P. (1951a) *C. R. Acad. Sci. Paris*, **232**, 1845.

Bastein, P. and Azou, P. (1951b) in *Proceedings of the First World Metallurgical Congress*, Cleveland, OH: ASM, p. 535.

Bastein, P., Azou, P. and Plusquellec, J. (1960) *J. Iron Steel Inst.*, **196**, 89.

Batte, A. D. and Murphy, M. C. (1979) *Full Metals Technol. (Lond.)*, **6**, 62.

Batte, A. D. and Murphy, M. C. (1980) *Phil. Trans. R. Soc. Lond. A*, **295** 293.

Baukloh, W. and Retziaff, W. (1937) *Arch. Eisenhüttenwes.*, **11**, 97.

Beck, T. R. (1971) in *The Theory of Stress Corrosion Cracking in Alloys*, ed. J. C. Scully, Brussels: NATO, p. 64.

Beere, W. (1980) *J. Mater. Sci.*, **15**, 657.

Beere, W. (1981) in *Cavities and Cracks in Creep and Fatigue*, ed. J. Gittus, Applied Science, p. 33.

Beere, W. and Roberts, G. (1982) *Acta Metall.*, **30**, 571.

Beevers, C. J. and Arrendu-Edmonds, W. (1968) *J. Less Common Metals*, **14**, 387.

Berg, O. (1958) *Corrosion*, **14**, 562.

Besnard, S. (1952) *Corros. and Anticorros.*, **10**, 182.

Besnard, S. and Talbot, J. (1959) *Rev. Metall.*, **56**, 163.

Bika, D., Pfaendtner, J. A., Menyhard, M. and McMahon, C. J. (1995) *Acta Metall.*, **43**, 1895.

Bika, D. and McMahon, C. J. (1993) *Mater. Res. Soc. Symp. Proc.*, **238**, 399.

Bika, D. and McMahon, C. J. (1995) *Acta Metall. Mater.*, **43**, 1909.

Blackburn, M. J., Feeney, J. A. and Beck, T. R. (1973) in *Advances in Corrosion Science and Technology*, New York: Plenum Press, Vol. 3, p. 67.

Blanchard, P. and Troiano, A. R. (1960) *Mem. Sci. Rev. Metall.*, **57**, 409.

Bodenstein, M. (1922) *Z. Electrochem.*, **23**, 517.

Boniszewski, T. (1982) *Weld. J.*, **161**, S164.

Bowen, P. and Hippsley, C. A. (1988) *Acta Metall.*, **36**, 425.

Bowen, P., Hippsley, C. A. and Knott, J. F. (1990) *Mater. Sci. Technol.*, **6**, 562.

Brian, J. and King, B. L. (1980) CEGB Report LM/MA/73/348.

Briant, C. L. (1980) *Corrosion*, **36**, 497.

Briant, C. L. and Banerji, S. K. (1978) *Int. Metals Rev.*, **4**, 232.

Briant, C. L. and Messmer, R. P. (1980) *Phil. Mag.*, **42**, 569.

Briant, C. L. and Messmer, R. P. (1982) *Acta Metall.*, **30**, 1811.

Briant, C. L. and Messmer, R. P. (1984) *Acta Metall.*, **32**, 2043.
Brown, B. F., Fujii, C. T. and Dahlberg, E. P. (1969) *J. Electrochem. Soc.*, **116**, 218.
Brown, L. M., Allen, G. C. and Flewitt, P. E. J. (1997) *Phys. World*, **45**, May.
Bruemmer, S. M. (1998) *Intergranular and Interphase Boundaries in Materials*, eds P. Lejček and V. Pandon, Trans Tech Publ (Switzerland), p. 75.
Bruemmer, S. M., Arey, B. W. and Charlot, L. A. (1992) *Corrosion*, **48**, 42.
Bruscato, R. (1970) *Weld. J. Res. Suppl.*, **49**, 148.
Bulloch, J. H. and Crowe, D. (1998) *Theor. Appl. Fracture Mech.*, **29**, 59.
Bulloch, J. H. and Wild, R. K. (1995) *Eng. Mater.*, **99**(1), 169.
Burton, B. (1976) *Vacancies '76*, eds R. E. Smallman and J. E. Harris, London: The Metals Society, p. 156.
Cannon, W. R. and Langdon, T. G. (1983) *J. Mater. Sci.*, **18**, 1.
Chang, K. M., Henry, M. F. and Benz, M. G. (1990) *JOM*, December, 29.
Chen, I.-W. and Argon, A. S. (1981) *Acta Metall.*, **29**, 1321.
Cheruvu, N. S. (1989) *Metall. Trans. A*, **20A**, 87.
Christoffel, R. J. (1962) *Weld. J. Res. Suppl.*, **11**, 251.
Chung, J. W. and Cahoon, J. R. (1979) *Metal Sci. J.*, **13**, 635.
Cocks, A. C. F. and Ashby, M. F. (1979) *Acta Metall.*, **27**, 1505.
Coehn, A. and Sperling, K. (1933) *Z. Phys.*, **83**, 291.
Cohron, J. W., Lui, Y., Zee, R. H. and George, E. P. (1998) *Acta Metall.*, **46**, 6245.
Cottrell, A. H. (1964) *The Mechanical Properties of Matter*, New York: Wiley.
Cottrell, A. H. (1989) *Mater. Sci. Technol.*, Ser. 5, 1165.
Cottrell, A. H. (1990a) *Mater. Sci. Technol.*, Ser. 6, 974.
Cottrell, A. H. (1990b) *Mater. Sci. Technol.*, Ser. 6, 807.
Cox, T. B. and Low, J. R. (1974) *Metall. Trans.*, **5**, 1457.
Crocker, A. G. and Flewitt, P. E. J. (1984) in *Reviews of Deformation Behaviour of Materials*, ed. P. Feltham, Tel Aviv: Freund Publ. House.
Crocker, A. G., Smith, G. E., Moskovic, R. and Flewitt, P. E. J. (1998) *Intergranular and Interphase Boundaries in Materials*, eds P. Lejcek and V. Pandon, Trans Tech Publ. (Switzerland), p. 673.
Crossman, F. W. and Ashby, M. F. (1975) *Acta Metall.*, **23**, 425.
Cui, Z., Zhong, W. and Wei, Q. (1994) *Scr. Metall. Mater.*, **30**, 123.
Darken, L. S. and Smith, R. (1949) *Corrosion*, **5**, 1.
Davidenkov, N. N. (1938) *Problems of the Impact in Materials Science*, Moscow: USSR Acad. Sci.
Davidenkov, N. N. (1981) *Dynamic Strength and Brittleness of Materials*, Kiev: Naukova Dumka, Vol. 1.
De Kazincky, F. (1959) *Acta Metall.*, **7**, 706.
Denk, J. (1994) Proc. Eurocorr 94 and UK Corrosion **2**, 31 October–4 November, Bournemouth, UK.
Dhooge, A. and Vinckier, A. (1982) Commission IX Int. Inst. Welding Doc. IX-T250-82, Ljubjanca.
Dhooge, A. and Vinckier, A. (1987) *Int. J. Pressure Vessels Piping*, **27**, 239.
Dickenson, J. H. S. (1917–19) *J. Auto Eng. Disc. Phil. Paper*, **12**(1), 342.
Dinda, S. and Warke, W. S. (1976) *Mater. Sci. Eng.*, **24**, 199.
Doig, P. and Flewitt, P. E. J. (1977) *Proc. R. Soc. (Lond.) A*, **A357**, 439.
Doig, P. and Flewitt, P. E. J. (1978) *Acta Metall.*, **26**, 1283.
Doig, P. and Flewitt, P. E. J. (1984) *Embrittlement by the Localised Crack Environment*, ed. R. P. Gangloff, New York: The Metallurgical Society of AIME, p. 304.
Doig, P. and Flewitt, P. E. J. (1985) *Solute-Defect Interaction Theory and Experiment*, eds. S. Samamoto, G. R. Pendry and G. V. Kidson. Toronto: Pergamon Press, p. 434.
Doig, P., Flewitt, P. E. J. and Edington, J. W. (1977) *Corrosion*, **33**, 217.

Doig, P., Lonsdale, D. and Flewitt, P. E. J. (1982) *Metal Sci.*, **16**, 335.

Dokko, P. C., Pask, J. A. and Mazdiyasni, K. S. (1977) *J. Am. Ceram. Soc.*, **60**, 150.

Donovan, J. A. (1976) *Metall. Trans. A*, **7A**, 145 and 1677.

Edwards, B. C., Gaze, G. and Eyre, B. L. (1980) *Acta Metall.*, **28**, 335.

Edwards, C. A. (1924) *J. Iron Inst.*, **110**, 9.

Emmer, L. G., Clauser, C. D. and Law, J. R. (1973) *Weld. Res. C Bull.*, **183**.

English, C. A., Murphy, S. M. and Perks, J. M. (1990) *Faraday Trans., J. Chem. Soc.*, **86**, 1263.

EPRI (1983) Report CS-3248, Project 559.

Erwin, W. E. and Kern, J. G. (1982) *Weld. Res. Coun. Bull.*, **275**.

Evans, R. W. and Wilshire, B. (1985) *Creep of Metals and Alloys*. London: Institute of Metals.

Exworthy, L. F., Ellis, B. J. C. and Flewitt, P. E. J. (1999) Boiler Shell Weld Repair, *I. Mech. E. Summer Pub.*, Prof. Eng. Pub. (London) p. 9.

Finnis, M. W. (1982) *Defects and Fracture. Proc. 1st Int. Symp. Tuczno, Poland, October 1980*, ed. Sih Zorski, The Hague/Boston/London: Martinus Nijhof, p. 253.

Foster, P., MacNabb, A. and Payne, C. (1965) *Trans. AIME*, **23**, 1022.

Frandsen, J. D., Paton, N. E. and Marcus, H. L. (1973) *Scr. Metall.*, **7**, 409.

Friedel, J. (1969) *The Physics of Metals*, ed. J. M. Ziman, Cambridge University Press, Section 8.2.1.

Gandhi, C. and Ashby, M. F. (1979) *Acta Metall.* **27**, 1565.

Gandhi, C. and Raj, R. (1991) *Acta Metall. Mater.*, **39**, 679.

George, E. P., Li, P. L., Pope, D. P. (1987) *Acta Metall.*, **35**, 2487.

Gittus, J. (1975) *Creep Viscoelasticity and Creep Fracture in Solids*, London: Applied Science.

Gittus, J. (1978) *Irradiation Effects in Crystalline Solids*, London: Applied Science.

Gleiter, H. (1989) *Prog. Mater. Sci.*, **33**, 223.

Grabke, H. J. (1989) *Surf. Interface Analysis*, **14**, 686.

Griffiths, A. A. (1920) *Phil. Trans. R. Soc. (Lond.) A*, **A221**, 163.

Guo, H., Chaturvedi, M. C., Richards, N. L. and McMahon, G. S. (1999) *Scr. Metall.*, **40**, 383.

Gupta, D. (1976) *Phil. Mag.*, **37**, 189.

Gupta, D. (1977) *Metall. Trans. A*, **8A**, 1431.

Hansworth, B., Bornszewski, T. and Eaton, N. F. (1969) *Br. Weld. J. Suppl.*, **1**(2), 5.

Harrison, R. P., Jones, D. de G. and Newman, J. F. (1977) *Int. Cong. on Stress Corrosion Cracking and Hydrogen Embrittlement of Iron Base Alloys*, eds R. W. Staehle, J. Hochmann and R. D. McCright. France: Ferning/Houston, TX: NACE, p 659.

Hart, E. W. (1967) *Acta Metall.*, **15**, 1545.

Hashimoto, M., Ishida, Y., Wakayama, S., Yamamoto, R., Doyama M. and Fujiwara, T. (1984) *Acta Metall.*, **32**, 13.

Herr, U., Jing, J., Gonser, U. and Gleitner, H. (1990) *Solid State Commun.*, **76**, 197.

Higashi, K., Ohnishi, T. and Nakatani, Y. (1985) *Scr. Metall.* **19**, 821.

Hippsley, C. A. (1981) *Metal Sci.*, **15**, 137.

Hippsley, C. A. (1985) *Mater. Sci. Technol.*, **1**, 475.

Hippsley, C. A. (1987) *Acta Metall.*, **35**, 2399.

Hippsley, C. A. and De Van, J. H. (1989) *Acta Metall.*, **37**, 1485.

Hippsley, C. A., Rauh, H. and Bullough, R. (1984) *Acta Metall.*, **32**, 1381.

Hippsley, C. A., Briant, C. L. and Edwards, B. C. (1986) *Mater. Sci. Technol.*, **2**, 386.

Hippsley, C. A., Buttle, D. J. and Scruby, C. B. (1988) *Acta Metall.*, **36**, 441.

Hirth, J. P. (1980) *Phil. Trans. R. Soc. (Lond.) A*, **295**, 139.

Holdsworth, S. R. and Thornton, J. V. (1993) *Microstructure and Mechanical Properties of Ageing Materials*, ed. P. K. Liaw, Waarrendale: Minerals, Metals and Materials Society, p. 83,.

Hondros, E. D. (1965) *Proc R. Soc. (Lond.) A*, **286**, 479.

Hondros, E. D. and McLean, D. (1974) *Phil. Mag.*, **29**, 271.
Hour, K. Y. and Stubbins, J. F. (1990) *Acta Metall.*, **38**, 1463.
Huang, X., Chanturvedi, M. C. and Richards, N. L. (1996) *Metall. Mater. Trans. A*, **27A**, 785.
Hull, D. and Rimmer, D. E. (1959) *Phil. Mag.*, **4**, 673.
Huntington, J. (1912) *Inst Metals*, **8**, 126.
Jaffee, R. I. (1986) *Metall. Trans. A*, **17A**, 755.
Jaffee, R. I., Machner, P., Meyer, W. and Steiner, J. E. (1986) *Iron and Steel Making*, **13**, 322.
Jenssen, A., Ljungberg, L. G., Walmsley, J. and Fisher, S. (1998) *Corrosion*, **54**, 48.
Jin Yu and McMahon, C. J. (1980) *J. Metall. Trans. A*, **11A**, 291.
Johnson, E. and Hill, M. (1960) *Trans. AIME*, **218**, 1104.
Johnson, H. H. (1969) in *Fundamental Aspects of Stress Corrosion Cracking*, eds R. W. Staehle *et al.*, Houston: NACE, p. 439.
Jokl, M. L., Vitek, V. and McMahon, C. J. (1980a) *Acta Metall.*, **28**, 1479.
Jokl, M. L., Kameda, J., McMahon, C. J. and Vitek, V. (1980b) *Metal Sci.*, **14**, 375.
Jones, R. H., Danielson, M. J., Bruemmer, S. M., Baer, D. R. and Thomas, M. T. (1984) *Embrittlement by the Localized Crack Environment*, ed. R. P. Gangloff. New York: ASM.
Joseph, B. (1998) Thesis, University Paris-XI, Orsay, France, 16 October.
Joseph, B., Picat, M. and Barbier, F. (1999) *Eur. Phys. J. Appl. Phys.*, **5**, 4341.
Joshi, Y. U. and Nix, W. D. (1983) *Metall. Trans. A*, **14A**, 2447.
Kameda, J. and McMahon, C. J. (1981) *Metall. Trans. A*, **12A**, 31.
Karch, J., Birringer, R. and Gleiter, H. (1987) *Nature*, **330**, 556.
Keeler, J. and Davis, H. (1953) *Trans. AIME*, **193**, 44.
Kelemen, D. G. (1955) Dissertation, University of Princetown, NY, Microfilm Publ. No. 10, Vol. 935, 135.
Kimura, A. and Birnbaum, H. K. (1988) *Acta Metall.* **36**, 757.
King, B. L. and Wigmore, G. (1973) *Metall. Trans. A*, **7A**, 1761.
Koterazawa, R. and Mori, T. (1980) *Engineering Aspects of Creep*, London: Institution of Mechanical Engineers, Vol. 1, 219.
Kupper, J., Erhart, H. and Grabke, H. J. (1981) *Corros. Sci.*, **21**, 227.
Kussmaul, K., Blind, D. and Ewald, V. (1977) *Int. J. Pressure Vessels Piping*, **5**, 159.
Lacombe, P. and Parkins, R. N. (1977) *Stress Corrosion Cracking and Hydrogen Embrittlement of Iron Base Alloys*, NACE, p. 521.
Latanision, R. M. and Staehle, R. W. (1969) *Fundamental Aspects of Stress Corrosion Cracking*, ed. R. W. Staehle, Hannover: NACE.
Latanision, R. M. and Opperhauser, H. (1974) *Metall. Trans.*, **5**, 483.
Lee, D. Y., Barrera E. V., Stark, W. P. and Marcus, H. L. (1984) *Metall. Trans. A*, **15A**, 1415.
Lewandowski, J. J., Hippsley, C. A. and Knott, J. F. (1987) *Acta Metall.*, **35**, 2081.
Lindley, T. C., McIntyre, P. and Trant, P. J. (1982) *Metals Technol.*, **9**, 135.
Lonsdale, D. and Flewitt, P. E. J. (1979) *Mater. Sci. Eng.*, **41**, 127.
Lovell, A. J. (1975) *Nucl. Technol.*, **26**, 297.
Lui, C. T. and White, C. L. (1987) *Acta Metall.*, **35**, 643.
Lyle Jr, F. F. and Burghard Jr, H. C. (1982) *Mater. Perform.*, **21**, 35.
McIntyre, P., Younes, C. M. and Chan, S. W. (1996) *Br. Corros. J.*, **31**, 133.
McLean, D. (1968) *Metall. Trans. Metals Soc. AIME*, **242**, 1193.
McMahon, C. J. (1975) *Grain Boundaries in Engineering Metals*, ed. J. L. Walton, Baton Rouge: Claitors, p. 525.
McMahon, C. J. (1984) in *Advances in Fracture Research*, Oxford: Pergamon Press, p. 143.
McMahon, C. J. and Vitek, V. (1979) *Acta Metall.*, **27**, 507.
McMinn, A., Lyle Jr, F. F. and Leverant, G. R. (1985) *Corrosion* **41**, 493.
Marek, M. and Hochman, R. F. (1970) *Corrosion*, **26**, 5.
Messmer, R. P. and Briant, C. L. (1982) *Acta. Metall.*, **30**, 457
Middleton, C. J. (1980) *Phil. Trans. R. Soc. (Lond.) A*, **295**, 265.

Middleton, C. J. (1981) *Metal Sci.*, **15**, 154.
Min, B. K. and Raj, R. (1978) *Acta Metall.*, **26**, 1007.
Mishra, R. S., Bieler, T. R. and Mukherjee, A. K. (1995) *Acta Metall. Mater.*, **43**, 877.
Misra, R. D. K. and Balasubramanian, T. V. (1990) *Acta Metall.* **38**, 1263.
Misra, R. D. K., McMahon, C. J. and Guha, A. (1994) *Scr. Metall. Mater.*, **31**, 1471.
Molonznik, K. L., Briant, C. L. and McMahon Jr, C. J. (1979) *Corrosion*, **35**, 331.
Morgan, T. S., Little, E. A., Faulkner, R. G. and Titchmarsh, J. M. (1992) *Effects of Radiation on Materials: 15th International Symposium*, eds R. E. Stoller, A. S. Kumar and D. S. Gelles, Philadelphia: ASTM, STP1125, p. 633.
Motta, A. T., Lemaignan, C. and Olander, D. R. (1992) *Effects of Radiation on Materials: 15th International Symposium*, eds R. E. Stoller, A. S. Kumar and D. S. Gelles, Philadelphia: ASTM, STP1125, p. 689.
Muller, D. A. (1996) PhD thesis, Cornell University.
Muraleedham, P., Schneider, P. and Mummer, K. (1999) *J. Nucl. Mater.*, **270**, 342.
Murr, L. E. (1975) *Interfacial Phenomena in Metals and Alloys*, Addison Wesley, Reading MA.
Murza, J. C. and McMahon, C. J. (1980) *J. Eng. Mater. Technol.*, **102**, 369.
Muthiah, R. C., Pfaendtner, J. A., Ishikawa, S. and McMahon, C. J. (1999) *Acta Metall.*, **47**, 2797.
Myers, J. (1977) *Metals Technol.*, **4**, 411.
Myers, J. (1980) *Phil. Trans. R. Soc. (Lond.) A*, **295**, 289.
Nabarro, F. R. N. and de Villier, H. L. (1995) *The Physics of Creep*, London: Taylor and Francis.
Nakamura, H., Naiki, T. and Okobayashi, H. (1970) *Trans. Jap. Weld. Soc.*, **60**, 210.
Nakamura, T., Shiroda, T. and Watanabe, H. (1979) *Trans. Jap. Iron Steel Ind.*, **19**, 365.
Nettleship, D. J. and Wild, R. K. (1992) *Effects of Radiation on Materials: 15th International Symposium*, eds R. E. Stoller, A. S. Kumar and D. S. Gelles, Philadelphia: ASTM, STP1125, p. 645.
Nixon, R. D., Chevacharoenkul, S., Davis, R. F. and Tiegs, T. N. (1990) *Ceram. Trans.*, **6**, 579.
Norris, D. I. R., Baker, C., Taylor, C. and Titchmarsh, J. M. (1992) *Effects of Radiation on Materials: 15th International Symposium*, eds R. E. Stoller, A. S. Kumar and D. S. Gelles, Philadelphia: ASTM, STP1125, p. 603.
O'Brian, T. J. and Wolstenholm, D. A. (1984) CEGB Report TPRD/M/1356/R84.
Okamoto, P. R., Santnanan, A. T., Wiedersich, H. and Taylor, A. (1974) *Nucl. Technol.*, **22**, 45.
Olefjord, I. (1978) *Int. Metals Rev.*, **231**, 149.
Ortner, S. R. and Hippsley, C. A. (1992) *Mater. Sci. Technol.*, **8**, 883.
Parker, J. D. and Wilshire, B. (1980) *Phil. Mag.*, **41A**, 665.
Petch, N. J. and Stables, P. (1952) *Nature*, **169**, 842 1952.
Petch, N. S. (1956) *Phil. Reg.*, **1**, 331.
Pfeil, L. B. (1926) *Proc. R. Soc. (Lond.) A*, **112**, 182.
Plusquellec, J. (1960) *Mem. Sci. Rev. Metall.*, **57**, 215.
Plusquellec, J., Azou, P. and Bastein, P. (1957) *C. R. Acad. Sci. Paris*, **244**, 1195.
Povich, M. D. (1978) *Corrosion*, **34**, 60.
Pugh, S. F. (1991) *An Introduction to Grain Boundary Fracture in Metals*, London: Institute of Metals.
Rauh, H., Hippsley, C. A. and Bullough, R. (1989) *Acta Metall.*, **37**, 269.
Rellick, J. R. and MacMahon, C. J. (1974) *Metall. Trans.*, **5**, 2439.
Rice, J. R. (1976) *Effect of Hydrogen on the Behaviour of Materials*, eds A. W. Thompson and I. M. Rennstein, Norfolk: AIME p. 455.
Rice, J. R. and Thomas, R. (1974) *Phil. Mag.*, **29**, 73.
Riedel, H. (1987) *Fracture at High Temperatures*, Berlin: Springer.
Ritter, J. C. and McPherson, R. (1974) *Metals Technol.*, **1**, 506.

Robinson, J. L. and Scott, M. H. (1980) *Phil. Trans. R. Soc. (Lond.) A*, **295**, 105.

Rosenthal, D. (1946) *Trans. ASME*, **68**, 849.

Rundgren, K., Elfving, P., Tabata, H. and Kanzaki, S. (1990) *Ceram. Trans.*, **6**, 553.

Savage, W. F. (1980) *Weld. World*, **18**, 89.

Schneider, H., Okada, K. and Pask, J. (1994) *Mullite and Mullite Ceramics*, Chichester: John Wiley & Sons.

Schulz, B. J. and McMahon, C. J. (1973) *Metall. Trans.*, **4**, 2485.

Seah, M. P. (1976) *Proc. R. Soc. (Lond.) A*, **349**, 535.

Seah, M. P. (1977) *Acta Metall.*, **25**, 345.

Seah, M. P. (1980a) *Phil. Trans. R. Soc. (Lond.) A*, **295**, 265.

Seah, M. P. (1980b) *Residuals and Additives and Materials Physics*, London: Royal Society.

Seah, M. P. (1980c) *Acta Metall.*, **28**, 955.

Sedriks, A. J., Green, J. A. S. and Novak, D. L. (1971) *Corrosion*, **27**, 198.

Sherby, O. D. and Wadsworth, J. (1989) *Progr. Mater. Sci.*, **33**, 169.

Shiga, H. and Katayama, K. (1991) *Ceramic Society of Japan*, **99**, 798.

Shin, J. and McMahon, C. J. (1984) *Acta Metall.*, **32**, 1535.

Shin, J., Kameda, J. and McMahon, C. J. (1982) in *Micro and Macro Mechanisms of Crack Growth*, eds K. Sadamanda, B. B. Rath and D. J. Michel, New York: The Metallurgical Society of AIME, p. 149.

Simonen, E. P. and Bruemmer, S. M. (1999) *Mater. Sci. Forum*, **294**, 755.

Simonen, E. P., Charlot, L. A. and Bruemmer, S. M. (1992) *Effects of Radiation on Materials: 15th International Symposium*, eds R. E. Stoller, A. S. Kumar and D. S. Gelles, Philadelphia: ASTM STP1125, p. 621.

Skelton, R. P. (1975) *Metal Sci.*, **9**, 192.

Smith, E. (1966) *Physical Basis of Yield and Fracture, Proceedings*, London Institute of Physics, p. 36.

Smith, G. E., Crocker, A. K., Flewitt, P. E. J. and Moscovic, R. (1997) *Damage and Failure of Interfaces*, ed. P. Rossmouth, Rotterdam: Balkemov, p. 229.

Smith, J. F., Reynolds, J. H. and Southworth, H. N. (1982) *Metal Sci.*, **16**, 431.

Speidel, M. O. (1974) in *Hydrogen in Metals*, eds I. M. Bernstein and A. W. Thompson, Metals Park, OH: ASM, p. 249.

Steigerwald, E. A., Schaller, F. W. and Troiano, A. R. (1959) *Trans. AIME*, **215**, 1048.

Svoboda, J. and Sklenicka, V. (1990) *Acta Metall.*, **38**, 1141.

Tagayanagi, K., Kondo, Y. and Ohnishi, H. (1999) *JEOL News*, **34E**, 20.

Teirlinck, D., Zok, F., Embury, J. D. and Ashby, M. F. (1988) *Acta Metall.*, **36**, 1213.

Thomas, R. D. (1984) *Weld. J.*, **63**, 24.

Thompson, A. W. (1973) *Metall. Trans.*, **4**, 2819.

Thompson, A. W. (1974) *Metall. Trans.*, **5**, 1855.

Thompson, A. W. and Bernstein, I. M. (1980) *Advances in Corrosion Science and Technology*, eds M. G. Fontana and R. W. Staehle, New York: Plenum, Vol. 7, p. 53.

Thompson, A. W. and Knott, J. F. (1993) *Metall. Trans.*, **24A**, 523.

Thompson, A. W. and Wilcox, B. A. (1972) *Scr. Metall.*, **6**, 689.

Thompson, R. G., Dobbs, J. R. and Mayo, D. E. (1986) *Weld. J.*, **65**, 299.

Tien, J. K., Richards, R. J., Buck, O. and Marcus, H. L. (1975) *Scr. Metall.*, **9**, 1097.

Troiano, A. R. (1960) *Trans. ASM*, **52**, 54.

Uhlemann, M. and Pound, B. G. (1998) *Corros. Sci.*, **40**, 645.

Vincent, R. (1985) *Acta Metall.*, **33**, 1205.

Vitovec, F. (1964) *Proc. Am. Petrol. Inst.*, **44**, 179.

Wakai, F. and Kato, H. (1988) *Advd. Ceram. Mater.*, **3**, 71.

Wang, C. Y. and Zhao, D. L. (1994) *Mater. Res. Soc. Symp. Proc.*, **318**, 571.

Was, G. and Latanision, R. M. (1981) *Metall. Trans.*, **12A**, 1397.

Weiss, B. and Stickler, R. (1972) *Metall. Trans.*, **3**, 851.

Werner, J. and Davis, H. M. (1961) *Trans. ASM*, **53**, 853.

Wigmore, G. (1973) CEGB Report SSD/SW/R172.

Wignarajah, S., Masumoto, I. and Hara, T. (1990) *ISI J. Int.*, **30**, 58.

Wilcox, B. A. and Clauer, A. H. (1972) *The Superalloys*, eds C. T. Sims and W. C. Hazel, New York: John Wiley & Sons, p. 197.

Wild, R. K. (1997) *Mater. World*, July, 389.

Wilkinson, D. S., Abiko, K., Thyagorajan, N. and Pope, D. P. (1980) *Metall. Trans. A*, **11A**, 1827.

Williams, D. P. and Nelson, H. G. (1987) *Metall. Trans.*, **2**, 1987.

Woo, O. T. and Carpenter, G. J. C. (1988) *J. Nucl. Mater.*, **159**, 397.

Wyatt, L. M. (1971) *J. Mater. Sci. Eng.*, **7**, 237.

Yamashita, M., Mimaki, T., Hashimoto, S. and Miura, S. (1991) *Phil. Mag.*, **A63**, 707.

Younes, C. M., McIntyre, P. and Jones, K. I. (1994) *Proc. Conf. UK Corr and Eurocorr '94*, Bournemouth, UK, October–November 1994, Institute of Materials, Vol. 2, p. 82.

Younes, C. M., Morrissey, F. H., Allen, G. C. and McIntyre, P. (1997) *Br. Corros. J.*, **32**, 185.

Yu, K. S., Joshi, Y. U. and Nix, W. D. (1983) *Met. Trans.* **14A**, 2447.

APPENDIX

Table A.1 Segregation at grain boundaries in iron-base alloys and steels (after Hondros and Seah, 1977)

Material solute	Solvent (wt%)	Element segregating	Monolayers segregated	Enrichment ratio, β_b	Heat treatment[a]
Fe	Sb (5.0)	Sb	2.1	95	SC
AISI 3340	Sb (0.065)	Sb	0.65	2150	SC
		Ni	0.75	22	
AISI 3340	Sb (0.062)	Sb	0.53	1880	SC
		Ni	1.31	38	
Fe–0.3C–4Ni–2.2Cr	P (0.02)	P	0.79	2180	SC
		Ni	0.39	11	
Fe–0.04C	Te (0.02)	Te	1.79	20000	FC
Fe	Sb (1.5)	Sb	1.00	150	AS
AISI 3340, 0.4Mo–0.12V	P (0.21), Sn (0.0128)	P	0.36	950	IT
		Sn	0.12	2000	
		Ni	0.87	24	
		Cr	0.45	22	
Fe–0.4C–3.5Ni–1.7Cr	Sb (0.062)	Sb	0.58	2039	SC
		Cr	0.36	20	
		Ni	0.44	13	
Fe–0.3C–3.7Ni	Sn (0.048)	Sn	0.19	820	SC
		Ni	0.05	2	
Fe–0.4C–3.5Ni	Sb (0.065)	Sb	0.33	1087	SC
		Ni	0.05	2	
Fe–0.24–4.4Ni–1.7Cr	Sb (0.040)	Cb	0.40	2160	SC
		Cr	0.30	16	
		Ni	0.79	19	
Fe–0.4C–1.7Cr	P (0.060)	P	0.58	536	SC
		Cr	0.23	13	
Fe–0.4C–3.5Ni	P (0.060)	P	0.34	314	SC
		Ni	0.17	5	
Fe–0.4C–3.5Ni–1.l7Cr	P (0.060)	P	0.64	590	SC
		Cr	0.35	19	
		Ni	0.66	20	

(*continued*)

Table A.1 (*continued*)

Material solute	Solvent (wt%)	Element segregating	Monolayers segregated	Enrichment ratio, β_b	Heat treatment[a]
Fe–0.4C–3.7Ni–1.7Cr	Sn (0.50)	Sn	0.81	3440	SC
		Cr	0.33	18	
		Ni	0.72	20	
Fe–0.4C–3.5Ni–1.7Cr	Sb (0.060)	Sb	0.56	2030	SC
		Cr	0.22	12	
		Ni	0.72	22	
		Sb	0.50	1823	IT
		Cr	0.25	14	
		Ni	0.61	18	
Fe–12Ni–6Mn	P (trace),	P	0.044	—	IT
	N (trace)	N	0.17	—	
		Mn	0.86	14	
Fe	S (0.005)	S	0.71	110	IT
Fe	P (0.2)	P	0.48	124	WQ
		P	1.15	319	FC
		P	1.20	333	WQ
Fe	P (0.9)	P	0.33	202	WQ
Fe	P (0.5)	P	0.19	211	FC
Fe	P (0.05),	P	0.18	201	WQ
	S (0.003)	S	0.38	7270	
Fe	P (0.02),	S	0.24	4650	FC
	S (0.003)	S	0.15	2960	WQ
Fe	P (0.02),	S	0.37	7170	FC
	S (0.003)	Sb	< 0.08	< 850	
Fe	Sb (0.06),	S	0.48	9300	FC
	S (0.003)	Sb	0.09	312	
Fe–0.1C AISI 1010	P (0.050)	P	0.12	130	WQ
		P	0.08	—	
Fe–0.4C–3.5Ni–1.6Cr	Sb (0.62)	Sb	0.31	1107	SC
		Ni	0.76	23	
		Sb	0.18	640	
		Ni	0.47	14	SC
Fe–3.5Ni–1.6Cr	P (0.50)	P	0.71	788	SC
		Ni	0.47	14	
		P	0.82	907	
		Ni	0.95	29	SC
Fe–0.4C–1.4Cr	P (0.50)	P	0.27	303	SC
Fe	Sn (0.25)	Sn	0.38	323	IT
Fe	Sn (1.0)	Sn	0.90	190	IT
Fe	Sn (4.0)	Sn	2.00	106	IT
Fe	S (0.003)	S	0.82	15 700	IT
Fe–02C–1.92Mn	Sb (0.12)	P	0.17	725	IT
	P (0.013)	Sb	0.28	510	
		Mn	1.09	56	
		P	0.02	85	IT
		Sb	0.09	165	
		Mn	1.07	55	

(*continued*)

Table A.1 (*continued*)

Material solute	Solvent (wt%)	Element segregating	Monolayers segregated	Enrichment ratio, β_b	Heat treatment[a]
Fe–0.12C–3.07Cr	P (0.02)	Cr	0.40	12	IT
		P	0.43	1192	
Fe–0.15C–2.98Cr	P (0.024)	P	0.49	1132	IT
Fe–0.09C–2.92Mn	P (0.020)	P	0.31	860	IT
Fe–0.10C–3.07Mn–0.39Mo	P (0.024)	P	0.30	693	IT
Fe–0.15C–0.24Si–1.5Mn	P (0.016)	P	0.25	867	IT
Fe–0.15C–2.98Cr	P (0.024)	P	0.13	300	WQ
Fe–0.09C–2.92Mn	P (0.020)	P	0.23	638	IT
Fe–0.10C–3.02Mn–0.39Mo	P (0.024)	P	0.18	416	IT
Fe–0.22C	P (0.028)	P	0.05	99	IT
AISI 3340	Sb (0.070)	Sb	0.37	1140	SC
		Cr	0.34	21	
		Ni	0.64	19	
		Sb	0.56	1744	
		Cr	0.34	21	IT
		Ni	0.88	26	
		Sb	0.40	1250	
		Cr	0.38	24	IT
		Ni	0.80	24	
Fe–0.008C–3.5Ni–1.7Cr	Sb (0.06)	Sb	1.43	5200	IT
		Ni	0.72	21	
Fe–0.008C–3.5Ni–1.7Cr–0.1Ti	Sb (0.06)	Sb	0.09	338	IT
		Ni	0.26	7	
AISI 3340	Sb (0.062)	Sb	0.96	3375	IT
	P (0.005)	P	0.07	780	
		Ni	0.50	14	
Fe–1.3Cr–1.2Mo–0.3V	P (0.25)	P	0.35	776	SC
	Sn (0.027)	Sn	0.24	1890	
		P	0.06	133	SC
		Sn	0.12	945	
Cast iron	P (0.070)	P	0.48	380	AS
		Sn	0.06	504	
Mild steel	P (0.022)	P	0.20	277	AS
En 24	P (0.016)	P	0.08	134	AS
	Sn (0.019)	Sn	0.012	482	
1Cr–Mo–V	P (0.022)	P	0.20	482	AS
3Cr–0.5Mo	P (0.03)	P	0.45	831	AS
2.25Cr 1 Mo	P (0.008)	P	0.21	1164	AS
5Ni–Cr–Mo–V	P (0.008)	P	0.27	1664	AS
	Sn (0.22)	Sn	0.17	1642	

[a] SC=step cooled; FC=furnace cooled; WQ=water quenched; IT=isothermal heat treatment; AS=as received.

Table A.2 Summary of results from heavy-iron irradiation and electron irradiation respectively at 823 K for ferritic steels (after Norris *et al.*, 1986)

Alloy	Heavy-ion irradiation	Electron irradiation
Fe–12G–0.5W–0.5Mn	Cr↓Ni↑	Cr↓Ni↑
Plus P	Cr↓Ni↑P↑ (Ni P rich pptn)	Cr↓Ni↑P–
Plus Si	Cr↓, Ni↑↑, Si↓	Cr↓, Ni–, Si–[a]
Plus B	Cr↓Ni↑V↓ (+$M_{23}C_6$ pptn)	Cr↓Ni–V (↑)
Plus Mo	Cr↓Ni↑Mo↓ (+$M_{23}C_6$ pptn)	Cr↓Ni↑Mo↓
Plus Mo+Si	Cr↓Mo↓Ni↑Si↑ (+$M_{23}C_6$ pptn)	Cr↓Ni↑Si↑Mo↓ ($M_{23}C_6$ pptn)

[a] Only measured at 723 K.

Table A.3 Misfit parameters, volume and radius, for various solutes in α-iron (after Norris *et al.*, 1986)

Element	Volume misfit, Ω_{sf} (%)	$(r_{sol}/r_{sol})^3 - 1$
P	−13.2	−32.3
Si	−7.0	−15.0
Ni	+4.0	+1.1
Cr	+4.4	+1.9
V	+10.5	+17.9
Mo	+27.5	+32.3
Mn	+4.9	+33.2

Table A.4 Summary of surface segregation data where the predicted segregation is derived from Burton and Machlin (1985) (after Wynblatt and Ku, 1977)

Alloy: Solvent (solute)	Segregating element (measured)	Segregating element (predicted)
Ag(Au)	Ag	Ag
Au(Ag)	Ag	Ag
Au(Ni)	Ni	Ni
Au(Pd)	Au	Au
Au(Sn)	Sn	Sn
Cu(Au)	Au	Au
Fe(Cr)	Cr	Cr
Fe(Sn)	Sn	Sn
Fe(Zr)	Sr	Zr
Ni(Au)	Au	Au
Ni(Cu)	Cu	Cu
Ni(Pd)	Pd	Pd
Pd(Ag)	Ag	Ag
Pd(Au)	Au	Au
Pt(Au)	Au	Au
Pt(Cr)	None	Cr
Pt(Fe)	None	Fe
Pt(Ni)	None	Ni
Pt(Sn)	Sn	Sn
Zr(Fe)	Fe	Fe

Table A.5 Experimental data of sulphur segregation to grain boundaries in Cu(Fe,S) and Cu(Ni,S) solid solutions (after Cabané and Cabané, 1991)

	Volume concentration $P^6 X_S$	Segregation coefficient, α	Coverage, θ (%)	Grain size compared with Pure Copper
Grain boundaries of polycrystals				
Cu	0.5	11 500	3.3	—
Cu–0.1 at.% Fe	0.5	51 000	14.5	Similar
Cu–0.2 at.% Fe	0.5	72 000	20.5	Similar
Twin grain boundaries of a bicrystal				
Cu	0.7	< 2000	—	—
Cu–0.1 at.% Fe	0.7	5100	—	—
Grain boundaries of polycrystals				
Cu–0.5 at.% Ni	0.74	8100	3.3	Large
Cu–1.1 at.% Ni	0.78	5100	2.2	Large
Cu–1.1 at.% Ni	0.78	11 700	6.4	Similar

Table A.6 Crystal structure and basic electrical properties of important semiconductor elements and compounds (after Grovenor, 1989)

	Crystal structure[a]	Lattice spacing (nm)	Band-gap width and type (at 300 K) (eV)	
Element				
Si	D	0.5431	1.12	Indirect
Ge	D	0.5646	0.66	Indirect
Group III to V compounds				
GaAs	S	0.5653	1.42	Direct
GaP	S	0.5451	2.26	Indirect
GaSb	S	0.6096	0.72	Direct
InAs	S	0.6058	0.36	Direct
InP	S	0.5869	1.35	Direct
InSb	S	0.6479	0.17	Direct
AlAs	S	0.5661	2.16	Indirect
AlSb	S	0.6136	1.58	Indirect
Group II to VI compounds				
CdS	S/W	$0.5832/a = 0.416$ $c = 0.6756$	2.42	Direct
CdSe	S	0.605	1.7	Direct
CdTe	S	0.6482	1.56	Direct
ZnS	SS/W	$0.542/a = 0.382$ $c = 0.626$	3.68	Direct
ZnSe	S	0.5669	2.7	Direct
ZnTe	S	0.6089	2.2	Direct
HgTe	S	0.644	0	—
Chalcopyrite $CuInSe_2$	S	$a = 0.5782$ $c = 1.1564$	1.04	Direct
Group IV to VI compounds				
PbS	N	0.594	0.41	Direct
PbSe	N	0.612	0.27	Direct
PbTe	N	0.646	0.31	Direct
SnTe	N	0.632	0.18	Direct

[a] D = diamond cubic; S = sphalerite; W = wurtzite; N = rock salt.

LIST OF SYMBOLS

a	lattice spacing
a_x	activity of solute atoms
A	area/atomic weight
A^*	effective Richardson coefficient
$\boldsymbol{b}$	Burgers vector
B	atom mobility
c	velocity of light
C_v	concentration
d	thickness/diameter/distance
d_g	grain size
d_{gb}	grain boundary width
D	diffusion coefficient
D_0	dispersion for the spectrometer
E	Young's modulus of elasticity
f	fraction
f_v	shape factor comparing the cavity volume with the cavity radius
F	force
g^*	point defect production rate
G	shear modulus
h	Planck's constant
H_I	enthalpies
I	intensity
j	interaction cross-section
J	current
J_e	electron flux/intensity
J_v	total vacancy flux
J_v, J_I and J_s	fluxes of vacancies, interstitials and solute atoms
k	Boltzmann constant
K	constant
l	crack length
L_v and L_I	dose rates for vacancies and interstitials to dislocation and/or grain boundary sinks
m	inverse size exponent
M	mobility of a grain boundary
n	integer
N_A	The Avogadro number
N_i, N_j	mole fractions
p	gas pressure
P	production rate for vacancies and interstitials
q, q_I and q_{II}	stress intensity factors
Q_A	activation energy
r	radius
R	gas constant
s_i	deviation parameter at the ith minimum
S	entropy
S_i and S_j	sensitivity factors for the ith and jth elements

LIST OF SYMBOLS (*continued*)

t	time
T	temperature
U_i	bias parameter
v	velocity
v	volume
V	voltage
V_α	applied potential
w	work function
W	work required to create unit area of surface A
x	distance
X_c	bulk concentration
X_{gb}	grain boundary concentration
Z	coordination number/atomic number
α	relaxation parameter
α_{ij}	interaction coefficients
β	value of the overlap integral controlling the splitting of states
γ	surface energy
γ_T	surface tension
Γ_A and Γ_B	surface excess concentrations of atoms A and B
Γ_S	segregation level
δ	grain boundary cross-section
ΔG	free energy terms
ΔH_j	enthalpy
ΔS	entropy
Δt	time interval
ΔV	relaxation volume
ε_{ij}	strain
$\dot{\varepsilon}$	creep rate
$\ddot{\varepsilon}$	steady-state creep rate
η	efficiency factor
θ	tilt angle
θ_D	Debye temperature for the solute atom rate parameter
θ_E	Einstein temperature
θ_m	coverage of boundaries
μ	chemical potential
ν	Poisson ratio
ν_E	Einstein frequency
ζ_g	extinction distance
σ	stress
τ_0	function of the relaxation time
τ_{eff}	effective shear stress on the dislocation pile-up $\cong \sigma_L/2\tau_f$
τ_f	'friction' stress which opposes dislocation motion
ϕ	angle between the feature and the foil surface, work function
ϕ_B	barrier height which is proportional to composition
Φ_A	ionisation cross-section
ω	maximum logarithmic decrement
Ω	atomic volume
∇	recombination coefficient

ACRONYMS FOR TECHNIQUES USED

AEM	Analytical electron microscopy
AES	Auger electron spectroscopy
APS	Appearance potential spectroscopy
Atom Probe	FIM plus time-of-flight MS
BSE	Backscattered electrons
CBD	Convergent beam diffraction
CL	Cathodoluminescence
EBSP	Electron backscattered spectroscopy
EDS	Energy dispersive (X-ray) spectrometry
EDX	Energy dispersive X-ray analysis
EELS	Electron energy loss spectroscopy
EPMA	Electron probe microanalysis
ESCA	Electron spectroscopy for chemical analysis
EXAFS	Extended (X-ray) absorption fine structure
FAB(MS)	Fast atom bombardment mass spectroscopy
FEG	Field emission gun
FEM	Field emission microscopy
FIM	Field ion microscopy
HVEM	High-voltage electron microscopy
ISS	Ion scattering spectroscopy
Kossel	Diffraction of electron-excited X-rays
LEED	Low-energy electron diffraction
LIMA	Laser-induced ion mass analysis
PEEM	Photoemission electron microscopy
RBS	Rutherford backscattering spectroscopy
RHEED	Reflection high-energy electron diffraction
SAM	Scanning Auger microscopy
SEM	Scanning electron microscopy
SIMS	Secondary ion mass spectroscopy
SNMS	Sputter neutral mass spectroscopy
STEM	Scanning transmission electron microscopy
STM	Scanning tunnelling microscopy
TEM	Transmission electron microscopy
UPS	Ultraviolet photoelectron spectroscopy
XPS	X-ray photoelectron spectroscopy
XRD	X-ray diffraction
XRF	X-ray fluorescence
ZAF	Atomic number, absorption and fluorescence corrections

COMMONLY USED CONVERSION FACTORS

1 Ångstrom (Å)	=	10^{-10} m		
1 dyne	=	10^{-5} N		
1 gauss (G)	=	10^{-4} tesla (T)		
0 °C	=	273.15 K		
1 curie (Ci)	=	3.7×10^{-10} s^{-1}		
10^7 erg	=	6.241×10^{18} eV	=	1 Joule (J)
1 eV	=	1.602×10^{-19} Joule (J)		
1 cal	=	4.184 Joule (J)		
e	=	2.718		
1 mol	=	2.24×10^{-2} m^3 (at STP)		
1 μm	=	10^{-6} m		
1 nm	=	10^{-9} m		

PHYSICAL CONSTANTS

Avogadro constant	N_A	6.022×10^{23} mol^{-1}
Bohr magneton	$\mu_B = e\hbar/2m_e$	9.274×10^{-24} J T^{-1}
Boltzmann constant (R/L_A)	k	1.381×10^{-23} J K^{-1}
Charge of an electron	e	-1.602×10^{-19} C
Compton wavelength of:		
electron $(h/m_e c)$	λ_c	2.426×10^{-12} m
proton $(h/m_p c)$	λ_{cp}	1.321×10^{-15} m
neutron $(h/m_n c)$	λ_{cn}	1.319×10^{-15} m
Electron radius	$r_e = \mu_0 e^2/(4\pi m_e)$	2.817×10^{-15}
Faraday constant	F	9.649×10^4 C mol^{-1}
Fine structure constant	$\alpha = \mu_0 e^2 c/2h$	7.297×10^{-3} ($\alpha^{-1} = 137.0$)
Gas constant	R	8.314 J *and*$^{-1}$ mol^{-1}
Gravitational constant	G	6.673×10^{-11} N m^2 kg^{-2}
Magnetic moment of electron	μ_e	9.284×10^{-24} J T^{-1}
Magnetic moment of proton	μ_p	1.410×10^{-26} J T^{-1}
Molecular gas constant	R	8.314 J K^{-1} mol^{-1}
Nuclear magneton	μ_N	5.051×10^{-27} J T^{-1}
Permeability of a vacuum	μ_0	$4\pi \times 10^{-7}$ H m^{-1}
Permittivity of a vacuum	ε_0	8.854×10^{-12} F m^{-1}
		($1/4\pi\varepsilon_0 = 8.988 \times 10^9$ m F^{-1})
Planck constant	h	6.626×10^{-34} J s
(Planck constant)/2π	$\hbar$	1.055×10^{-34} J s
		$= 6.582 \times 10^{-16}$ eV s
Rest mass of electron	m_e	9.11×10^{-31} kg
		$= 0.511$ MeV/c^2
Rest mass of proton	m_p	1.673×10^{-27} kg
		$= 938.3$ MeV/c^2
Rydberg constant	$R_\infty = \mu_0^2 m_e e^4 c^3/8h^3$	1.097×10^7 m^{-1}
Speed of light in a vacuum	c	2.998×10^8 m s^{-1}
Stefan–Boltzmann constant	$\sigma = 2\pi^5 k^4/15h^3 c^2$	5.670×10^{-8} W m^{-2} K^{-4}
Unified atomic mass unit (^{12}C)	u	1.661×10^{-27} kg
		$= 931.5$ MeV/c^2
Wavelength of a 1 eV photon		1.243×10^{-6}

Index